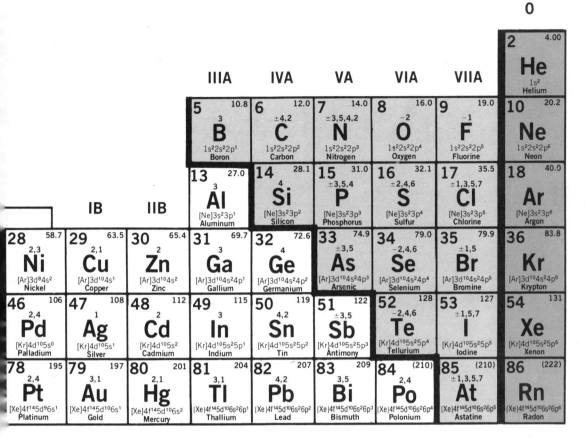

*Roe's principles
of chemistry*

Roe's principles of chemistry

Alice Laughlin, B.S., M.S., Ed.D.

Professor and Former Chairman, Department
of Chemistry, Jersey City State College,
Jersey City, New Jersey

with 122 illustrations

TWELFTH EDITION

The C. V. Mosby Company

Saint Louis   1976

TWELFTH EDITION

Copyright © 1976 by The C. V. Mosby Company

All rights reserved. No part of this book may be reproduced in any manner without written permission of the publisher.

Previous editions copyrighted 1927, 1929, 1932, 1936, 1939, 1944, 1950, 1956, 1963, 1967, 1972

Printed in the United States of America

Distributed in Great Britain by Henry Kimpton, London

Library of Congress Cataloging in Publication Data

Roe, Joseph Hyram, 1892-1967.
    Roe's Principles of chemistry.

    Includes index.
    1. Chemistry. I. Laughlin, Alice, 1918-
II. Title: Principles of chemistry. [DNLM:
1. Chemistry—Nursing texts. QD33 R698p]
QD33.R6   1976   540   75-37565
ISBN 0-8016-1470-8

GW/CB/B 9 8 7 6 5 4 3 2 1

*An introductory textbook of
inorganic, organic, and
physiological chemistry for students
in the allied health fields*

Preface

For almost two generations, instructors and students of nursing, medical technology, and other allied health fields have found *Roe's Principles of Chemistry* to be a highly teachable and readable book. Its strengths have been its simple explanations, its ability to relate principles to practice, and its brief, but thorough, coverage of the essential areas of inorganic and organic chemistry and biochemistry.

In this new revision, although the basic forms and emphasis remain the same, new developments and new points of view have prompted many changes. Greater emphasis has been placed on the metric system, molecular and atomic structure, and recent discoveries in biochemistry.

Important additions are five new tables and 35 new illustrations by Robert Wiethop; a new appendix, containing a discussion of logarithms and a table of four-place logarithms; and a new chapter, "The Physical States of Matter." In all areas the material has been presented in as simple and uncomplicated manner as possible without distorting the true picture revealed through scientific investigation.

Alice Laughlin

Contents

Introduction Chemistry as a science, 1

Science, 1
Chemistry, 1
  History of chemistry, 2
  Divisions of chemistry, 2
  Importance of chemistry, 2

1 Matter—introduction to chemical concepts, 4

Matter, 4
  Inertia, 4
  Mass, 4
  Weight, 4
  Energy, 5
Measurement of matter and energy, 5
  Length, 5
  Volume, 6
  Weight, 6
  Pressure, 6
  Temperature, 8
  Heat, 8
Chemical concepts, 8
  Elements, 8
  Symbol, 9
  Compound and mixture, 9
  Formula, 10
  The atom, 10
  Atomic mass, 11
  The molecule, 11
  Molecular mass, 11
Physical and chemical changes, 12
  Equation, 13
  Types of chemical reaction, 13
  Reversible reactions—equilibrium, 14
  Catalyst, 15
  Laws of conservation of mass and energy, 15
  Law of definite composition, 16
  Combining masses, 17
  Law of multiple proportions, 18

2 The atom, periodic classification, and chemical reactivity, 20

The atom, 20
  Subatomic particles, 20
Atomic structure, 24
  Rules governing atomic structure, 24
  The Bohr atom, 27
  Atomic structure according to quantum theory, 28
Periodic classification of the elements, 36
  Atomic numbers, 38
  Natural families, 38
  Discovery of new elements, 38
Chemical reactivity, 38
  Noble gases—relation to chemical theory, 39
  Chemical changes involving several of the simplest elements, 41
  Ions, 41
Metals, nonmetals, and metalloids, 43
Valence, 43
  Elemental, 44
  Radical, 45
  Mechanisms underlying valence, 47
Bonding, 47
  Formation of electrovalent bonds, 47
  Formation of covalent bonds, 48
  Polar bonds, 50
Naming of binary compounds, 50

3 Physical states of matter, 54

Solid state, 54
   Molecular crystals, 55
   Ionic crystals, 55
   Metallic solids, 55
Liquid state, 56
Gaseous state, 59
Allotropism, 62

4 Some important metals and nonmetals, 63

Metals, 63
   Sodium and potassium, 63
   Lithium, 63
   Calcium, 63
   Magnesium, 64
   Iron, 64
   Mercury, 64
   Bismuth, 65
   Electromotive series, 66
Nonmetals, 66
   Nitrogen, 66
   Nitric acid, 67
   Ammonia and ammonium hydroxide, 67
   Sulfur, 68
   Phosphorus, 70
   Arsenic, 71
   Boron, 71
   The halogen family, 71

5 Hydrogen, oxygen, oxidation-reduction, and equilibrium constant, 75

Hydrogen, 75
Oxygen, 78
   Ozone, 84
Oxidation and reduction, 85
Equilibrium constant, 88

6 Water, hydrogen peroxide, solutions, standard solutions, and milliequivalents, 91

Water, 91
Hydrogen peroxide, 97
Solutions, 98
   Solubility, 98
   Concentration, 99
Milliequivalent, 103
   Summary, 104

7 Electrolytes, ionization acids, bases, salts, and pH, 106

Electrolytes and nonelectrolytes, 106
Ionization, 106
   Nonaqueous ionizations, 108
The effect of solute on solution, 108
   Calculations involving critical temperatures, 109
Acids, 110
   Naming of inorganic acids, 113
Bases, 114
   Naming of bases, 115
   Neutralization, 115
   Titration, 116
Salts, 117
   Naming of salts, 118
   Insoluble salts, 120
Antidotes for wounds produced by acid or base, 121
pH, 122
Buffers, 123

8 Crystalloids, colloids, diffusion, dialysis, and osmosis, 127

Crystalloids and colloids, 127
Diffusion, 130
Dialysis, 130
Osmosis, 131

9 Radiochemistry, 137

Radioacvitity, 137
Radium, 137
X-rays, 142
Atomic fission, 144
Atomic fusion, 146
Uses of radioactive isotopes in biology and medicine, 148
Units of measurement of radioactivity, 149

10 The chemistry of carbon compounds, 151

Carbon, 151
Some inorganic carbon compounds, 153
   Carbon monoxide, 153
   Carbon dioxide, 154
Organic chemistry, 156
   Some fundamental principles, 157
   Classification of organic compounds, 159
   Naming organic compounds, 160

11 Organic chemistry—hydrocarbons, halogen compounds, alcohols, aldehydes, ketones, 162

Hydrocarbons, R-H, 162
   Saturated hydrocarbons—alkanes, 162
   Unsaturated hydrocarbons—alkenes and alkynes, 164
Halogen compounds—halides, R-X, 165
Alcohols, R-OH, 166
   Some important alcohols, 168
Aldehydes, RCHO, 171
   Some important aldehydes, 172
Ketones, RCOR, 173

12 Organic chemistry—organic acids, esters, ethers, amines, 176

Organic acids, RCOOH, 176
   Some important organic acids, 177
Esters, RCOOR, 179
   Some important esters, 179
Ethers, ROR, 181
Amines, RNH_2, 182

13 Organic chemistry—aromatic compounds, 186

Aromatic hydrocarbons, Ar-H, 186
Aromatic halogen compounds, Ar-X, 190
Phenols and aromatic alcohols, Ar-OH, 191
Aromatic aldehydes, Ar-CHO, 192
Aromatic acids, Ar-COOH, 193
Aromatic esters, Ar-COOR, 194
Aromatic amines, Ar-NH_2, 195
Polynuclear hydrocarbons, 197
Heterocyclic compounds, 198
Carbocyclic compounds, 199

14 Carbohydrates, 201

Individual occurrence, importance, and uses of principal carbohydrates, 205
   Monosaccharides, 205
   Disaccharides, 206
   Polysaccharides, 206

15 Proteins, 209

16 Lipids, 222

Fats, 223
Sterols, 229
Phospholipids, 230
Glycolipids, 231
Waxes, 231

17 Nucleoproteins, nucleic acids, and genetics, 233

Nucleoproteins, 233
Nucleic acid, 234
   Components, 234
      Pyrimidines, 234
      Purines, 235
      Sugars in nucleic acid, 235
      Nucleosides, 236
      Nucleotides, 236
   Structure, 237
Genetics, 240
   Storage of genetic information, 240
   Replication, 241
   Transmission of genetic information, 241
   The genetic code, 242

18 Enzymes, 246

19 Hormones, 251

Thyroid gland, 251
   Cretinism, 253
   Simple goiter, 254
   Hyperthyroidism, 254
Parathyroid glands, 255
Pancreas, 255
Pituitary gland, 257
   Physiological effects of secretion of the anterior pituitary gland, 257
   Secretions of the posterior pituitary gland, 260
Adrenal glands, 261
   Secretions of the adrenal cortex, 261
   Secretions of the adrenal medulla, 263
Thymus, 263
Female sex hormones, 264
   Menstrual cycle, 265
   Pregnancy tests, 267
   Synthetic progestins—antiovulatory drugs, 267
Male sex hormones, 267
Gastrointestinal hormones, 268

20 Digestion, 272

Salivary digestion, 274
Gastric digestion, 275
Digestion in the intestinal tract, 277
Supplementary action of enzymes in the digestive tract, 278
Bile, 279

21 Metabolism of carbohydrates, fats, proteins, and inorganic salts, 282

Metabolism of carbohydrates, 282
Metabolism of fats, 290
Metabolism of proteins, 295
Metabolism of inorganic salts, 298

22 Nutrition, 304

Proteins, 304
Carbohydrates, 307
Fats, 308
The balanced diet, 309
Milk, 314
Energy metabolism, 315

23 Vitamins, 322

Fat-soluble vitamins, 323
  Vitamin A (retinol), 323
  Vitamin D (calciferol), 325
  Vitamin E (α-tocopherol), 327
  Vitamin K (menadione), 328
Water-soluble vitamins, 330
  Ascorbic acid (vitamin C), 330
  Thiamine (vitamin B_1), 332
Vitamin B complex, 333
  Niacin—the antipellagra vitamin, 334
  Riboflavin (vitamin B_2), 335
  Pyridoxine, 336
  Choline, 337
  Pantothenic acid, 338
  Biotin, 339
  Folic acid, 340
  Vitamin B_{12} (cyanocobalamin), 341
Bacterial synthesis of vitamins in the alimentary tract, 342

24 The blood, 344

Respiration, 349

25 The urine, electrolyte balance, water balance, and acid-base balance, 354

Urine, 354
  Composition, 354
  Secretion of urine, 355
    Structure of the nephron, 355
    Mechanism of secretion, 356
    Summary, 357
  Qualitative analysis, 357
Electrolyte balance, 360
  Electrolyte composition, 360
  Balance between anions and cations in extracellular fluid, 361
Water balance, 361
Acid-base balance, 364
  The $BHCO_3/H_2CO_3$ system, 364
  Causes of acid-base imbalance, 365
    Disturbances in the alimentary tract, 366
    Imbalance arising in body tissues, 366
    Imbalance due to abnormal respiration, 366
    Imbalance due to failure in the kidneys, 366
  Laboratory findings, 367

Appendix, 369

Logarithms, 369
  The use of log tables, 370
Four-place logarithms, 372

Glossary, 374

*Roe's principles
of chemistry*

Introduction *Chemistry as a science*

SCIENCE

For thousands of years man has made observations of the things about him and passed on to succeeding generations his thoughts regarding them. Man's early ideas concerning the universe and the substances of which it is composed were naturally very simple. These ideas appeared first as myths, which were often interwoven with religion, and later as theories proposed by the philosophers—all of which were imaginative and have literary charm to the present-day reader but were lacking in exactness because they were not based on man's greatest test of truth, that is, experimentation. As time passed, the earlier observations were found by experience to be inexact, and the old concepts gradually gave way to new ones that contained more of the truth because they were based on observation and the tests of experimentation. Thus a mass of ideas about matter and the universe has gradually accumulated and has been recorded in a systematic form.

This accumulation of knowledge of the universe constitutes the physical sciences. *A science is the systematic arrangement of observations and the conclusions derived from them.* The sciences of today contain much that is true because in them are observations that have been made thousands of times by thousands of different observers. When an occurrence happens many times in the same manner, the observation of this similarity of result establishes a fact that the observer can feel very sure is true. Out of such often repeated observations have developed the principles or laws that are the basis of the sciences. In science *a law is a statement of a fact that has been demonstrated to be true by often repeated tests or experiments* and that one believes would be verified at any future time one chooses to repeat these tests.

CHEMISTRY

Chemistry is a science that deals with the structure of matter and the changes that matter undergoes. Its purpose is to find out what are the simplest parts of matter; how these parts can be separated in a pure form from crude matter, put together, or rearranged to produce new forms of matter; and what energy is liberated or absorbed in making these rearrangements of matter. The energy changes of matter are considered more especially, however, by the science called *physics*. Physics and chemistry of necessity overlap in their treatment of material because both deal with the investigation of matter. *Biology* is a science that studies the structure and changes that take place in matter in living organisms. Chemistry and biology also have areas that overlap.

History of chemistry

The earliest recorded investigators in the field of chemistry were known as *alchemists,* and the principles that they developed and followed were called *alchemy*. The great purpose of the alchemists was to convert base metals, such as lead, into gold. In seeking a method to accomplish this dream, they produced little of value to the world, but as pioneer experimenters with matter, they initiated the science of chemistry, and in this respect they made a great contribution to civilization. The earliest records of alchemy date back to the beginning of the Christian Era. It probably originated with the Alexandrian Greeks and later spread to Egypt, Rome, and the nations of western Europe.

Just when the modern age of chemistry began is a somewhat indefinite date, but it can be conservatively said that practically all the chemistry known at the present time is a development of the past 200 years.

Divisions of chemistry

Chemistry is divided into two major branches—*inorganic* and *organic*. Inorganic chemistry deals especially with inanimate or lifeless matter; it includes a study of all the elementary substances, but the element carbon is treated only to a minor extent in this division of chemistry. Organic chemistry considers matter of which living things are composed or that is closely associated with living things; it is essentially a study of the compounds of carbon. Both these great divisions have many subdivisions, such as analytical, synthetical, physical, and physiological. The division of chemistry into inorganic and organic is not a clear-cut separation and is to be regarded as a convenience for purposes of study.

Importance of chemistry

Before the advent of chemistry, knowledge of the universe was clouded with mystery and superstition, and the human race lacked most of its present-day comforts. Chemistry has fulfilled a great mission in solving many of the mysteries of nature and revealing the wonderful properties of matter. Perhaps no other subject has contributed so much to the development of human knowledge, and certainly no other science has contributed so much to human happiness by utilizing the products and forces of nature and creating new products for human use. The significance of chemistry in the production of articles for human comfort and happiness can be seen readily by noting the articles surrounding us, wherever we may be. It is practically impossible in modern times to observe an article of commerce that has not been created, improved, or influenced in some manner by an application of the principles of chemistry. It will be an interesting and instructive exercise for the student to try to note an exception to this statement. Today chemists, along with others, recognize their responsibility to help combat the threat of pollution that may destroy our environment, to increase the production of food, to conserve energy, and to find new sources of energy.

Questions for study

1. Define science.
2. What is a law in science?
3. What is chemistry? What other sciences are closely related to chemistry?
4. State the two major divisions of chemistry. What is meant by inorganic chemistry? By organic chemistry?
5. Discuss the importance of chemistry.

1 Matter—introduction to chemical concepts

MATTER

Matter is anything that possesses inertia, has mass, and occupies space. It includes all the substances that we can see, feel, or otherwise perceive by means of our senses. The physical properties of matter are those characteristics that can be recognized by the senses, for example, color, size, taste, and odor. These properties of matter can be changed without greatly altering the chemical properties of the matter. The chemical properties of matter are those characteristics that have to do with the activity of the substance in a reaction, the ease or difficulty with which it enters into chemical change.

Inertia

Inertia is described by Newton's first law of motion, which states that matter possesses the ability to remain at rest until it is acted on by some outside force. *Force* is the cause of motion, or the cause of a change in motion, of matter. Newton's law goes on to state that, once set in motion, matter has the ability to remain in motion in a straight path at a constant momentum until it is acted on by an external force, or forces, to deflect its path, slow it down, or speed it up.

Mass

Mass is a quantitative measure of the inertia of an object. The mass of an object determines how difficult it is to accelerate its motion. The rate at which movement occurs with respect to time is called *velocity* and is described by the distance an object moves in a given period of time, for example a car moving at the velocity of 50 miles per hour. *Acceleration* can be described as the rate at which the velocity changes with time. Force and mass are related through the equation, $F = ma$, where F is the force causing the acceleration (a) of the mass (m).

Weight

Weight is caused in an object because the mass of that object is being accelerated by the force of the gravitational attraction of a large object. In everyday experience that large object is the earth. From the beginning of time, mass and weight were almost synonymous because mankind was earthbound. Scientists theorized that an object in space could reach points in that space at which the gravitational attractions of stellar bodies would cancel each other and the object would have mass but would be weightless. They also realized that an object on another celestial body would weigh more or less than it did on earth, depending on the size of that celestial body with respect to the earth. The larger the body, the more gravitational force is associated with it. It was not until the dawn of the space age in the late 1950s that man could experience the truth of these theories.

Energy

Energy is the capacity to do work. *Kinetic energy* is that energy that matter exerts because of its motion and mass. When a force *(F)* operates on an object through a distance *(d)*, work *(W)* is done, $W = Fd$. Kinetic energy is equal to one-half the mass of the object *(m)* times the square of its velocity *(v)*, $E = \frac{1}{2} mv^2$.

Potential energy is the energy that an object possesses due to its position with respect to another object. A raised hammer has the ability to do work because of its position above the earth. Gravity will add to the muscular activity of the person wielding the hammer and increase the force of the blow. As the hammer falls, it does work, and the potential energy is converted to kinetic energy.

MEASUREMENT OF MATTER AND ENERGY
Length

Early in history, measurements were taken in a rather unscientific manner. A foot was the length of the foot of some leader in the community, the lord of the manor, or the tribal chief. It can be realized that the length of a foot varied from place to place. Soon it became clear that measurements needed to be standardized. Today we have the English measurements, inches, feet, and yards, and the metric measurements, millimeters, centimeters, and meters (Fig. 1-1). For many years the meter was standardized by a meter stick made of a platinum-iridium alloy kept at a constant temperature at the International Bureau of Weights and Measures in a suburb of Paris, France. This meter was one ten-millionth of the distance from the North Pole to the equator.

To make the standardization of the length of the meter more universally available, it was necessary to define the length as 1,650,763.73 times the wavelength of the orange-red line in the spectrum of the element krypton-86.* The metric system has an advantage over the English system in the fact that the metric system is based on units of ten.

        1 kilometer (km) = 1000.0 m
        1 meter (m) = 1.0 m
        1 decimeter (dm) = 0.1 m
        1 centimeter (cm) = 0.01 m
        1 millimeter (mm) = 0.001 m
        1 micron (μ) = 10^{-6} m
        1 millimicron (mμ) (nanometer, nm) = 10^{-9} m
        1 angstrom (Å) = 10^{-10} m

---

*An isotope of the element krypton.

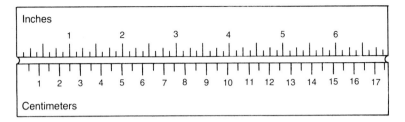

Fig. 1-1. Units of measurement.

Table 1-1. Some equivalent values

Metric system	English system
1 meter	39.37 inches
2.54 centimeters	1.0 inch
1 liter	1.0567 quarts
1 kilogram	2.205 pounds
453.6 grams	1 pound
28.35 grams	1 ounce
3.785 liters	1 gallon
1.60933 kilometers	1 mile

Volume

The English system of measuring volume uses the values of pints, quarts, and gallons, whereas the metric system uses the cubic millimeter, cubic centimeter, and the liter (l). The milliliter (ml) is 1/1000 of a liter.

Weight

The English system of weight consists of ounces, pounds, and tons. The metric system uses the gram (g) as the basic unit of weight. A gram is defined as the weight of 1 ml of water at 4° C. A kilogram (kg) is 1000 g, a milligram (mg) is 0.001 g, a microgram (μg) is 0.000001 g (10^{-6} g). The Greek letter μ is mu. The microgram is also called by the Greek letter γ, gamma.

Pressure

Air above the earth exerts a pressure on the surface of the earth. The higher areas of the earth have less atmosphere over them and, therefore, have less weight of atmospheric pressure than areas near sea level. It is usual to speak of air pressure at sea level as being 1 atmosphere of pressure. An instrument used to measure air pressure is the barometer. A barometer consists of a tube sealed at one end. A vacuum exists in the tube, and the open end of the tube rests in a well of mercury. The air pressure over the well of mercury forces the mercury into the tube to a level that corresponds to the amount of air pressure existing over the open well of mercury. It is usual to report barometric readings in either inches or millimeters, corresponding to the height of the mercury in the tube. One atmo-

Matter—introduction to chemical concepts 7

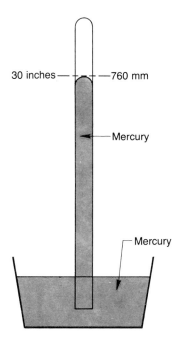

Fig. 1-2. A simple barometer.

Fig. 1-3. The three temperature scales.

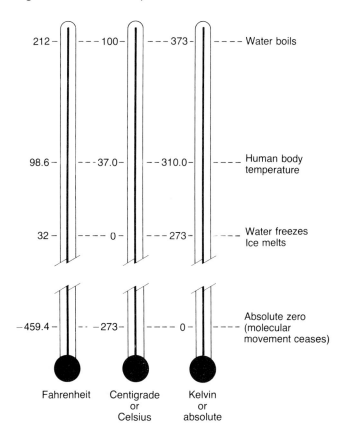

sphere of pressure is the pressure necessary to elevate the mercury in the tube to the level of 29.9 inches or 760 mm (Fig. 1-2). (A new term, the *torr,* is frequently being substituted for millimeters of mercury in modern scientific literature, 760 torr being equivalent to 760 mm of mercury.)

Temperature

There are three types of temperature scales in use (Fig. 1-3). The Fahrenheit scale is the one most generally used in the United States. According to this scale, at 1 atmosphere of pressure, water boils at 212° and freezes at 32°. The centigrade or Celsius scale shows water under the same atmospheric pressure boiling at 100° and freezing at 0°. The third scale is used in scientific work at low temperatures and is called the Kelvin or absolute temperature scale. It has the advantage of showing no negative values. The zero point on the centigrade scale is equal to 273.16° (usually rounded to 273°) on the Kelvin scale, whereas the 100 mark of the centigrade scale is equivalent to 373° on the Kelvin scale.

To convert from Fahrenheit to centigrade and back again, use the following equation: 1.8° C = °F − 32. For example, 14° F corresponds to what reading on the centigrade scale?

$$°C = \frac{14 - 32}{1.8} = -10° C$$

To convert centigrade to Kelvin, add 273° to the centigrade figures. To convert Kelvin to centigrade, subtract 273° from the Kelvin figures.

Heat

Heat energy is measured in calories. A calorie (cal) is the quantity of heat required to raise 1 g of water 1° C at 1 atmosphere of pressure. A calorie can be converted to units of electrical energy, 1 cal being equal to 4.1840 joules (J).* A kilocalorie (kcal or Cal) is the heat equal to 1000 cal.

CHEMICAL CONCEPTS
Elements

The early philosophers of Greece, Egypt, and other nations of antiquity believed that there were four primary substances—earth, water, air, and fire—that composed all the forms of matter in the universe. Of these four substances, earth was typical of solids, water stood for liquids, air represented the winds and the breath, and fire was a mystic essence that existed in greater or lesser amounts in all things capable of burning. Until about 200 years ago, this four-element concept of matter was the basis of all attempts to classify the many forms of matter in the world. It finally disappeared when investigators

---

*The particles that make up a substance offer resistance to the flow of electricity. A unit of resistance, the ohm, is that resistance to the flow of electricity offered by a thread of mercury containing 1 g of mercury in a length of 106.3 cm at 0° C. The coulomb is the amount of electricity needed to plate out 0.001118 g of silver from a solution of silver ions. The potential needed to cause the flow of 1 coulomb (C) through the thread of mercury in 1 second is a volt. The energy produced by the flow of 1 C at a potential of 1 volt (V) is a joule.

found that simpler substances could be made from earth, water, or air and that fire is a manifestation of energy released in the form of heat and light.

We now have knowledge of over 100 primary substances in matter that have distinctive, characterizing properties. These substances are called elements. (See the table of elements on the inside cover of this book.) *An element is a basic form of matter incapable of being decomposed by chemical means into more simple substances, having distinct chemical and physical characteristics, and occupying a specific place in the periodic table of elements.* Each element has characteristics or properties that make it distinct from every other element. Examples of elements are iron, gold, silver, lead, carbon, copper, sulfur, oxygen, and hydrogen. The element iron, for example, has properties that distinguish it from all other elements, and the same is true of each of the elements. Up to the time of the publication of this edition, 106 elements have been identified. Since 1940, new elements, with atomic numbers ranging from 93 to 106, have been synthesized by scientists.

Symbol

A chemical symbol is an abbreviation of the name of an element. It may be considered a unit of chemical shorthand, its purpose being to save the extra work of writing the whole name of the element. It usually is the first letter in the name, but since there are over 100 elements, it is obvious that the initial letter cannot be used as the abbreviation in every case. Where there are two or more elements whose names begin with the same letter, the practice is to use the first letter of the name for one element and to use for the others two significant letters of their name, usually the first two. In some instances a symbol derived from the Latin name given to the element by the alchemists is used. For example, the names and symbols of the seven elements having the initial letter A are as follows: actinium, Ac; aluminum, Al; americium, Am; antimony, Sb; argon, Ar; arsenic, As; and astatine, At. Six of these symbols need no explanation; the symbol for antimony, Sb, is derived from stibium, the Latin word for this element. Symbols are thus used for convenience and economy of expression in chemistry.

Compound and mixture

A compound is a combination of two or more elements chemically united in a definite proportion by mass. A mixture is a mass of two or more substances that are not chemically combined and need not be in any definite proportion. A compound differs from a mixture in the following ways:
1. In a compound the elements are definitely combined with each other; in a mixture the components are not combined but are in loose contact with each other.
2. The elements of a compound are always present in the same proportion; the components of a mixture may be present in any proportion.
3. A mixture will exhibit the properties of each of its constituents; the properties of a compound will be different from those of its individual constituents.
4. A compound can be separated only by chemical means; mixtures may be separated by physical methods.

To illustrate these distinctions let us take the two elements sulfur and iron, grind each into very fine particles, and then mix them intimately. This mass of intermingled sulfur and iron particles is a mixture. The iron and sulfur are not firmly combined and can be separated easily by holding a magnet near the mass; a magnet will pick out the iron particles but will not attract the sulfur. Furthermore, the sulfur and the iron may be present in any proportion. But if we thoroughly mix 28 parts of iron and 16 parts of sulfur by mass and heat the mixture to a high temperature, the iron and sulfur will combine to form a compound called iron sulfide. This compound will have quite different properties from either iron or sulfur; the iron and sulfur cannot be separated by a magnet, and they are always present in the proportion of 28 parts of iron to 16 parts of sulfur by mass.

Formula

A formula is an abbreviated expression for a compound. It includes the symbols of all the elements in the compound, with subscripts to show the number of atoms of each element present. For example, H_2O is the formula for water; this designation shows that a molecule of water contains 2 atoms of hydrogen and 1 atom of oxygen. To represent sulfuric acid, we use the formula H_2SO_4, which is an abbreviated expression showing that the compound contains 2 atoms of hydrogen, 1 atom of sulfur, and 4 atoms of oxygen in its molecule. The formula for cane sugar is $C_{12}H_{22}O_{11}$. The symbols C, H, and O show that cane sugar contains the elements carbon, hydrogen, and oxygen, and each subscript (12, 22, and 11) indicates the number of atoms of the element immediately preceding it that are in a molecule of cane sugar.

From these examples it is clear that the use of formulas decreases work, in that the time and effort that would be required for writing out in full the names of compounds is greatly reduced, as well as is the space consumed in writing or printing chemical literature. Formulas are therefore a great convenience in chemistry. The student will also find them an important aid in studying the many substances encountered in chemistry.

The atom

When different kinds of matter react with each other and produce a chemical change, the process is called a *chemical reaction*. Chemical reactions are orderly rearrangements of the extremely minute particles of matter called atoms. *An atom is the smallest fundamental unit of an element maintaining the distinguishing features of that element that is able to take part in chemical change.* When a chemical reaction takes place, the atoms are not decomposed but are torn away from or made to combine with atoms of other elements. For example, wood is a substance composed principally of many atoms of carbon, hydrogen, and oxygen. When wood is burned, its atoms are torn apart and combined with atoms of oxygen from the air, and new substances are formed. The combustion of wood is a disruption of its substance down to the stage of the atom, but the atoms are not broken up. They are merely rearranged to form other substances—ashes and certain gases. The atom is thus seen to be a stable unit of matter; that is, it enters into chemical reactions involving moderate energy changes and the formation or decomposition of compounds but is not itself decomposed.

Matter—introduction to chemical concepts

This preliminary description of the atom has been introduced here to prepare the student for a discussion of some basic concepts. A more fundamental discussion of the structure and behavior of the atom is presented in Chapter 2.

Atomic mass

Atomic masses are relative average masses of atoms of the elements. An atomic mass of an element is a value that represents the mass of an atom of that element when compared with the mass of an atom of carbon, which has been arbitrarily set at 12. For example, the atomic mass of hydrogen is approximately 1; of carbon, 12; of iron, 55. These numbers have a relative significance. They mean that 1 atom of carbon is 12 times heavier than 1 atom of hydrogen, that 1 atom of iron is 55 times heavier than 1 atom of hydrogen, and that the ratio of the mass of 1 atom of iron to 1 atom of carbon is 55:12. Again, uranium, the heaviest of all the naturally occurring elements, has an atomic mass of 238. This figure means that the ratio of the mass of 1 atom of uranium to 1 atom of iron is 238:55 and of 1 atom of uranium to 1 atom of hydrogen is 238:1. The atomic masses of the elements are given in the table on the inside cover.

The molecule

A molecule is the smallest electrically neutral particle of any substance that can maintain a stable existence. In some instances the molecule consists of a single atom. The atoms of many elements, however, have a strong tendency to combine, and in doing so they form molecules that consist of more than 1 atom. Molecules may be composed of atoms of the same element or atoms of different elements. If composed of atoms of different elements, the molecule is a structural unit of the compound. The molecules of compounds are as a rule more complex than the molecules composed of atoms of a single element. The molecules of compounds may contain from 2 to thousands of atoms.

Molecular mass

Molecular mass is the average of the masses of the molecules of a substance according to the atomic mass scale. Molecular masses are relative masses. A molecular mass is a value that represents the mass of a molecule of a substance as compared with the mass of an atom of carbon-12. The molecular mass of a substance is the sum of the atomic masses of the elements in its molecule. The hydrogen molecule (H_2) contains 2 atoms; therefore, the molecular mass of hydrogen is twice the atomic mass (1.008×2) or 2.016. Since the atomic mass of oxygen is 16 and the oxygen molecule contains 2 atoms, the molecular mass of oxygen is 32 (16×2). If we know the formula of a compound, we can calculate its molecular mass by taking the sum of the atomic masses of all the elements in the molecule. This is illustrated by the following three examples:

For water (H_2O)
Atomic mass of   2H = 2.016
Atomic mass of   O = 16.000
Molecular mass of H_2O = 18.016

For ammonia (NH_3)
Atomic mass of	N = 14.0
Atomic mass of	3H = <u>3.024</u>
Molecular mass of	NH_3 = 17.024

For sulfuric acid (H_2SO_4)
Atomic mass of	2H = 2.016
Atomic mass of	S = 32.06
Atomic mass of	4O = <u>64.000</u>
Molecular mass of	H_2SO_4 = 98.076

In cases of a crystalline compound in which individual molecules exist only in the gaseous state, the *formula mass* is used instead of the molecular mass. For example, the formula mass of NaCl is 22.99 + 35.45 = 58.44.

PHYSICAL AND CHEMICAL CHANGES

The changes that take place in matter under ordinary conditions are of two distinct kinds; they are either physical or chemical. *A chemical change is one in which a new substance is formed.* In a chemical change the substances involved lose their identity and change into other substances that have different properties. *A physical change is one in which a new substance is not formed and the substance at the end of the change has the same chemical properties as at the beginning of the process.* If we tear a piece of paper into small bits, we are carrying out a physical change; we still have paper at the end of the process, although in a finer state of division. If we set fire to paper, however, a chemical change takes place; new substances, that is, ashes and certain gaseous products of burning, are formed. In burning, the paper loses its identity, and new substances with different properties are formed. We may heat a platinum wire in a flame until it becomes red-hot, but as soon as it is removed, it cools, and we have unchanged platinum. The heating and cooling of platinum is a physical change. Now if we treat the platinum wire with an appropriate mixture of acids, it will dissolve, and a new substance will be formed. This change will be a chemical change.

In nature, matter is constantly undergoing both physical and chemical changes. The evaporation, condensation, and freezing of water; the erosion of the land by the waters of small streams and rivers; and the crumbling of rocks as a result of the wear of rainfall and the freezing and thawing of water in their pores are examples of the physical changes continually occurring in nature. The many changes taking place in plants and animals are interesting examples of the chemical changes in nature. Thousands of chemical changes are taking place in the human body at any instant of time. These changes are involved in the processes of the digestion and absorption of foods, the building up of tissues, the burning of foods in the body, the wear and destruction of tissues, respiration, and elimination of waste products. Many similar chemical changes are constantly taking place in plants, some resulting in the production of foods, which are stored in the seeds, roots, and fruits of the plant.

A chemical change will later be referred to as a chemical reaction. In chemistry the words "change" and "reaction" are used as synonyms, meaning essentially the same thing.

Equation

An equation is an abbreviated expression for a chemical change. It is a further application of the shorthand system of the chemist. Just as the symbol is an abbreviation for an element and the formula is an abbreviation for a compound, so the equation may be considered a shorthand method of indicating a complete chemical reaction in which symbols and formulas are used. For example, the reaction between zinc and hydrochloric acid may be represented briefly as follows:

$$Zn + 2HCl \rightarrow ZnCl_2 + H_2 + Energy$$
Zinc   Hydrochloric   Zinc   Hydrogen
**   acid   chloride**

This notation is a completed chemical equation. It shows that when zinc and hydrochloric acid are mixed, zinc chloride and hydrogen are produced, and energy is released. One atom of zinc reacts with 2 molecules of hydrochloric acid to form 1 molecule of zinc chloride and 1 molecule of hydrogen. The number of atoms of zinc, hydrogen, and chlorine on one side is exactly equal to the number of the same atoms on the other side. The sum of the masses of the zinc and hydrochloric acid on the left is equal to the sum of the masses of the zinc chloride and hydrogen on the right. Reading this long dissertation on the reaction of zinc with hydrochloric acid is unnecessary for the trained student, since the student can get the same thoughts from the equation, which explains the reaction with greater clearness and rapidity.

It is important to note that an equation always stands for some reaction that has previously been carried out and proved by experiment. We do not form equations with our imagination but on the basis of knowledge first obtained in the laboratory. We treat certain substances chemically, and from our observations we write out equations to represent the results as we find them. *The laboratory experiment is always the basis of a chemical equation.*

Types of chemical reaction

Practically all chemical reactions will come under one of four types: combination or synthesis, decomposition or analysis, double decomposition or double displacement, and single replacement or substitution.

Combination or synthesis. Two or more substances combine to form a single product.

$$C + O_2 \rightarrow CO_2\uparrow \qquad (1)$$
Carbon dioxide

This requires little explanation. Carbon and oxygen combine to form the compound carbon dioxide.

Decomposition or analysis. A single substance breaks down into two or more simpler substances.

$$2KClO_3 \xrightarrow{Heat} 2KCl + 3O_2\uparrow \qquad (2)$$

Potassium   **Potassium**   **Oxygen**
chlorate   **chloride**

In this reaction the force of chemical attraction between the atoms of potassium chlorate is overcome by heating to a high temperature, and the result is a decomposition, with the liberation of oxygen and the formation of potassium chloride.

Double decomposition or double displacement. This is similar to the action that takes place during a "change partners" call in a square dance.

$$AgNO_3 + HCl \rightarrow HNO_3 + AgCl\downarrow \qquad (3)$$

Silver   Hydrochloric   Nitric   Silver
nitrate   acid   acid   chloride

Single replacement or substitution. One element takes the place of another in a compound.

$$2HCl + Zn \rightarrow ZnCl_2 + H_2\uparrow \qquad (4)$$

Hydrochloric   Zinc   Zinc   Hydrogen
acid   chloride

$$Cl_2 + MgI_2 \rightarrow MgCl_2 + I_2\downarrow \qquad (5)$$

Chlorine   Magnesium   Magnesium   Iodine
              iodide   chloride

The arrows pointing downward denote an insoluble product that precipitates to the bottom of the reaction vessel. The arrows pointing upward denote a gaseous product that bubbles out of the reaction vessel. In either case, these products are no longer free to be an active part of the reaction.

Reversible reactions—equilibrium

In the previous discussion of types of chemical reactions, the reaction symbol ($\rightarrow$) was made to point to the right, indicating that only the products on the right are formed. This is correct, and the reaction goes to completion toward the right because one of the products is removed from the field of reaction. Thus in equation 3 the reaction goes to the right because the AgCl is highly insoluble and after its formation does not react appreciably with the HNO_3 to form the products on the left of the reaction symbol.

Again, in equation 4 the substances on the right of the arrow are the only products of the reaction because hydrogen is a gas and escapes from the presence of the $ZnCl_2$. It is satisfactory to write this equation with the arrow pointing to the right. By convention, reactions to the right are called *forward reactions*. Similarly, reactions proceeding to the left are called *reverse reactions*.

These reactions are examples of one-way reactions. Not all chemical reactions are like this. For example, in the following reaction

$$2NaCl + H_2SO_4 \rightarrow 2HCl + Na_2SO_4$$

it is indicated that the substances on the left of the arrow react with each other to produce the substances on the right. However, it is true that the substances on the right of the arrow also react to a certain extent with each other to give the substances on the left. To indicate the facts fully, it is necessary to write this equation with arrows pointing in both directions.

$$2NaCl + H_2SO_4 \rightleftarrows 2HCl + Na_2SO_4$$

This reaction is called a *reversible reaction*. After allowing adequate time, we find that the rate of the forward reaction and the rate of the reverse reaction become equal to each other. We then can say that a state of *equilibrium* exists. In the equilibrium state, two opposing changes are taking place simultaneously, at the same rate.

For those reactions that go one way, we write the reaction symbol pointing toward the products formed; for those reactions that are reversible, we write the reaction symbols pointing both ways.

Catalyst

A catalyst is a substance that enters into a chemical reaction altering the rate of that reaction but remaining unchanged when the reaction is ended. A catalyst that speeds up the rate of a reaction is called a *positive catalyst,* one that slows down the rate of a reaction is called a *negative catalyst.* Since it is not itself changed in the reaction, a catalyst is needed in only small amounts in any reaction.

Laws of conservation of mass and energy

When beginning the study of chemistry, the student may at first get the impression that matter can be destroyed. When we pour acid over a lump of marble, the marble disappears, and it may seem, because the marble has disappeared from view, that some matter has been destroyed. But this is not true. What happens when the marble disappears is that its particles are separated and rearranged to form new substances, some of which are gaseous, some of which are in solution, and none of which are visible. If all the new substances formed were collected and weighed, the total weight would be exactly equal to the weight of the original substances before they were mixed.

Let us refer to the diagrams in Fig. 1-4. A chemical reaction, triggered by a tiny current from a battery, causes the light from a flash bulb to appear. When the flash is over, the used flash bulb weighs the same as it did prior to the reaction that caused the light. There has been no loss in mass, only a rearrangement of the atoms making up the components of the chemicals responsible for generating the flash of light.

This experiment illustrates the law of conservation of mass: *in an ordinary chemical reaction, matter is neither created nor destroyed; the total mass of the reacting substances is equal to the total mass of the products formed.* In an ordinary chemical reaction in which there is a decomposition of matter with the formation of a gas, the matter is only broken up into smaller particles that may disappear from sight but are not destroyed; such widely dispersed, minute gas particles are not perceptible to the senses.

Another law of importance in this connection is the law of conservation of energy,

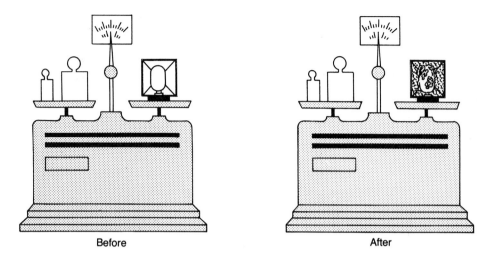

Fig. 1-4. Experiment demonstrating law of conservation of mass.

which states that *under ordinary conditions energy can neither be created nor destroyed.* The energy used in doing work is not lost but is transformed into other forms. The mechanical energy of running water may be transformed into electrical energy by the devices of the modern electric power plant, and the electrical energy may be used for running an elevator, lighting a city, or producing heat by means of an electric stove. The original energy of the falling water is successively transformed into several different forms of energy, which are finally dissipated and made nonutilizable after doing work but are not destroyed.

These laws apply to the conservation of mass and energy under the usual conditions that prevail around us. Under ordinary conditions the laws of the conservation of mass and energy are experimentally verifiable, and demonstrations of their validity are reproducible. However, they are not true under all circumstances. *Under extraordinary conditions, such as the explosion of an atomic bomb or with matter that is radioactive and undergoing disintegration, mass is converted into energy. The reverse is also true; that is, energy is convertible into matter.* Thus scientists now know that mass and energy are interconvertible in certain cases or under certain circumstances. The conditions of these interconversions are extraordinary, however. Since under ordinary conditions the laws of the conservation of mass and energy hold good, they are useful concepts.

Law of definite composition

It has been noted previously that one of the essential differences between a compound and a mixture is that the elements of a compound are always combined in definite proportions, whereas the mixture may have elements in any proportion. The fact that the elements in forming a compound always unite in the same manner and in the same ratios is the basis of a fundamental law, the law of definite composition. This law states that *a compound always contains the same elements united in the same manner and in the same proportions by mass.* Thus 2 parts by mass of hydrogen always combine with 16 parts by mass of oxygen to form 18 parts by mass of water. In the formation of water from hydrogen and oxygen, we cannot combine 2 parts of hydrogen with 10 parts of oxygen or 2 parts of hydrogen with 20 parts of oxygen or use any other combination than that of 2:16 by mass. Water always has the same ratio of hydrogen to oxygen under any conditions anywhere in the universe. The same law applies to all other compounds. We may change the proportions of the atoms in the molecule of a compound and produce a new compound, but for the same compound the proportions are always fixed and definite.

Combining masses

When elements combine with each other, a definite mass of one element always combines with a definite mass of another element. By chemical analysis the combining mass relationships between the elements may be determined, and it is important to have such information. To have combining mass values of the greatest usefulness, it is necessary to adopt some element as a standard and to set a definite amount of this element to be known as the standard for combining masses. This has been done. The element classically selected for this purpose is oxygen, and the amount of oxygen adopted as a standard for combining masses is 8 g. The combining mass of an element is the number of grams of that element that will combine with 8 g of oxygen or with an amount of another element that is chemically equivalent to 8 g of oxygen. The purpose of the combining masses is to provide established and definite quantitative values that will stand for the combining ability of every element.

If we want to know how many grams of one element will combine with a certain amount of another element, the answer can be obtained by a simple calculation from the combining masses of the elements concerned. The designation of 8 g of oxygen as the standard for combining masses is arbitrary. Other standard values might have been chosen, but the use of 8 g of oxygen as the standard has certain advantages. This value for oxygen fixes the combining mass of hydrogen, the lightest element, at 1.008; this is convenient because it avoids having values below 1 for any of the elements. The use of oxygen as a standard is important also because so many elements combine with oxygen, thereby affording a direct experimental method of determining the combining masses of these elements.

The combining mass relationships of the elements can be explained best by taking up a few examples. Let us consider the following data:

> In water (H_2O), 1.008 g of hydrogen combine with 8 g of oxygen.
> In chlorine monoxide (Cl_2O), 35.45 g of chlorine combine with 8 g of oxygen.
> In sodium oxide (Na_2O), 23 g of sodium combine with 8 g of oxygen.
> In calcium oxide (CaO), 20 g of calcium combine with 8 g of oxygen.
> In carbon dioxide (CO_2), 3 g of carbon combine with 8 g of oxygen.

The amounts of hydrogen, chlorine, sodium, calcium, and carbon that will combine with 8 g of oxygen, which by definition are the combining masses of these elements, are indicated. The importance of such figures is that they not only show the combining values of certain elements for oxygen but also indicate the amounts of these elements that will be used when they combine with each other or with the combining mass of any other

Table 1-2. Examples of compounds in which multiple proportions are shown

Compounds	Combining proportions by mass
Cupric oxide (CuO)	Cu:O = 63:16
Cuprous oxide (Cu_2O)	Cu:O = 126:16
Mercuric oxide (HgO)	Hg:O = 200:16
Mercurous oxide (Hg_2O)	Hg:O = 400:16
Nitrous oxide (N_2O)	N:O = 14: 8
Nitric oxide (NO)	N:O = 14:16
Nitrogen trioxide (N_2O_3)	N:O = 14:24
Nitrogen dioxide (NO_2)	N:O = 14:32
Nitrogen pentoxide (N_2O_5)	N:O = 14:40

element. For example, the amount of chlorine combining with 8 g of oxygen is 35.45 g; this amount of chlorine will also combine with 1.008 g of hydrogen, 23 g of sodium, 20 g of calcium, or 3 g of carbon.

Law of multiple proportions

Under one set of conditions two elements may unite to form one compound. Under other conditions the same two elements may unite to form an entirely different compound. When they do this, different proportions by mass of the elements entering into the formation of the compounds are involved. Examples are given in Table 1-2.

Inspection of the data in Table 1-2 shows an interesting relationship. Consider first the two oxides of copper. The difference in these two compounds is that the first one contains by mass 63 parts of copper to 16 parts of oxygen, and the second one contains by mass 126 parts of copper to 16 parts of oxygen. The ratio of the amount of copper in the first compound to the amount of copper in the second compound is 63:126, or 1:2. The change in the amounts of the varying element, copper, is thus represented by whole numbers.

A similar relationship is shown by the two oxides of mercury. In mercuric oxide the ratio by mass of mercury to oxygen is 200:16; in mercurous oxide the ratio by mass of mercury to oxygen is 400:16. In these compounds the variation in the amount of mercury present is 200:400, or 1:2, which again is a whole-number ratio.

In the group of compounds containing nitrogen and oxygen there are still more interesting examples of whole-number variations in one of the elements. In the column showing the amount of oxygen, it will be seen that the proportions of oxygen by mass in the second column to 14 parts by mass of nitrogen are, respectively, 8, 16, 24, 32, and 40 (or 1, 2, 3, 4, and 5). Here again we find a simple whole-number relationship in the amounts of the varying element (oxygen) in these five compounds.

With these data in mind the student should begin to understand the law of multiple proportions. This law states that *when two elements form more than one compound with each other, a simple whole-number relationship exists in the proportions of the element whose amounts are varied.*

The student will at once see that the law of multiple proportions is a very obvious and reasonable principle. If a variation in the elements of a compound occurs, this will not be in terms of one-half of an atom, one-fourth of an atom, or any other fraction of an atom; whatever change occurs must necessarily be in terms of whole atoms. Therefore, when proportions that exist between the elements of a compound are changed, the ratio between the amounts of the varying elements must be a ratio of whole numbers.

Questions for study

1. Define matter. Give ten examples of matter.
2. Define energy. In what different forms is energy manifested?
3. Distinguish between mass and weight.
4. What is an element? How many elements exist in the matter composing the earth?
5. What is a symbol? Of what value is it?
6. What is a compound? Distinguish between a compound and a mixture.
7. What is a formula? Illustrate by examples. What is the purpose of formulas?
8. What is an atom? What part does the atom play in chemical changes?
9. Define atomic mass.
10. What is a molecule? What is the relation of molecules to atoms?
11. Define molecular mass.
12. Calculate the molecular mass of the following compounds using the table of elements on the inside of the back cover: hydrochloric acid, HCl; sodium hydroxide, $NaOH$; potassium nitrate, KNO_3; calcium sulfate, $CaSO_4$; magnesium sulfate, $MgSO_4$.
13. Distinguish between physical and chemical changes.
14. What is an equation? What is its purpose?
15. What are the four types of chemical reaction? Explain each.
16. What is a reversible reaction?
17. What is meant by equilibrium in a chemical reaction?
18. Explain the laws of the conservation of mass and energy. Are there conditions under which these laws are not applicable? Under what conditions are these laws applicable?
19. Explain the law of definite composition.
20. Explain the law of multiple proportions.
21. Convert 5 feet to centimeters, 6 km to miles, 40° C to degrees Fahrenheit and to degrees Kelvin, and 1 quart to milliliters.
22. What is a catalyst?
23. What is meant by the term *combining mass?*

2 The atom, periodic classification, and chemical reactivity

THE ATOM

The atom is of great importance to the structure of matter and chemical reaction. Proportional to that importance has been the interest of scientists in trying to understand the structure of the atom. About 400 BC, the early Greek philosopher, Democritus, theorized that the atom was the smallest particle of matter. However, his theory could not be tested by scientific experimentation because the means for such experimentation were not available until much later in history. In 1802, his ideas were given scientific verification in the work of John Dalton. Dalton's theories were based on such facts as the laws of conservation of mass and energy, definite composition, and multiple proportions and the observation of combining mass ratios among the elements. Since simple, whole-number ratios held true when two elements combined to form more than one compound with each other, it was necessary to assume that in each mass of element combining, there must be tiny units of that element reacting, for example, on a 1:1, 1:2, 1:3, or 2:1 ratio with tiny units of the second element.

With substantiation by experimental data, Dalton published his now famous atomic theory, which can be summarized as follows:
1. All elements consist of small particles—atoms.
2. These atoms cannot be created, destroyed, or decomposed into simpler units.
3. All the atoms of one element are identical to one another in chemical and physical properties and differ from the atoms of all other elements in these same properties.
4. Atoms of the same or different elements can unite in definite proportions by mass to form molecules. The mass of these atoms combine in the proportion of small whole numbers.

During the following 100 years, Dalton's theory was the basis of many important experimental discoveries. With minor corrections, it is still valid today.

Subatomic particles

One of the most valid indications that the atom was not the ultimate particle of matter came from the study of radioactivity (see p. 137). Certain elements are naturally radioactive; that is, the atoms of that element disintegrate at a fixed rate with time, giving rise to elements with less atomic mass, energy, and assorted particles that are subatomic in size. Some of these particles, like the electron, had been known prior to the study of radioactivity.

Electron. The first evidence that a particle smaller than the atom existed came from the study of the flow of electricity. Electrical discharges such as those that appear in thunderstorms had intrigued the mind of man from early times. The phenomenon of static electricity had also been known for a long time, but the electron was studied in detail by using the flow of electricity that takes place when an electrical potential exists in a nearly evacuated glass tube. An electrode is placed in either end of the tube, and the gas in the tube (such as air, neon, or mercury vapor) is pumped from the tube. When the electrodes are attached to a source of electricity and the voltage is increased, an electric current runs from the negative electrode, *cathode,* to the positive electrode, *anode.* The remaining gas in the tube glows as streams of electrons pass through the near vacuum in the tube. Neon signs, fluorescent lights, and mercury-vapor lamps are commercial examples of this glowing tube. The stream of electrons is called a *cathode ray,* and the evacuated tube is called the *cathode-ray tube* (Fig. 2-1).

The cathode ray has been shown to bend toward the positive pole in an electric field suggesting that it is made of particles having a negative charge, since opposite charges attract. Determination of the mass and charge of the particles proved them to have a mass of 1/1838 of a hydrogen atom with a negative charge equal to 1.6×10^{-19} coulomb

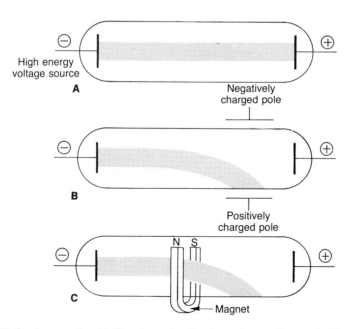

Fig. 2-1. A, Cathode-ray tube. **B,** Bending of cathode ray by an electric field. **C,** Bending of a cathode ray by a magnetic field.

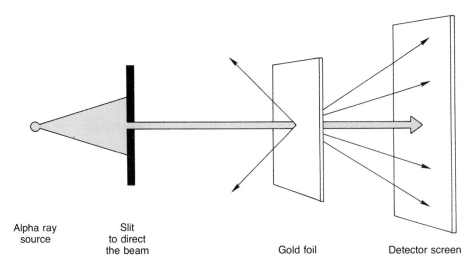

Fig. 2-2. Alpha particles directed toward gold foil.

(0.00000000000000000016 C). The mass of the electron is so small that, for all practical purposes, it is assigned a mass of 0 atomic mass units (amu). The electrical charge on the electron is the smallest charge that exists on a particle of matter and has been assigned the value of -1.

Proton. The presence of positively charged particles in the atom was studied extensively by Ernest Rutherford and his associates beginning in 1911 and ending with the identification of the proton in 1919. Working with naturally occurring radioactive elements, they found that such elements emitted rays of negatively charged particles. Such rays were called β (beta) rays and were proved to be streams of fast moving electrons. Besides these β or electron rays, radioactive sources gave off another ray called α (alpha) rays. These rays were made up of particles that proved to have the mass of four times the mass of the hydrogen atom but with a positive charge twice the magnitude of the charge on the electron. These particles were called alpha particles.

When gold foil was bombarded with alpha rays, it was found that most of the alpha particles passed through the foil but that a significant number of the particles were deflected or rebounded back from the foil. From this it was theorized that the atoms making up the gold foil had an outside area of low density but that the center of the atoms were small, dense, and positively charged. The alpha particles were deflected or repulsed whenever they approached a nucleus of a gold atom because like charges repel each other. The alpha particles passing through the less dense outer layers of the gold atoms went through unimpeded (Figs. 2-2 and 2-3). This experiment indicated that the outer areas of the atom make up the greater area of the atom. The area of the nucleus is very tiny in comparison. These conclusions were based on the fact that a greater number of alpha particles got through the gold foil without deflection; only an occasional alpha particle came near enough to a nucleus to be deflected, and even fewer made direct contact with a nucleus and were repelled in a backward path. Since the nucleus of the atom

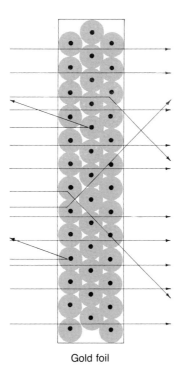

Gold foil

Fig. 2-3. Schematic description of the deflection of alpha particles by gold foil.

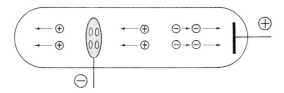

Fig. 2-4. The formation of positively charged particles in the cathode-ray tube.

was proved to be positively charged, the outer areas must contain the negatively charged electrons to comprise a particle of matter that is electrically neutral.

The use of the cathode-ray tube made the discovery of the proton possible. When a cathode penetrated with holes was used, a stream of charged particles appeared and passed through the holes in the cathode. The lightest of these particles was formed when the residual gas in the tube was hydrogen. The cathode ray stripped the negative charges from the hydrogen atoms, leaving the hydrogen nuclei (protons), which were attracted to the electrical pole having a charge opposite to their own, the cathode (Fig. 2-4). The proton was found to have a mass almost the same as the mass of the hydrogen atom, approximately 1 amu. Its charge was found to be equal to the charge of the electron but of the opposite sign. It was assigned the arbitrary charge of $+1$.

Atomic number. Concurrent with the work on the discovery of the proton, it was found that the anode of the cathode-ray tube emitted shortwave rays when struck by the

Table 2-1. Subatomic particles

Particle	Charge	Mass in amu (carbon-12 = 12.000)	Atomic location
Electron	−1	Actual 0.00055 Approximate 0	Outside nucleus
Proton	+1	Actual 1.00727 Approximate 1	In nucleus
Neutron	0	Actual 1.00866 Approximate 1	In nucleus

cathode ray. These shortwaves were called *x-rays* (see p. 143). Using a modified cathode-ray tube, H. G. J. Moseley discovered that different elements gave characteristic x-rays when these elements were used as anodes. He found that the length of the x-rays emitted decreased slightly as the atomic mass of the element used as the anode increased. From the data thus obtained, he was able to state that the number of positive charges in the nucleus of the atom increases from element to element by a single electronic unit. The positive charges in the nucleus are the result of the protons present in the nucleus. The number of positive charges in the nucleus of the atom of an element is equal to the number of protons in that nucleus. This number of positive charges is called the *atomic number of the element*. In the neutral atom, the atomic number also indicates the number of electrons in the outer areas of that atom, since the positive charges in the nucleus must be balanced by negative charges to form an electrically neutral atom.

Neutron. The masses of atoms were determined by use of the mass spectroscope. It was discovered that except for one form of hydrogen, the atoms of the respective elements all weighed more than could be accounted for by adding the masses of the protons and electrons necessary to fulfill the concept of the neutral atom. If the atomic number of an element such as iron is 26, this means that the nucleus of the iron atom contains 26 positively charged protons and that the outer areas of the atom contain 26 negatively charged electrons. This would account for a mass of slightly over 26 amu for the iron atom. In actual fact, the mass of the iron atom is 55.847, a fractional number more than twice the mass that can be accounted for by the mass of the electrons and protons. This data and research conducted by Rutherford indicated, as early as 1920, that the existence of at least one more subatomic particle with mass but no electrical charge could be suspected. In 1932, Chadwick discovered and described the neutron, a particle with approximately the mass of the proton but with no electrical charge. The atomic location of the neutron was fixed in the nucleus, since ample evidence existed showing that the mass of the atom was concentrated in its center. Table 2-1 summarizes the data on the subatomic particles.

ATOMIC STRUCTURE
Rules governing atomic structure

Rules regarding the structure of the atoms of the different elements may be summarized as follows:

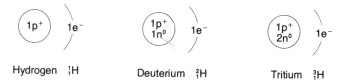

Fig. 2-5. Diagrams showing the difference in the three isotopes of element 1, hydrogen.

1. The number of protons in the nucleus of an atom is equal to the atomic number of that atom.
2. The number of neutrons in the nucleus of an atom is equal to the difference between the atomic mass and the atomic number.
3. The number of electrons in the shells of a neutral atom is equal to the number of protons in the nucleus.

Isotopes. The recognition of the neutron offered one explanation for the fact that atomic masses are fractional figures. The reason for these fractional values is twofold. One reason stems from the fact that the atomic masses are relative values obtained when the mass of a certain amount of one element is compared to the mass of the same amount of another element, the most common reference element being carbon-12. Because of this, the masses of the electron, the proton, and the neutron are really fractional numbers, since they are also based on a relative mass relationship with carbon-12. Adding fractional numbers usually results in a fractional sum. Another reason for fractional atomic masses is that the presence or absence of neutrons in an atom of an element will not alter the chemical properties of that atom but will make it weigh more or less than other atoms of the same element. The chemical reactivity of an atom is related to its atomic number, which in turn is determined by the number of protons present in the nucleus. All atoms of an element must have the same atomic number or their chemical reactivity would not be similar. However, all the atoms of an element do not have to be identical with respect to the number of neutrons found in the nucleus. Many elements have two or more species of *atoms containing different numbers of neutrons* (Fig. 2-5). *A species of an element having a different atomic mass than the other species of that same element is called an isotope of that element.* The atomic mass of an element is an average value of all of the masses of all of the isotopes of that element. Average values often are fractional figures. The atomic mass of an atom can be calculated with fair accuracy by adding the number of protons to the number of neutrons present in that atom. Because of their small weight, the electrons can be ignored without causing grave error in the calculation. *The average atomic mass of an element is called the atomic mass number.* Isotopes of an element also have atomic mass numbers (Table 2-2). The atomic masses of the three isotopes of carbon are shown to be

	Carbon-12	Carbon-13	Carbon-14
Electrons	6	6	6
Protons	6	6	6
Neutrons	6	7	8
Atomic mass number	12	13	14

26 Roe's principles of chemistry

Table 2-2. Isotopes of some elements

Element	Average atomic mass of isotopes in naturally occurring mixture	Atomic mass of isotopes
Magnesium	24.31	24, 25, 26
Chlorine	35.45	35, 37
Potassium	39.10	39, 41
Iron	55.84	54, 56
Silver	107.87	107, 109

Isotopes are identified by their atomic mass number and often by both their atomic number *and* their atomic mass number.

Importance of isotopes. One application of isotopes that is of great interest is their use in studying the changes that take place in living things. Isotopes may be introduced into compounds to serve as labels. Such compounds may then be fed to an animal or introduced into the tissues of a plant. These compounds, labeled by the presence of isotopes in their molecules, may then be traced by the investigator through an animal's body or in the tissues of a plant. The use of isotopes for labeling compounds has already resulted in many important scientific discoveries and offers great possibilities for the future.

From all of the study of the atom a newer concept of the atomic theory has been developed. Essentially the Daltonian atomic theory has been validated. The atom is the smallest particle of an element that enters into chemical reactions. It is also the smallest part of an element that still retains the chemical and physical properties of that element. It is now realized that the atom can be decomposed, but this decomposition does not take place during chemical change. Decomposition of the atom results in the loss of the chemical and physical properties that associated the atom with its original element. Atomic disintegration gives rise to energy. Energy is viewed as a form of matter, so the laws of conservation of mass and energy are still valid. Atoms of an element are all similar with the exception of mass; isotopes of an element exhibit different masses. Atoms of an element are different from the atoms of all other elements, again, with the exception of the masses of isotopic forms.

Energy levels or orbits. To this point, we have a generalized picture of the atom as having a dense, positively charged nucleus in which are found the protons and the neutrons surrounded by an area of less density in which the electrons are found. Clarifying the exact location of the electrons has been the goal of a great amount of experimental and mathematical endeavor.

The atom, periodic classification, and chemical reactivity

Using the evidence gained from the discrete lines appearing on the emission spectra of the elements, it has been determined that electrons exist in certain discrete states outside the nucleus of the atom. These states are reached by gaining or losing distinct amounts of energy and are called *energy levels or orbits*. For an electron to appear in a certain location around its nucleus, it must gain or lose the amount of energy necessary for it to move from one energy level to another. This gaining or losing of a certain amount of energy is called the *quantization* of the energy of the electron. To move farther away from the nucleus, the electron needs to gain energy; if it loses energy, it moves closer to the nucleus, until it reaches the closest area in which it can exist without being drawn into the nucleus itself. This final stage is called the *ground state* of the electron. Higher energy states farther from the nucleus are called *excited* states of the electron.

It is a well-known principle in electricity that unlike, charged particles attract each other. There is an electrical attraction between the electrons in the outer parts of the atom and the nucleus because the electrons have a negative charge and the nucleus contains protons, which are positively charged. The force of attraction that tends to pull the electron toward the nucleus is balanced by the opposing momentum of the moving electron. Thus the atom is kept stable. The electron does not fall into the nucleus because it has a momentum that pulls in a direction away from the nucleus; and unless highly energized, it does not leave its orbit and go off into space because it is held by the balancing force of attraction of the oppositely charged nucleus.

The Bohr atom

In 1913, Niels Bohr, a Danish physicist, proposed an atomic structure that featured electrons moving in greater and greater concentric circles around a central nucleus. Each successive circle or orbit has its fixed energy state and the maximum number of electrons that it can accommodate at any one time. Fig. 2-6 shows a drawing of the Bohr atom. The first orbit outside the nucleus is the lowest energy state at which an electron can exist,

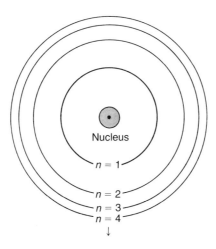

Fig. 2-6. The Bohr atom.

$n = 1$. Orbits farther out hold electrons with greater energy, such as $n = 2, n = 3$, and $n = 4$. An older system uses such letters as K, L, and M to indicate energy levels. The first energy level can accommodate only 2 electrons, the second, 8, the third, 18, the fourth, 32.

Atomic structure according to quantum theory

Bohr did his work utilizing the hydrogen atom with its single electron energized into various orbits. His model did not answer the problems posed by the multielectron atom. Schrödinger presented a mathematical equation describing the three dimensional space about the nucleus of an atom where electrons might be found. This equation has its limitations, especially when its solution involves a calculation for more than one electron. However, this theoretical description of the three dimensional atom does give an idea as to the possible position of the electrons around the nucleus. The application of quantum mechanics to the structure of the atom utilizes the concept of quantized energy levels for the electron. Describing a three dimensional model, it uses the three quantum numbers n, ℓ, and m, plus the spin of the electron, quantum number s.

In the quantum theory, as in the Bohr model, the principle energy levels are designated by $n = 1$, $n = 2$, and so on up to $n = 7$. Beginning with $n = 2$, the principle energy levels are subdivided into subenergy levels or suborbits. The $n = 1$ energy level has only one energy area, a sphere-shaped space surrounding the nucleus. The $n = 2$ energy level has two suborbits, the s and the p orbits. They are designated as the $2s$ and the $2p$ (pi) suborbits. With $n = 3$, another suborbit appears, the d; and third energy level electrons exist in the $3s$, $3p$, and $3d$ suborbital areas. When $n = 4$, four suborbital energy areas exist, the $4s$, $4p$, $4d$, and $4f$ (Fig. 2-7).

Each orbital can accommodate 2 electrons if their spins counterbalance each other. Evidence exists that the electrons spin on their axes roughly similar to the spinning of a top. To counterbalance each other, 1 electron spins in a clockwise ($s = +½$), and the other in a counterclockwise ($s = -½$), manner. Fig. 2-8 shows a schematic expression of the various energy levels, each square represents an orbital. In the outer reaches of space around the nucleus, the orbits of one energy level overlap the orbits of the next energy level. An example of this can be seen in the fact that $4s$ is at a lower energy level than $3d$. The $4s$ orbital usually accepts 2 electrons before $3d$ begins to accept electrons. The common pattern for filling the orbits is as follows, $1s$, $2s$, $2p$, $3s$, $3p$, $4s$, $3d$, $4p$, $5s$, $4d$, $5p$, $6s$, $4f$, $5d$, $6p$, $7s$, $5f$, $6d$, with the first named orbits filling first. Variations to this pattern exist in the larger atoms.

To properly visualize these spatial areas of probability for electron accommodation around the nucleus, consideration must be given to the work of W. Heisenberg, which shows that limits exist for the degree of precision with which a simultaneous calculation can be made for the position and the momentum of a body. The electron in the atom has continuous momentum, and it is realized that the exact location of the electron is impossible to ascertain at any one time. It is only possible to locate areas of space where an electron has a high probability of existing. All of these areas are calculated as though only the atom and the electron in question existed. The interaction of electron with electron or

The atom, periodic classification, and chemical reactivity 29

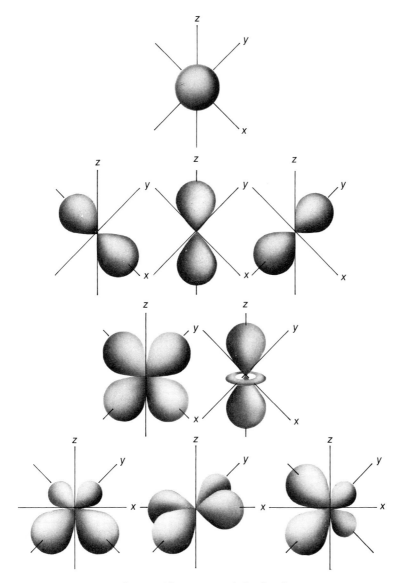

Fig. 2-7. The s, p, and d orbitals.

30   Roe's principles of chemistry

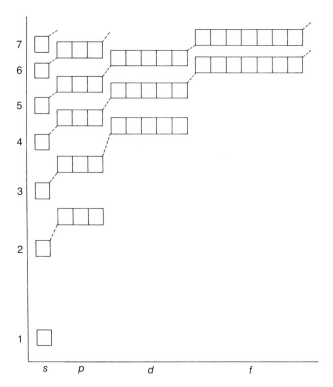

Fig. 2-8. Electron energy levels.

atom with atom no doubt alters these theoretical designations in space. As nearly as the shape of molecules can be determined, they tend to reinforce this model of the atom.

The nucleus of the atom is still under study. There is evidence that various kinds of particles exist in the nucleus. It is sufficient to visualize the neutron and the proton, held together by some force not completely understood. It has been suggested that each kind of particle occupies close-packed concentric spheres inside the nucleus, with a sphere of protons alternating with a sphere of neutrons.

The number of electrons in an orbital can be indicated by use of a superscript placed above the orbital in question. For example, an atom having 7 electrons would have 2 electrons in the 1s orbital, 2 electrons in the 2s orbital, and 3 electrons in the 2p orbitals. The 1s orbital would be indicated in the following manner:

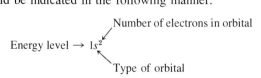

The total electron distribution of the atom with 7 electrons would be $1s^2$, $2s^2$, $2p^3$.

Table 2-3 gives the electron distribution, atomic number, atomic mass, protons, and neutrons of the ten smallest atoms.

The atom, periodic classification, and chemical reactivity 31

Table 2-3. Data showing structure of group of the simplest elements

			Structure				
			Nucleus		Outer portion		
					Number of electrons in shells		
Element	Atomic number	Atomic mass	Number of protons	Number of neutrons	1s	2s	2p
H	1	1.0079	1		1		
He	2	4.0026	2	2	2		
Li	3	6.9390	3	4	2	1	
Be	4	9.0122	4	5	2	2	
B	5	10.8110	5	6	2	2	1
C	6	12.0111	6	6	2	2	2
N	7	14.0067	7	7	2	2	3
O	8	15.9994	8	8	2	2	4
F	9	18.9984	9	10	2	2	5
Ne	10	20.1830	10	10	2	2	6

Using the information developed in Table 2-3, it can be seen that the atomic number of hydrogen is 1 and that the atomic mass of this element is essentially the same (1.0079). Hydrogen contains a single proton in its nucleus and 1 electron in its only orbit. The structure of hydrogen is shown in Fig. 2-9.*

Helium in its most abundant isotope has an atomic mass of approximately 4 and an atomic number of 2. The number of protons in the nucleus of helium is the same as its atomic number, 2. The number of neutrons in the nucleus is equal to the difference between the atomic weight and the atomic number, 2. The number of electrons is the same as the number of protons, 2. Thus helium has an atom containing 2 protons and 2 neutrons in its nucleus and 2 electrons in a single orbital, the first orbit or first main energy level, around the nucleus. In Fig. 2-9 the structure of helium is shown. Helium is an element that is chemically inert. Therefore a structure such as helium, in which there are 2 electrons in the first energy level, is considered to be a very stable, or inert, atomic configuration.

The atomic number of lithium is 3. Its nucleus therefore contains 3 protons. Lithium has an atomic mass of approximately 7; hence, the number of neutrons in its nucleus is 7 − 3, or 4. Since the atomic number or number of protons is 3, the number of electrons is 3. Lithium is an element whose atom consists of a nucleus in which there are 3 protons and 4 neutrons and an outer portion in which there are 2 electrons in the 1s orbital and 1 electron in the 2s orbital (Fig. 2-9).

The element beryllium has 4 protons, 4 electrons, and 5 neutrons, whereas boron has

---

*The drawings illustrating the structure of the atoms of elements in this book are merely diagrammatic representations of the constituents present and their approximate location. The size of the nucleus is greatly exaggerated to enable the artist to indicate the number of protons and neutrons present. The electrons are placed in arcs or squares to indicate the areas of space in which they move. The electrons are at much greater distances from the nucleus than the drawings suggest. The whole structure is spherical in shape and not a flat disk as the two-dimensional drawings might indicate.

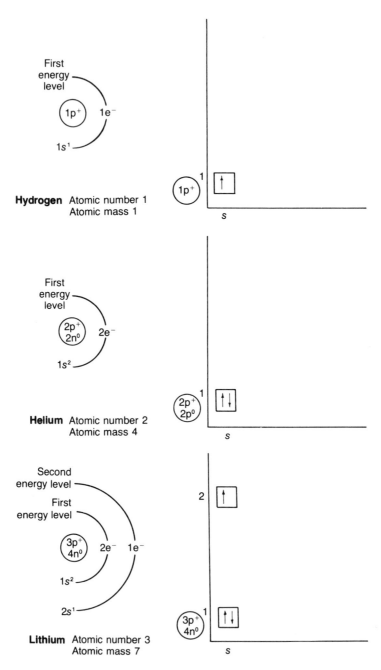

Fig. 2-9. Diagrams showing structures of the atoms of hydrogen, helium, and lithium. p^+ represents protons, n^0 represents neutrons, and the symbol e^- indicates electrons. Arrows show the opposite spins of the electrons.

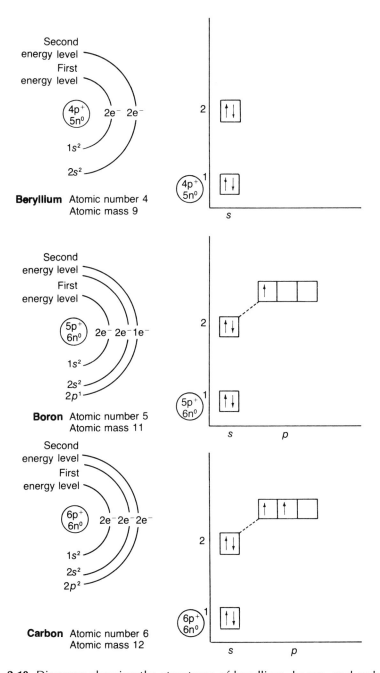

Fig. 2-10. Diagrams showing the structures of beryllium, boron, and carbon.

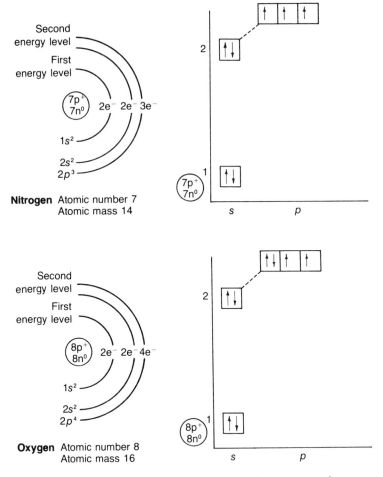

Fig. 2-11. Diagrams showing the structures of nitrogen and oxygen.

5 protons, 5 electrons, and 6 neutrons. Carbon exists with 6 protons, 6 electrons, and 6 neutrons (Fig. 2-10).

Nitrogen has 7 electrons, 7 protons, and 7 neutrons, giving it a mass of approximately 14 amu. Its electron distribution can be indicated by the following: $1s^2$, $2s^2$, $2p^3$ (Fig. 2-11).

Oxygen has an atomic number of 8 and an atomic mass of 16. The nucleus of the oxygen atom contains 8 protons and 8 neutrons, and there are 8 electrons in the orbits with the following distribution, $1s^2$, $2s^2$, $2p^4$. Fluorine has 9 protons, 10 neutrons and 9 electrons. Fluorine has an electron distribution of $1s^2$, $2s^2$, $2p^5$.

Neon has 10 protons and 10 neutrons in its nucleus. Having 10 electrons in the following electron distribution, $1s^2$, $2s^2$, $2p^6$, it displays an *outer pi orbit filled with electrons. This type of electron distribution is typical of a chemically inert element* (Fig. 2-12).

Uranium has an atomic number of 92 and an atomic mass of 238. It is the heaviest of

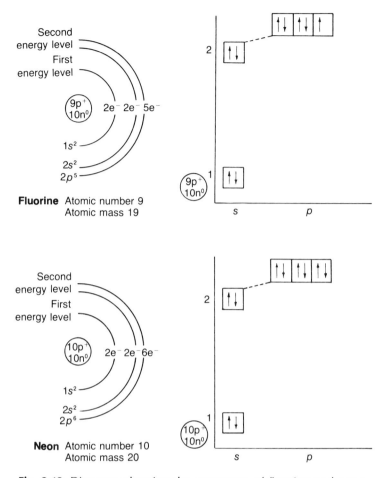

Fig. 2-12. Diagrams showing the structures of fluorine and neon.

the naturally occurring elements. In the neutral atom, there are 92 protons in the nucleus, electrically balanced by 92 electrons in the orbits. Subtracting 92 from 238 gives the number of neutrons, 146. In this manner, given the atomic number and the atomic mass, one can arrive at the number of protons, neutrons, and electrons in any atom.

The outer electrons of the atom determine the chemical reactivity of the elements. These electrons are called the *valence electrons*. More than one element has atoms with similar configurations in the outer or valence electrons. Lithium and sodium are examples of this. Lithium has the electron configuration of $1s^2$, $2s^1$, with 2 electrons in the first orbit and 1 electron in the 2s orbital. Sodium has the electron configuration of $1s^2$, $2s^2$, $2p^6$, $3s^1$, ending up with 1 electron in the 3s orbital. They are both similar in the outer electron configuration, $2s^1$ and $3s^1$. A single electron is in the outer orbit of both atoms. This similarity of outer electron configuration is the reason that there are also similarities among groups of elements with respect to their chemical properties.

PERIODIC CLASSIFICATION OF THE ELEMENTS

This relationship of properties within groups of the elements was first clearly announced in 1869, independently, by Mendeleev in Russia and by Meyer in Germany. The periodic classifications of the elements proposed by those authors have been modified since their appearance, as a result of an increased knowledge of the elements.

The great advantage of such an arrangement as the periodic table is that it places the

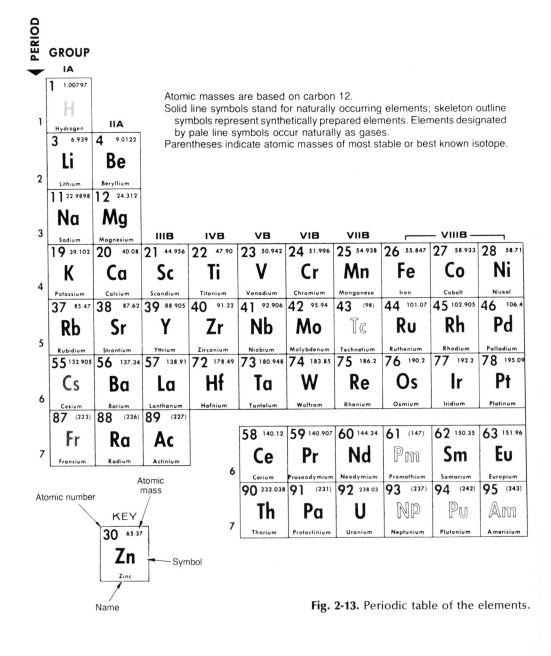

Fig. 2-13. Periodic table of the elements.

elements in groups that have many similarities with respect to chemical and physical properties. These related groups are the ones in the vertical columns formed by drawing lines between each vertical row of elements. The periodic table is thus an arrangement of all the elements in related groups or families. This arrangement shows a striking relationship of each of the elements to the place it occupies in the table (Fig. 2-13). In fact, the periodic table is so significant an arrangement that it is now recognized that each place in it defines the properties of the element that belongs to that place.

		IIIA	IVA	VA	VIA	VIIA	NOBLE GASES
							2 4.0026 He Helium
		5 10.811 B Boron	6 12.01115 C Carbon	7 14.0067 N Nitrogen	8 15.9994 O Oxygen	9 18.9984 F Fluorine	10 20.183 Ne Neon
IB	IIB	13 26.9815 Al Aluminum	14 28.086 Si Silicon	15 30.9738 P Phosphorus	16 32.064 S Sulfur	17 35.453 Cl Chlorine	18 39.948 Ar Argon
29 63.54 Cu Copper	30 65.37 Zn Zinc	31 69.72 Ga Gallium	32 72.59 Ge Germanium	33 74.922 As Arsenic	34 78.96 Se Selenium	35 79.909 Br Bromine	36 83.80 Kr Krypton
47 107.870 Ag Silver	48 112.40 Cd Cadmium	49 114.82 In Indium	50 118.69 Sn Tin	51 121.75 Sb Antimony	52 127.60 Te Tellurium	53 126.904 I Iodine	54 131.30 Xe Xenon
79 196.967 Au Gold	80 200.59 Hg Mercury	81 204.37 Tl Thallium	82 207.19 Pb Lead	83 208.980 Bi Bismuth	84 (210) Po Polonium	85 (210) At Astatine	86 (222) Rn Radon

64 157.25 Gd Gadolinium	65 158.924 Tb Terbium	66 162.50 Dy Dysprosium	67 164.930 Ho Holmium	68 167.26 Er Erbium	69 168.934 Tm Thulium	70 173.04 Yb Ytterbium	71 174.97 Lu Lutetium
96 (247) Cm Curium	97 (247) Bk Berkelium	98 (249) Cf Californium	99 (254) Es Einsteinium	100 (253) Fm Fermium	101 (256) Md Mendelevium	102 (254) No Nobelium	103 (257) Lw Lawrencium

Atomic numbers

In the periodic arrangement, hydrogen, the lightest element, is element No. 1, helium is element No. 2, lithium is element No. 3, and so on up through the heaviest elements. *Each element is given a serial number corresponding to the place in the periodic table that it should occupy according to its properties. This serial number given to each element is the same as the atomic number.*

The study of x-rays mentioned previously showed that the atomic number actually represents a fundamental quantity in the atom. It was found that when electric discharges in a vacuum tube are directed on different solid elements as targets, x-rays are emitted whose wavelengths vary according to the atomic numbers of the elements. The behavior of the elements with respect to x-rays was shown to vary according to an atomic number arrangement. The periodic law is stated as follows: *The ordinary physical and chemical properties of the elements are periodic functions of their atomic numbers.*

Natural families

Grouping the elements in periods places them in an arrangement in which the vertical columns are natural families. In the first column (Group IA) on the left side of Fig. 2-13 we find (besides hydrogen) the elements of the alkali family—lithium, sodium, potassium, rubidium, cesium, and francium; these elements have similar characteristics and form a natural family. In Group VIIA we find the halogen family—fluorine, chlorine, bromine, and iodine. Similarly, in each of the vertical columns of the periodic table we find families of elements related in properties. This division of the elements into natural families is of great value in becoming familiar with the characteristics of the elements.

Discovery of new elements

The periodic table has pointed the way to the discovery of new elements. When Mendeleev and Meyer first arranged the elements in a periodic table there were a number of gaps or vacant spaces in the table. Mendeleev rightly claimed that these gaps represented places for undiscovered elements and actually predicted the discovery of three new elements, now called scandium, gallium, and germanium. Furthermore, he predicted with amazing accuracy the properties of these three elements from his knowledge of the periodic table. Mendeleev's announcement was a stimulus for research that eventually brought about the discovery of unknown elements. Now all the places in the periodic table left vacant by Mendeleev have been filled.

CHEMICAL REACTIVITY

Chemical reactions consist of changes involving either transfer of electrons from one atom to another atom or sharing of electrons between atoms. An atom entering into a chemical reaction either takes from, shares with, or gives to another atom 1 or more electrons. The chemical activity of an element is thus determined by the ability of its atoms to give up, share, or gain electrons. This is dependent on the number of electrons in the outer orbit or orbits of an atom and the distance of these electrons from the nucleus. *There is no change in the nucleus of the atom during ordinary chemical reactions.*

The atom, periodic classification, and chemical reactivity 39

Table 2-4. Electron structure of the noble gases

Element	Atomic number	Number of electrons in orbits						
		1	2	3	4	5	6	7
Helium	2	2						
Neon	10	2	8					
Argon	18	2	8	8				
Krypton	36	2	8	18	8			
Xenon	54	2	8	18	18	8		
Radon	86	2	8	18	32	18	8	

What happens when a chemical reaction takes place has been outlined briefly. It now becomes important to consider why these changes occur. *Chemical reactions are a result of the tendency of matter to reach a more stable state under the existing conditions.*

An *energy change* is involved in all chemical reactions. Energy may be absorbed when a chemical reaction occurs or liberated as a result of the reaction. When energy in the form of heat is absorbed, the reaction is characterized as *endothermic,* and when energy in the form of heat is given off, the reaction is called *exothermic.*

The fundamental change in a chemical reaction is a filling of the outer orbitals of atoms that contain unpaired electrons. An orbital of an atom is "satisfied," and therefore stable, when it contains 2 paired electrons.

Noble gases—relation to chemical theory

The group of elements known as the noble gases have had a special significance in the development of chemical theory. The outer electron orbit in each of these elements, except helium, contains 8 electrons (see Table 2-4). Since these elements do not react with each other and for a long time no one was able to make them react with other elements, they were called the "inert gases." Except for helium, which is inert with the filling of the $1s$ orbital, the filling of an outer p orbit results in chemical inertness. Compare the atomic structure of neon versus that of argon in Fig. 2-14. On account of this behavior, chemists were led to believe that the 8-electron outer orbit (s^2, p^6) is a completely stable configuration, which the atoms of all other elements assume by undergoing chemical reactions.

In 1962, chemists in several laboratories prepared compounds by reacting one of these elements, xenon (Xe), with fluorine (F) to make such compounds as xenon difluoride (XeF_2), xenon tetrafluoride (XeF_4), and xenon hexafluoride (XeF_6). In making these compounds, a relatively simple procedure was used: heating a mixture of xenon and fluorine gases in a closed vessel at 400° C. This process resulted in energizing a $5p$ electron into the $6s$ orbital. The valence electron was now that single electron in the $6s$ orbital not the 6 electrons in the p orbit as had previously existed. An electron is easily lost from an s orbital, and xenon could react with fluorine. As more and more $5p$ electrons were energized, the xenon was able to react with more and more fluorine atoms. Fluorides of

40 Roe's principles of chemistry

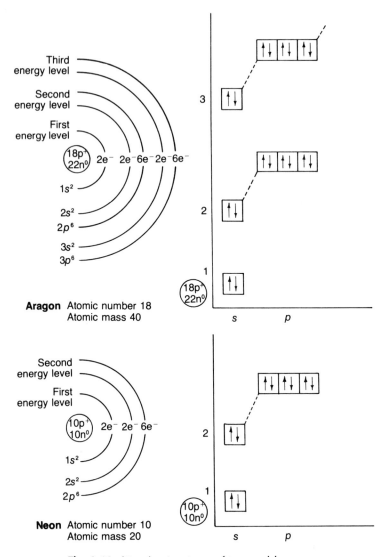

Fig. 2-14. Atomic structure of two noble gases.

radon and krypton have also been prepared. The fluorides of krypton are extremely unstable, however.

The preparation of compounds from xenon and fluorine and, later, the reacting of krypton or radon with fluorine completely changed the dogma of the chemical inertness of the noble gases. But it did not alter the importance of the chemical theory that had been built up by noting the high degree of stability and unreactivity of these gases. The principle that chemical reactions are changes that make the outer electron orbits of atoms contain 8 electrons, therefore conferring stability on the resulting compounds, is applicable in many chemical reactions. In most chemical reactions there are formed

The atom, periodic classification, and chemical reactivity 41

compounds whose atoms have outer electron shells containing 8 electrons arranged in 4 orbitals (one *s* and three pi orbitals), each orbital having a satisfied electron pair. This outer structure is called the *octet arrangement*.

Chemical changes involving several of the simplest elements

Hydrogen. The hydrogen atom contains 1 proton in its nucleus and a single electron in an orbital near the nucleus (Fig. 2-9). This is not a stable arrangement; consequently, atoms of hydrogen have a strong tendency to combine with each other or with the atoms of other elements. Hydrogen atoms in the presence of each other form stable molecules consisting of 2 atoms; a stable combination is formed by 2 atoms sharing a pair of electrons. Hydrogen may also form stable combinations with other elements by giving up, taking in, or sharing a single electron. In this manner hydrogen helps to fill unsatisfied orbitals in the outer energy levels of atoms of other elements, making the other elements stable; at the same time it is stabilized itself by sharing an electron of the other atom.

Lithium. Lithium has 2 electrons on its first energy level and 1 electron on its second energy level (Fig. 2-9). To become stable, lithium may lose 1 electron from its outer orbit and have an arrangement like helium. The only other possibility for lithium to assume a stable form is to fill out its second energy level by taking on 7 electrons. This is obviously very difficult and would create an unbalanced situation, since there are only three positive charges in the nucleus. Hence it does not occur. Lithium enters into chemical reactions when it gives up the 1 electron of its second energy level to some element that needs an electron to make it stable.

Oxygen. Oxygen has 6 electrons on its outermost energy level. To assume a stable outer configuration, oxygen might lose the 6 electrons of its outer orbit and have an outer arrangement like that of helium or add 2 electrons to its outer orbit and have an outer shell containing 8 electrons like that of neon. The latter process is the one that usually occurs. There are 2 unpaired electrons in unsatisfied orbitals in the oxygen atom (Fig. 2-11). The addition of 1 electron to each of its unsatisfied orbitals forms a stable outer shell of 8 electrons. Oxygen can also share electrons.

Sodium. Sodium has 2 electrons in its first orbit, 8 electrons in its second orbit, and 1 electron in its third orbit. In chemical reactions this element assumes a more stable form by losing the 1 electron of its outer orbit, resulting in an outer shell (the number 2 orbit) of 8 electrons.

Ions

From what has been said it is obvious that all chemical reactions are divided into two types. These involve either (1) a transfer of electrons or (2) a sharing of electrons. *An ion is an atom or group of atoms that has gained or lost electrons and formed a negatively or a positively charged particle. Ions are different from atoms.* For example, consider sodium chloride, which exists as ions in the solid state.

$$Na^+ - Cl^-$$

The sodium ion has properties that are quite different from those of the sodium atom.

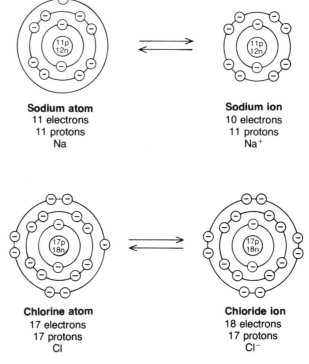

Fig. 2-15. Diagram showing differences between the atom and the ion of sodium and chlorine.

Sodium atoms decompose water, liberating hydrogen; sodium ions exist in water and do not attack it. Chlorine in the atomic state is a greenish gas with a poisonous action on living matter, and when dissolved in water, it is an excellent disinfectant; the chloride ion, however, is a harmless particle in solution and is in fact an essential constituent of the fluids that bathe living tissues.

Such differences in behavior are quite marked and therefore must be due to fundamental differences in the nature of the ion and the atom. To understand the differences in the behavior of ions and atoms, one must consider the electron structure of these two particles of matter. The differences in the structure of the atoms and ions of sodium and chlorine are shown in Fig. 2-15. The nuclei of the sodium atom and the sodium ion are the same. The differences in these two particles are to be found in their electrons. The sodium atom contains 11 electrons; the sodium ion contains 10 electrons. The sodium ion and the sodium atom are therefore different substances structurally; hence, one would expect them to manifest different properties.

The chlorine atom contains 17 electrons. The negatively charged chloride ion contains 18 electrons. Here again we find a difference in structure. This difference in structure is a fundamental one and accounts for the difference in properties of the chlorine atom and the chloride ion.

The diagrams of Fig. 2-15 also show why the sodium ion is positive and the chloride ion is negative. The neutral sodium atom contains 11 protons in its nucleus. This number of positive particles is balanced by the 11 electrons in its orbits. (Verify these figures by counting the protons and electrons in the diagrams.) This balance of protons or positive charges by electrons or negative charges accounts for the neutrality of the sodium atom. The sodium ion contains 11 protons and 10 electrons. The difference between the number of protons and electrons in the sodium ion is 1 and is in favor of the protons. Thus there is 1 unbalanced proton in the nucleus of the sodium ion, which gives it a positive charge of 1.

The neutral chlorine atom contains 17 protons in its nucleus and 17 electrons in its orbits. (Again, verify these figures by using the diagrams in Fig. 2-15.) Here again we find a balanced condition. The number of protons in the neutral chlorine atom is equal to the number of its electrons; hence, the chlorine atom is neutral. The negative chloride ion contains 17 protons in its nucleus and 18 electrons in its orbits. The difference in the number of charges is one and is in favor of the electrons; hence, the chloride ion has a negative charge of 1.

Radical. A radical is a group of *atoms held together by covalent bonds* and having a positive or negative charge; it is therefore able to undergo further chemical reaction as a unit. The following are examples of radicals:

SO_4^{-2} group in $ZnSO_4$ (zinc sulfate)
NO_3^{-1} group in KNO_3 (potassium nitrate)
OH^{-1} group in $NaOH$ (sodium hydroxide)
NH_4^+ group in NH_4OH (ammonium hydroxide)

METALS, NONMETALS, AND METALLOIDS

In chemical reactions, *metallic elements* generally lose electrons and are often called *electron donors*. When a metal loses electrons, it becomes a positively charged ion. The majority of elements are electron donors and thus are considered to be metals. They are found on the left side of the periodic table (Fig. 2-13).

The *nonmetals* are usually *electron acceptors*. They gain electrons and then form negative ions. These elements are found on the right side of the table. The most reactive metals are found in the lower left portion of the table, and the most active nonmetals, in the upper right.

Finally, those elements along the borderline between the metal and nonmetal elements (the heavy steplike line in the table) have some of the properties of both and are often referred to as *metalloids*.

VALENCE

The word "valence" comes from a Latin word meaning strength. In chemistry it signifies the combining strength or combining requirement of an element. More specifically, *valence is a number that tells how many atoms of one element are required to combine with an atom or atoms of another element when a compound containing both* elements is formed. For each of the elements there is a number that represents the combining requirement of that element when uniting with other elements, and this number is known as the valence of the element.

Elemental

A considerable number of elements have a valence of 1. Among these elements is hydrogen. A single atom of hydrogen will require only 1 atom of another element with a valence of 1 to form a compound. Reasoning from this, we find that in any compound in which hydrogen is combined with another element, if only 1 atom of hydrogen is required to unite with 1 atom of the second element, the valence of the second element is 1. To illustrate, consider the compound HCl. It is known that the valence of hydrogen is 1. Therefore, since only 1 atom of hydrogen is found combined with 1 atom of chlorine, the valence of chlorine is 1. *Since hydrogen has a valence of 1 and combines with many other elements, it has been found desirable to use this element as a reference standard of valence.*

Oxygen is a good example of an element with a valence of 2. When oxygen combines with hydrogen, it forms water (H_2O). In endeavoring to establish the valence of oxygen, the student needs only to observe that in water 1 atom of oxygen is combined with 2 atoms of hydrogen. Since the valence of hydrogen is 1, it follows that the valence of oxygen must be 2, because oxygen requires 2 hydrogen atoms to satisfy its combining requirement. To carry the analysis further, What is the valence of mercury in the compound HgO? We know from what has just been stated that oxygen has a valence of 2. Therefore the valence of mercury is 2, since 1 atom of oxygen is combined with 1 atom of mercury.

When the element nitrogen combines with hydrogen, it forms the compound ammonia (NH_3). The student will readily observe from the formula for ammonia that, in this case, the valence of nitrogen is 3 because it combines with 3 atoms of hydrogen.

Carbon is an element whose valence is 4. In entering into chemical combination with hydrogen, 1 atom of carbon unites with 4 atoms of hydrogen to form the compound methane (CH_4). This can be seen more easily if an equation is written showing both of the elements with a dash or "bond" to represent each valence.

$$-\underset{|}{\overset{|}{C}}- \quad + \quad 4H- \quad \rightarrow \quad H-\underset{\underset{H}{|}}{\overset{\overset{H}{|}}{C}}-H$$

Carbon   Hydrogen   Methane

Again, let us use carbon for an exercise in the study of valence. What should be the formula of a compound containing carbon and oxygen? Since carbon has a valence of 4 and oxygen a valence of 2, 2 atoms of oxygen will be required to unite with 1 atom of carbon, forming the compound carbon dioxide (CO_2). The equation for this chemical reaction is as follows:

$$-\overset{|}{\underset{|}{C}}- \ + \ 2-O- \ \rightarrow \ O=C=O$$

Carbon   Oxygen   Carbon dioxide

Now let us undertake a slightly more difficult application of the principles of valence. What is the formula for a compound of aluminum (Al) and oxygen (O)? Here the student will have to examine the formulas of some of the simpler known compounds of these elements. Looking for simple compounds of aluminum, the student will come upon the substance aluminum chloride ($AlCl_3$). It is at once obvious that the valence of Al is 3, since Al is combined with 3 chlorine atoms and, as noted, chlorine has a valence of 1 because it has been found to combine with hydrogen in a 1:1 ratio (HCl). In a similar manner it was ascertained that oxygen has a valence of 2 (see discussion of oxygen in H_2O). Al and O cannot be put together in a 1:1 ratio since their valences, 3 and 2, respectively, do not balance each other. Trying other combinations, we soon hit on the formula Al_2O_3, which is correct; but let us prove its correctness by the following deductions:

3 (the valence of Al) × 2 (atoms of Al) = 6 (total valences of 2 Al)
2 (the valence of O) × 3 (atoms of O) = 6 (total valences of 3 O)

The formula Al_2O_3 is thus seen to be correct since the total valences of the two elements of the molecule are the same (6) and since each element satisfies the combining requirement of the other element.

Valences of 5, 6, 7, and 8 are exhibited by other elements. But these need not be discussed here since the principles involved are the same as those just mentioned.

Radical

The principles of valence outlined for individual elements apply also to radicals. Take, for example, these three compounds: HNO_3 (nitric acid), Na_2SO_4 (sodium sulfate), and K_3PO_4 (potassium phosphate). In the first compound the valence of the NO_3 radical is 1 because it needs 1 hydrogen electron to satisfy its orbital requirement. Similarly, in the compound Na_2SO_4, the valence of the SO_4 radical is 2 because 2 atoms of the element sodium, of which the valence is 1, enter into combination with this radical. In K_3PO_4 the valence of the PO_4 radical is seen to be 3 since this radical is combined with 3 atoms of potassium, a monovalent element.

Some elements have more than one valence. Mercury, for example, is an element having two valences. Under certain conditions the mercury atom assumes a valence of 2, whereas under other conditions the mercury atom has a valence of 1. Mercury may therefore form two compounds with chlorine. Exhibiting a valence of 2, mercury forms the compound mercuric chloride ($HgCl_2$), and manifesting a valence of 1, mercury forms the compound mercurous chloride (HgCl).

The student need not let variations in valence be disturbing. The conditions determining valence lie in the atomic orbital configuration and are easily understood when a more advanced knowledge of chemistry has been obtained. Whenever an element mani-

Table 2-5. Valence and its application to formula writing

Elements	Valence	Compounds	Graphic representation
Hydrogen	1	HCl	H—Cl
Chlorine	1	Hydrochloric acid	
Hydrogen	1	H_2O	H—O—H
Oxygen	2	Water	
Hydrogen	1	NH_3	H—N(—H)(—H)
Nitrogen	3	Ammonia	
Hydrogen	1	CH_4	H—C(—H)(—H)—H
Carbon	4	Methane	
Mercury	1	HgCl	Hg—Cl
Chlorine	1	Mercurous chloride	
Mercury	2	$HgCl_2$	Cl—Hg—Cl
Chlorine	1	Mercuric chloride	
Oxygen	2		
		HgO	Hg=O
		Mercuric oxide	
		$FeCl_3$	Fe(—Cl)(—Cl)(—Cl)
		Ferric chloride	
Iron	3		
Chlorine	1		
Oxygen	2	Fe_2O_3	Fe—O—Fe (with O above and below)
		Ferric oxide	

fests another valence, it is always due to a change in the chemical conditions under which that element is placed. *For any given set of conditions the valence of an element is constant, and when we know these conditions, we can apply our knowledge to writing formulas for compounds with absolute certainty.*

From the preceding discussion it is evident that the practical value of a knowledge of valence is that it enables one to write the formulas of compounds and work chemical equations. The application of a knowledge of a valence in writing formulas for compounds is shown in Table 2-5.

Mechanisms underlying valence

Valence has previously been defined as the combining requirement of an element entering into a chemical reaction. We shall now consider the mechanisms underlying this characteristic of the elements. Valence is a property determined by the electrons in the outer portion of the atom of an element, the valence electrons. It was shown that in a chemical reaction there is a shifting of electrons; electrons are either transferred from the atoms of one element to the atoms of another element or are shared by 2 atoms.

BONDING

In the formation of a compound, the transfer or sharing of electrons establishes a point of chemical union designated as a chemical bond. *There are two types of chemical bonds: (1) electrovalent or ionic, which is produced by a transfer of electrons, and (2) covalent, in which 2 electrons are shared.* In the covalent bond, 1 electron is donated by each of the reacting atoms. There is a modification of covalence, however, called coordinate covalence; in the coordinate covalent bond both the shared electrons are contributed by the same atom. Ammonium ion, a radical, is a coordinate covalent molecule. In the fourth nitrogen to hydrogen bond, both binding electrons come from the nitrogen atom.

$$\left[\begin{array}{c} H \\ H:N:H \\ H \end{array} \right]^{+}$$

The radical is placed in brackets to indicate the fact that the positive charge is diffused throughout the whole molecule making up the ion and is not located in any one place.

Formation of electrovalent bonds

It is the practice to designate valence by placing a dot near the symbol of the element to represent a valence electron. Using this system, we will discuss the following elements:

$$Li\cdot \qquad \cdot Be\cdot \qquad \cdot \overset{\cdot}{B}\cdot$$

     Lithium   Beryllium   Boron

Lithium has 2 electrons in its $1s$ orbital and 1 in its $2s$ orbital. The latter is a valence electron. When lithium reacts chemically, its 1 valence electron unites with a single electron in the outer orbit of an atom of the element with which it reacts, thereby forming a stable pair of electrons and completing the orbit of the other reacting atom. The reaction of lithium with fluorine is as follows:

$$Li\cdot \;+\; \cdot \ddot{\underset{\cdot\cdot}{F}}: \;\rightarrow\; Li^{+} \;:\ddot{\underset{\cdot\cdot}{F}}:^{-}$$

   Lithium   Fluorine   Lithium
                                fluoride

In this reaction, lithium loses 1 electron and fluorine gains 1 electron. Lithium becomes a positive ion because it now contains more positively charged particles (protons) in its nucleus than negatively charged particles (electrons) in its outer portion. Fluorine becomes a negative ion because it now contains more negative particles (electrons) in its outer portion than positive particles (protons) in its nucleus. This is an example of the changes that occur in forming an electrovalent bond.

Beryllium and boron have valences of 2 and 3, respectively. When beryllium and boron react chemically, the valence electrons of their atoms are transferred to the outer orbits of the atoms of the elements with which they react, thereby forming stable pairs of electrons and completing the outer orbits of the other atoms. Beryllium and boron frequently lose electrons when they react chemically, and their atoms become positive ions; the atoms of the elements with which they react gain electrons and become negative ions.

Formation of covalent bonds

Three examples of covalence are discussed.

Carbon and hydrogen in methane. The carbon atom has 4 single or unpaired valence electrons. When an atom of carbon reacts chemically, each of its 4 single electrons combines with a single electron belonging to an atom of another element or another carbon atom, forming 4 stable pairs of electrons or 8 electrons in its outer orbit. Thus in a chemical compound, carbon is a copossessor of 4 stable pairs of electrons, each pair being made up of 1 electron belonging to carbon and 1 electron belonging to an atom of the element or elements with which carbon reacts. The combination of carbon with hydrogen is shown below.

$$\cdot \overset{\cdot}{\underset{\cdot}{C}} \cdot \;+\; 4H\cdot \;\rightarrow\; H\overset{H}{\underset{H}{:\overset{\cdot\cdot}{\underset{\cdot\cdot}{C}}:}}H$$

Carbon　　Hydrogen　　Methane

Thus carbon and hydrogen share 4 covalent bonds in the compound methane. Methane is stable because carbon has an outer electron configuration in which there are 4 stable pairs of electrons in satisfied orbitals.

Nitrogen and hydrogen in ammonia. Nitrogen possesses 1 stable pair of electrons and 3 single unpaired electrons in its outer orbital. In its usual chemical reaction the 3 unpaired electrons of the nitrogen atom are made to combine with 3 unpaired electrons held by atoms of other elements; thus 3 stable pairs of electrons are formed. As a result of this chemical reaction, the nitrogen atom in the compound formed has 4 stable pairs or 8 electrons in its outer orbit. The reaction of nitrogen with hydrogen is as follows:

$$:\overset{\cdot}{\underset{\cdot}{N}}\cdot \;+\; 3H\cdot \;\rightarrow\; :\overset{H}{\underset{H}{N}}:H$$

Nitrogen　　Hydrogen　　Ammonia

Oxygen and hydrogen in water. The outer orbit of the oxygen atom consists of 2 stable pairs of electrons and 2 single unpaired electrons. When oxygen reacts chemically, the 2 single electrons in its outer orbit become paired with 2 electrons in the outer orbits of other atoms. Thus in compounds of oxygen the oxygen atom has an outer orbit that contains 4 stable pairs of electrons. The reaction of oxygen with hydrogen is as follows:

$$:\overset{..}{\underset{..}{O}}\cdot\ +\ 2H\cdot\ \rightarrow\ :\overset{..}{\underset{..}{O}}:\overset{H}{H}$$

Oxygen   Hydrogen   Water

Models of covalent molecules can be found in Fig. 2-16.

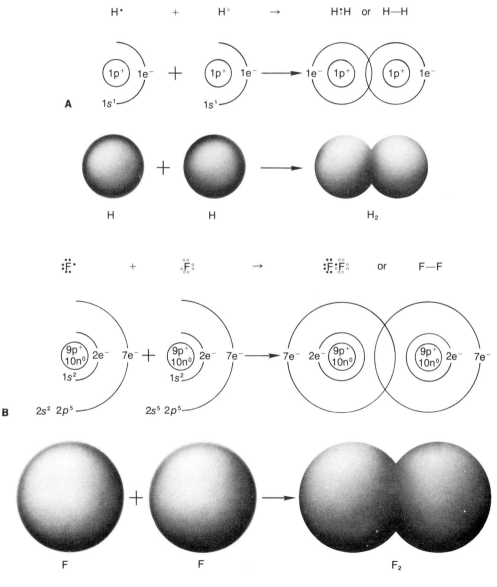

Fig. 2-16. A and **B,** Some molecular models.

Continued.

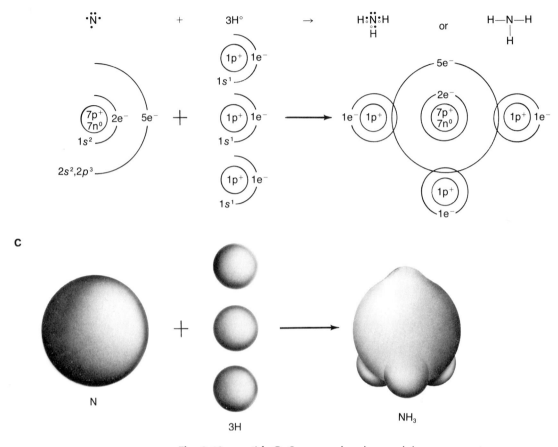

Fig. 2-16, cont'd. C, Some molecular models.

Polar bonds

The term covalent indicates that the two atoms involved share the bonding electrons equally. Some atoms form covalent bonds that do not have equal attractions for the bonding electrons. These electrons are not lost to either atom but are drawn to one atom more than to the other. Thus one atom has a partial negative electric charge, whereas the other has a partial positive electric charge. This partial charge results in a polar molecule and is indicated by the Greek letter δ (delta). Hydrogen chloride has this type of molecule.

$$\overset{\delta^+}{H} \quad \overset{\delta^-}{:\!\ddot{C}\!l\!:}$$

NAMING OF BINARY COMPOUNDS

The word "binary" means two. Binary compounds contain two elements. A binary compound is given a name that includes the names of both of its elements. The practice is to place the name of the metallic element first and to follow it with the name of the nonmetallic element, which is modified by changing its last syllable to *ide*. Examples:

Table 2-6. Valence of some elements and radicals

	Valence number	Designation of ion
Element		
Hydrogen	+1	H^+
Lithium	+1	Li^+
Sodium	+1	Na^+
Potassium	+1	K^+
Silver	+1	Ag^+
Calcium	+2	Ca^{++}
Magnesium	+2	Mg^{++}
Barium	+2	Ba^{++}
Zinc	+2	Zn^{++}
Mercuric mercury	+2	Hg^{++}
Mercurous mercury	+1	Hg^+
Cupric copper	+2	Cu^{++}
Cuprous copper	+1	Cu^+
Aluminum	+3	Al^{+++}
Ferric iron	+3	Fe^{+++}
Ferrous iron	+2	Fe^{++}
Fluorine	−1	F^-
Chlorine	−1	Cl^-
Bromine	−1	Br^-
Iodine	−1	I^-
Oxygen	−2	$O^=$
Sulfide sulfur	−2	$S^=$
Radical		
Ammonium	+1	NH_4^+
Hydroxyl	−1	OH^-
Nitrate	−1	NO_3^-
Bicarbonate or hydrogen carbonate	−1	HCO_3^-
Sulfate	−2	$SO_4^=$
Carbonate	−2	$CO_3^=$
Phosphate	−3	$PO_4^=$
Nitrite	−1	NO_2^-
Sulfite	−2	$SO_3^=$
Chlorate	−1	ClO_3^-

Elements in compound	Name of compound formed
Sodium and chlorine	Sodium chloride
Iron and sulfur	Iron sulfide
Potassium and iodine	Potassium iodide
Calcium and oxygen	Calcium oxide

When two elements form more than one compound with each other, it becomes necessary to have distinguishing names for the different compounds. One rule to take care of this situation is to change the suffix of the name of the varying element to "*ous*" in the compound in which this element has its lower valence and to "*ic*" in the compound in which it has the higher valence. Examples:

HgCl	Mercur*ous* chloride	(valence of mercury = 1)
HgCl$_2$	Mercur*ic* chloride	(valence of mercury = 2)

Cu_2O	Cup*rous* oxide	(valence of copper = 1)
CuO	Cup*ric* oxide	(valence of copper = 2)
$FeCl_2$	Fer*rous* chloride	(valence of iron = 2)
$FeCl_3$	Fer*ric* chloride	(valence of iron = 3)

Another practice for distinguishing between compounds of the same two elements is to use the prefixes such as *mon, di, tri, tetr,* and *penta,* which mean 1, 2, 3, 4, and 5, respectively, the prefix used being appropriate to designate the number of atoms of the varying element in the molecule. Examples:

CO	Carbon *mon*oxide
CO_2	Carbon *di*oxide
NO	Nitrogen *mon*oxide
NO_2	Nitrogen *di*oxide
PCl_3	Phosphorus *tri*chloride
PCl_5	Phosphorus *penta*chloride

A more modern system of naming compounds is that suggested by Stock, a chemist. Roman numbers put in parentheses are included in the name and indicate the oxidation state of the more positive element. For example, some of the compounds named in the preceding sets are named in the Stock system as follows:

$HgCl$	Mercury (I) chloride	CO	Carbon (II) oxide
$HgCl_2$	Mercury (II) chloride	CO_2	Carbon (IV) oxide
Cu_2O	Copper (I) oxide	NO	Nitrogen (II) oxide
CuO	Copper (II) oxide	NO_2	Nitrogen (IV) oxide
$FeCl_2$	Iron (II) chloride	PCl_3	Phosphorus (III) chloride
$FeCl_3$	Iron (III) chloride	PCl_5	Phosphorus (V) chloride

Questions for study

1. What is an electron? What is the unit that carries an electric current?
2. What is a proton? A neutron?
3. Describe the structure of the atom.
4. What value of the atom corresponds to the number of protons in the nucleus?
5. How is the number of neutrons in the nucleus of the atom ascertained?
6. What atomic value corresponds to the number of electrons in the shells of a neutral atom?
7. How many protons and neutrons are in the nucleus of an atom of helium? Lithium? Oxygen? Sodium? How many electrons are in the outer shell of an atom of each of these elements?
8. What are isotopes?
9. What is deuterium? Tritium?
10. Give an exact definition of an element.
11. What is the importance of isotopes?
12. What is the periodic law of the elements as expressed by Mendeleev and Meyer?
13. Show how the periodic table is developed.
14. What are atomic numbers?
15. Mention an experimental demonstration to justify the use of atomic numbers as a basis for the present arrangement of the elements in the periodic table.
16. State the periodic law in its modern form.
17. Show how the elements of the periodic table are arranged in groups of natural families.
18. How are the elements in the periodic table grouped with respect to metals and nonmetals?

The atom, periodic classification, and chemical reactivity 53

19. In what way has the periodic table led to the discovery of new elements?
20. What is a chemical change defined in terms of electrons?
21. Does the nucleus of the atom take part in ordinary chemical reactions?
22. Why do chemical reactions occur?
23. What is the fundamental change in a chemical reaction?
24. Describe the changes that occur when atoms of the following elements enter into chemical reactions: lithium, oxygen, and sodium.
25. What happens when lithium reacts with fluorine? When fluorine atoms react with each other?
26. What is valence?
27. What is the value of a knowledge of valence?
28. What is the relation of valence to the electron structure of the atom?
29. What is a valence electron?
30. Distinguish between transfer and sharing of electrons.
31. What is an electrovalent bond? A covalent bond?
32. What makes an ion positive? Negative?
33. Give a summary of valence and chemical changes in terms of electrons.
34. What is a polar molecule?
35. What is the general rule for naming binary compounds?
36. When two elements form more than one compound with each other, how are the compounds given distinguishing names?

3 Physical states of matter

Matter exists in three physical states: the solid state, the liquid state, and the gaseous state. All molecules of matter possess kinetic energy; that is, they are in continuous motion. The amount of motion depends on the amount of heat energy that is applied to the matter. The greater the temperature, the more the motion; the less the temperature, the less the motion. The zero point on the Kelvin temperature scale was chosen as the temperature at which all molecular motion would cease at 1 atmosphere of pressure (760 mm Hg). The pressure exerted on matter has an effect on the volume of that matter. This effect is greater on matter in the gaseous state, less on matter in the liquid state, and very little on matter in the solid state.

SOLID STATE

Matter in the solid state is characterized by a rigid, hard structure. Solids have a fixed volume and retain their shape. Great pressure is needed to compress a solid; indeed, some solids resist pressure and will break and crumble under high pressure but will not be compressed into a smaller volume. Usually the molecules of a solid are either closely packed or in a crystal formation. Closely packed molecules and crystal structures resist compression.

The solid state has two main forms, the *amorphous* and the *crystalline*. Amorphous matter has no fixed internal structure. The particles of an amorphous solid can be close packed or a finely divided powder. Tar and glass are examples of amorphous substances. The crystalline form of matter has a structural arrangement. The units of the solid (molecules, atoms, or ions) have geometric configurations in the form of crystals. Two substances having the same crystalline form and crystallizing together in all proportions are said to be *isomorphs* of each other.

Similarities among crystals suggests similarities among the molecules making up those crystals. This is the basis for the chemically analytical tool called *crystallography,* the study of crystal shapes. Early work with crystals was limited due to the technology then available. Later, x-ray defraction allowed studies in which the association of the wavelengths of reflected x-rays and the vertical distances between parallel reflecting planes of crystals were explored. Some types of crystal configuration are shown in Fig. 3-1.

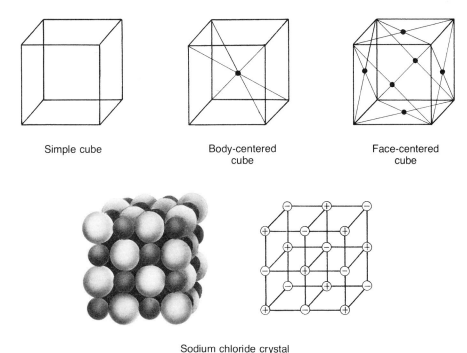

Fig. 3-1. Types of crystals.

Molecular crystals

Molecular crystals are held together by attractive forces between molecules. These forces are the weakest forces found in crystals. The molecular crystal tends to be soft and have a low melting point. Ice is an example of a molecular crystal.

Ionic crystals

Oppositely charged ions make up ionic crystals. Each positive ion is surrounded by negative ions and each negative ion is surrounded by positive ions. The number of oppositely charged ions surrounding any one ion is called the *coordination number*. The attractive forces between ions are strong, and these crystals are hard with high melting points. Common table salt (NaCl) is an example of this type of crystal. The usual practice of writing the formula of sodium chloride as NaCl results from using the emperical (simplest) formula in place of the molecular formula. When ionic solids are changed to the gaseous state, they have small molecules. In the solid, the crystal itself, no matter how large, is the molecule.

Metallic solids

The atoms of a metal form a close-packed crystal lattice. Neighboring atoms touch each other, and the electrons in the outer shells can move about the lattice structure. Owing to this easy migration of electrons, metals are good conductors of electricity. Metal crystals vary with respect to their hardness and their melting points, some being very hard with high melting points, like iron, others soft with low melting points, such as lead.

The molecules of a solid diffuse (mix together) very slowly. If two pieces of metal are bound together for a long time, some of the atoms of one metal will diffuse into the other metal and vice versa. A solid is characterized as having its own volume and shape. The molecules of a solid are in a low kinetic energy state with their movement chiefly limited to oscillation around a fixed point.

If heat energy is applied to a solid, the molecules making up that solid begin to move with greater and greater velocity. In some cases a few of the molecules can gain a great deal of energy and may leave the surface of the solid in the form of gas particles. Many solids, especially those with weak intermolecular attractive forces, such as ice, exist with a gas phase over them. This can be especially true if the pressure over the solid is low. Some solids always have their molecules move directly into the gaseous state. Iodine is one of these solids, and this conversion of a solid state directly into a gaseous state is called *sublimation*.

It is usual to have a solid melt into a liquid as the temperature increases. With an increase in temperature there is an increase in velocity of the molecules making up the solid. The amorphous solid slowly becomes liquid. With increased molecular velocity, the intermolecular attractive bonds that serve to hold molecules in crystals are broken. At this point the temperature remains constant. This constant temperature is the *melting point* of the solid. The melting point of the solid state is the same temperature as the *freezing point* of the corresponding liquid state. The heat in calories per gram of solid necessary to maintain a solid at the melting (freezing) point is called the *heat of fusion*. Passing from the solid to the liquid state is called *liquefaction*. Cooling a liquid causes molecular motion to slow, and the liquid becomes solid (freezes) at the freezing point.

LIQUID STATE

In the liquid, the molecules are still closely associated. In a majority of cases, the liquid will occupy a larger volume than the same weight of solid. Sometimes the liquid that melts from a crystalline solid takes up less volume than the same weight of the crystal. An example of this is water from melting ice. Because of its crystal lattice, ice expands as it freezes causing "heaving" of the ground in winter and the breaking of pipes holding freezing water.

Liquids have no definite shape but take the shape of the vessel into which they are placed. Liquids have definite volume and can only fill a vessel to the level of that volume. Increasing the pressure over a liquid reduces its volume slightly.

The molecules in a liquid have greater kinetic energy and greater motion than molecules of a solid. This is demonstrated by the phenomenon known as *Brownian movement*, named after Robert Brown, a Scottish botanist. In examining a water suspension of pollen grains, Brown noticed that the particles moved back and forth with a quivering, rotating motion. This continuous movement of particles suspended in a liquid is explained as being due to the bombardment of the particles by the molecules of the liquid. The invisible molecules of the liquid strike the surface of the visible suspended particles, and since the blows cannot be uniform on all sides, these particles are made to move with a spinning, rotating, zigzag motion. This movement of visible particles in suspension (Brownian movement) may be readily observed under the microscope. It is a good demonstration of the kinetic energy of the molecules of a liquid.

In liquids, diffusion brings about the intermingling of the molecules of one substance with those of another. Diffusion is readily demonstrated in the laboratory by placing a colored substance in a liquid. The colored substance rapidly diffuses and in a short time is uniformly distributed throughout the liquid. It is the impact of the moving molecules of the liquid on the molecules of the suspended or dissolved substance, as well as, to some extent, the kinetic energy of the molecules of the suspended or dissolved substance itself, that brings about the uniform distribution of the molecules of the suspended substance. Thus diffusion in a liquid is brought about by molecular motion because of the kinetic energy of molecules in the liquid state.

The attraction between the molecules of a liquid cause it to exhibit a physical characteristic, *viscosity*. All liquids tend to flow; that is, the molecules of the liquid slip over one another. However, if these molecules have attractions, one to another, the flowing characteristic is hampered in proportion to the strength of the attractive forces. Heating a liquid tends to increase the kinetic energy of the molecules, increasing their movement and decreasing the viscosity. Attractive forces between molecules of a liquid account for the *surface tension* of a liquid. In the center of a volume of liquid, the molecules are attracted in all directions by other particles. At the surface of the liquid, the molecules are attracted inward and sideward only. For this reason, a small amount of liquid will form a sphere—a drop—having the smallest surface area per unit of volume. When drawn by the force of gravity, a drop forms an egg-shaped form called "tear shaped."

When a liquid is heated, its molecules move with ever-increasing velocity. The faster moving molecules migrate to the surface and, when they have sufficient energy, break through the surface tension and form gas molecules. A liquid exists with its gas molecules over it. This movement of particles from the liquid to the gas state is called *evaporation*. If left open to the atmosphere, the gas particles escape; the migration of liquid particles to gas particles is one-directional. In a closed container, gas particles lose energy and return to the surface of the liquid. At a fixed temperature and pressure, an equilibrium is set up with molecules moving from the liquid to the gas as fast as molecules move from the gas to the liquid. The pressure of a gas on the surface of its liquid is called the *vapor pressure*. This pressure is caused by the mass of the gas molecules reacting to the force of gravity resulting in pressure on the surface of the liquid.

When a liquid is heated to the point that the vapor pressure of the liquid equals the atmospheric pressure, boiling occurs. At this point the temperature remains constant, and all of the heat energy is converted into kinetic energy. This constant temperature is the *boiling point* of the liquid. It is also the *condensation point* of the gas that is being formed from the liquid (Fig. 3-2). If heat energy is added at the boiling point, all of the liquid will soon be converted into a gas. If the atmospheric pressure is low, the boiling point of a liquid is reached at a lower temperature. In some higher points of the world, the

58 Roe's principles of chemistry

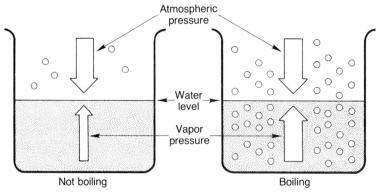

Fig. 3-2. Vapor pressure: boiling.

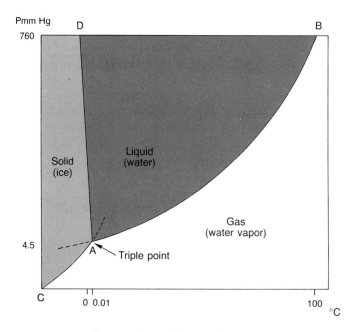

Fig. 3-3. Phase diagram for water.

boiling point of liquids are much lower than they are at sea level. Increasing the pressure allows for higher temperatures to be reached before boiling occurs. This is the idea behind the autoclave and the pressure cooker. The heat in calories per gram of liquid necessary to maintain a liquid at the boiling point is called the *heat of vaporization*. If heat energy is removed from a gas at the condensation (boiling) point, the gas will condense into a liquid.

The freezing point of a liquid is that point at which the vapor pressure of the liquid and the solid are equal. The boiling point is that point at which the vapor pressure of the liquid equals atmospheric pressure. The freezing point and the boiling point are called *critical temperatures* and are characteristic for each liquid.

The relationship existing between the solid, the liquid, and the gaseous state of a substance can be shown in a phase diagram. Each of the states of the substance can be called a *phase*. Fig. 3-3 shows the phase diagram for water. The coordinates are in pressure versus temperature. Line AB is the vapor pressure of water, whereas the line AC is the vapor pressure for ice. All the area below the line CAB represents the gas phase, water vapor. The line AD marks the melting-freezing point between water and ice. At A (temperature 0.01° C and pressure 4.58 mm), all three phases, solid (ice), liquid (water), and gas (water vapor), can exist in equilibrium with each other. To the left of this point, ice and water vapor exist; to the right, water and water vapor. At the temperatures of 0.01° C to 0.00° C, water and ice are in equilibrium at pressures greater than 4.58 mm Hg.

GASEOUS STATE

Molecules exhibit their greatest freedom of motion in the gaseous state of matter. The behaviour of the molecules of a gas is explained by the kinetic theory of gases. According to this theory, the molecules of a gas are in constant rapid motion. The individual gas molecule moves in a straight line; its course, however, is continually being changed by collision with other gas molecules or with the walls of the container. The molecules of a gas have elasticity, and they continue in motion. The collision of gas molecules with the walls of their container gives rise to the pressure exerted by a gas. Since gas molecules have elasticity, the pressure of a gas remains constant if the temperature and volume remain the same. The average velocity of gas molecules at 0° C and 760 mm Hg is about 1 mile per second.

It is the molecular motion of a gas that causes its rapid diffusion through space. If a gas is prepared in a closed container, the molecules quickly spread uniformly throughout the container. If a gas is released in an unconfined region, its molecules diffuse into space indefinitely. The rapid diffusibility of a gas is readily observed when a small amount of hydrogen sulfide, an ill-smelling gas, is prepared in the laboratory. In such an experiment the odor of hydrogen sulfide is quickly noticed in all parts of the laboratory, even though a very small amount of this substance is released.

The inert gases have monoatomic molecules—the atom is the molecule, such as He, Ar, and Ne. All other elements and compounds have gaseous molecules with two or more atoms held by covalent bonds, for example, CO_2, NO_2, and NH_3. It is usual for the non-inert elements to exist as diatomic in the gaseous state, for example, O_2, N_2, and H_2.

Gases are transparent, compressible, and expandable, and they mix with one another easily. The molecules of a gas are widely separated from each other and have only slight attraction to each other. The gaseous state is the least dense of the states of matter. Particles of a gas can be pictured as having very small volumes when this volume is compared to the intermolecular space around them. Because of their elasticity, the particles of a gas can expand into larger volumes. Because of their ability to expand, they completely fill any container into which they are placed.

If the temperature is lowered, the kinetic energy of the gas particles is lowered and the volume of the gas will decrease. This is the basis of *Charles' law*, which states that the volume of a gas varies directly with the temperature. The greater the temperature, the greater the volume.

$$V \propto T \text{ or } \frac{V_1}{V_2} = \frac{T_1}{T_2}$$

Because the gas particles are so far apart, the volume of a gas can be compressed more easily than can the volume of a liquid or a solid. *Boyle's law* states that the volume of a gas varies inversely as to the amount of pressure placed on it. The greater the pressure, the less the volume.

$$V \propto \frac{1}{P} \text{ or } \frac{V_1}{V_2} = \frac{P_2}{P_1}$$

Put together, these two laws give us the Boyle-Charles gas law, which states that *the volume of a gas varies inversely as to the pressure and directly as to the temperature*.

$$V \propto \frac{T}{P} \text{ or } \frac{V_1}{V_2} = \frac{T_1}{T_2} \times \frac{P_2}{P_1}$$

This law can be used to adjust the volume of a gas as it moves from one condition of temperature and pressure to another. For example: A sample of gas occupies a volume of 25 ml at 800 mm Hg and 25° C. What will its volume be a 0° C and 760 mm Hg?

$V_1 = 25$ ml
$V_2 = X$ ml
$P_1 = 800$ mm Hg
$P_2 = 760$ mm Hg
$T_1 = 25°$ C or 298° K   (The temperature of a gas must be expressed
$T_2 = 0°$ C or 273° K   in Kelvin degrees for proper calculations.)

$$\frac{V_1}{V_2} = \frac{T_1}{T_2} \times \frac{P_2}{P_1}$$

$$X = 25 \text{ ml} \times \frac{273}{298} \times \frac{800}{760} = 24.37 \text{ ml}$$

The conditions 0° C and 760 mm Hg (1 atmosphere of pressure) are called *standard conditions of temperature and pressure*, or *STP*.

The internal pressure of a gas on the walls of its container is dependent on the number (n) of gas molecules present. For two different gases

$$P_1 \propto \frac{n_1 T_1}{V_1} \text{ and } P_2 \propto \frac{n_2 T_2}{V_2}$$

If the pressure, temperature, and volume are kept constant, each sample of gas will have the same number of molecules. This is the basis for *Avogadro's law*, which states that *equal volumes of gases at the same temperature and pressure contain the same number of molecules*. Experience showed that these equal volumes did not weigh the same. The reason for this being that the molecular mass of one gas does not equal the molecular mass of another gas unless the two gases are identical. Carbon dioxide (CO_2) has the molecular mass of 44 amu, whereas ammonia (NH_3) has the molecular mass of 17 amu. At a certain temperature and pressure, a specific volume of CO_2 will contain a fixed number of CO_2 molecules. At the same temperature and pressure, an equal volume of NH_3 will contain the same number of NH_3 molecules. No matter how many molecules are in the two identical volumes, the weight of CO_2 will always be greater than the weight of NH_3 by a 44:17 ratio.

This fact offered a convenient analytical tool to determine the molecular mass of any substance that could be converted to the gaseous state under reasonable conditions of temperature and pressure. Research developed that 22.4 liters of any gas at STP weighed the same number of grams as the actual gas molecule was calculated to weigh in atomic mass units. At STP, 22.4 liters of any gas contains 6.02×10^{23} molecules. This amount is called an *Avogadro's number* of molecules. A *molecular mass* expressed in grams is called a *mole*, or a *gram-molecule*. A mole contains an Avogadro's number of molecules.

A full 22.4 liters volume need not be used nor do the conditions have to be STP when the gas is weighed. Using the Boyle-Charles law, it is possible to adjust the volumes to STP after the weighing is effected. A simple ratio calculation can give the weight of 22.4 liters of gas from any volume that is convenient to weigh (Fig. 3-4).

The experimentally discovered weight can be substituted for mass since mass and weight are the same on earth. For example: 100 ml of a gas weighs 0.1302 g at 20° C and 750 mm Hg. What is its molecular weight?

Adjusting for the temperature and pressure

$$X = 100 \times \frac{273}{293} \times \frac{750}{760} = 91.14 \text{ ml at STP}$$

Correcting for 22.4 liters (22,400 ml)

$$\frac{22,400}{91.14} = \frac{X}{0.1302}$$

$$X = 32 \text{ g (probably } O_2\text{)}$$

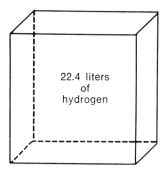
Absolute weight = 2.016 g
Molecular mass of hydrogen = 2.016

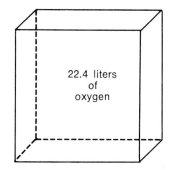
Absolute weight = 32 g
Molecular mass of oxygen = 32

Fig. 3-4. Diagram illustrating method of determining molecular mass.

62 Roe's principles of chemistry

The total pressure of a mixture of gases is equal to the sum of the pressures of each of the gases. This observation is the basis for the *law of partial pressures*. At sea level the pressure of air is 760 mm Hg. Air is approximately 20% oxygen and 80% nitrogen. Therefore, oxygen has 760 × 0.2, or 152, mm Hg, and nitrogen, 760 × 0.8 = 608 mm Hg at sea level.

ALLOTROPISM

Some elements occur in more than one form while still in the same physical state. This phenomenon is called *allotropy,* and the forms in which the element appears are called *allotropes*. Allotropy may result in more than one elementary molecular formula or differences in crystalline forms in the solid state.

Questions for study

1. Name the physical states of matter.
2. Describe the solid state of matter.
3. Name three types of crystals.
4. What are the two main forms of the solid state?
5. What causes metals to be good conductors of electricity?
6. What is sublimation? Liquefaction? Viscosity?
7. What relationship exists between the melting point and the freezing point of the same substance?
8. What is the heat of fusion?
9. Describe the liquid state of matter.
10. What is Brownian movement? What causes it?
11. What relationship exists between the boiling point of a liquid and the condensation point of its gas?
12. What is the heat of vaporization?
13. Describe the gaseous state of matter.
14. State the Boyle-Charles gas law.
15. In speaking of a gas, what is meant by STP?
16. A sample of gas occupies a volume of 1 liter at STP. What will be its volume if the temperature is raised to 20° C at SP? What will be its volume if the pressure is raised to 780 mm Hg at ST? What will be its volume at 25° C and 790 mm Hg?
17. What is Avogadro's law? Avogadro's number?
18. What is a mole (gram-molecule)? How many molecules are in a mole?
19. At 25° C and 755 mm Hg, 234 ml of a gas weighs 0.25 gm. What is the molecular weight of the gas? What is a possible molecular formula for the gas?
20. State the law of partial pressures.

4 Some important metals and nonmetals

METALS
Sodium and potassium

Sodium and potassium belong to a closely related chemical family called the *alkali metals* (included in Group IA of the periodic table), of which lithium, rubidium, cesium, and francium are the other members. Sodium and potassium are somewhat similar in properties, potassium being the more active chemically. They are soft, silvery white solids. They combine readily with most nonmetals. They react violently with water, liberating hydrogen and forming sodium or potassium hydroxides, which are strong bases. Sodium and potassium react so readily with water and oxygen that they have to be stored under kerosene. Sodium salts give a yellow color to the Bunsen flame. Pure potassium compounds impart a violet color to a flame. The salts of sodium and potassium occur rather abundantly in nature and are important constituents of the human body and of plant tissues. Oxides of metals react with water to form bases.

$$Na_2O + H_2O \rightarrow 2NaOH$$
$$\text{Sodium oxide} \quad \text{Water} \quad \text{Sodium hydroxide}$$

Lithium

Lithium, an alkali metal, occurs in a number of minerals, its ores being tryphilite ($LiFePO_4$) and silicates such as petalite and lepidolite. It is important industrially, and the compounds lithium aluminum hydride and lithium boron hydride are of commercial importance. Lithium salts inhibit the release of norepinephrine (see p. 263) and serotonin from the brain. Many cases of mania and similar mental disorders related to disturbances in amine metabolism (see p. 296) can be effectively treated with lithium salts.

Calcium

Calcium belongs to a chemical family called the *alkaline earth metals*, which falls in Group IIA of the periodic table. The members of this family, in addition to calcium, are beryllium, magnesium, strontium, barium, and radium. Calcium is a silvery white metal that is soft and easily cut or rolled. It reacts with water, liberating hydrogen and forming calcium hydroxide. It also combines with many nonmetals when heat is applied. It exists in nature in the form of some important compounds, chief of which are calcium carbonate, calcium phosphate, and gypsum, or calcium sulfate. Calcium salts are important constituents of the tissues of the body, especially the bones and teeth. Dissolved calcium ions in water are responsible for "hard" water.

Magnesium

Magnesium is a silvery white metal produced commercially from seawater. It has a tenacity that permits it to be rolled into a ribbon. It will react with water when the water is boiling hot, liberating hydrogen. It combines with most nonmetals when heat is applied, and it oxidizes but very slightly at ordinary temperatures and may therefore be kept exposed to dry air. When burned, it produces a brilliant white light; hence magnesium is used in the preparation of "flashlight" powder for night photography, for illuminating military objectives, and for the preparation of flares for signaling. Magnesium salts are constituents of the tissues of plants and animals. Magnesium ions in water create "hard" water.

Iron

Pure iron is a silvery lustrous metal that is malleable and ductile. It is not found in nature as a free element, its chief occurrence being in the form of iron oxides and iron carbonate. It is a constituent of hemoglobin, the oxygen-carrying compound of the blood. The great industrial importance of iron is its extensive use in manufacturing structural materials such as cast iron, wrought iron, and steel.

Mercury

Mercury was known to the early alchemists, who ascribed to it many mysterious properties. It is a silvery white liquid. It is the only metal that exists in the liquid state at ordinary temperatures. It has a high density, being 13.6 times heavier than an equal volume of water. Under appropriate conditions it combines with nonmetals, forming many compounds. Mercury vapors and the soluble compounds of mercury are severe poisons. Mercury is used in the construction of thermometers, barometers, and apparatus for measuring the volume of a gas. Some mercurial compounds are used as drugs.

Mercuric and mercurous chlorides. Mercuric and mercurous chlorides are used extensively in medicine. They have widely differing properties and will cause disaster if confused with each other.

Mercuric chloride ($HgCl_2$), also called bichloride of mercury, is highly poisonous and is a valuable disinfectant. Mercurous chloride ($HgCl$), or calomel, is not poisonous and is used as a purgative. Calomel is not poisonous because it is not soluble in water.

Large quantities of mercury have evidently been polluting the waterways and oceans of the world. Toxic amounts of mercury have been detected in certain fish. Greater care in the disposal of poisonous heavy metals must be exercised in the future if man is not to poison the environment.

Bismuth

Bismuth is a grayish white metal. It is used principally in the preparation of alloys with low melting points, such as Wood's metal, which melts at 60° C. Its chief importance in medicine is as a constituent of bismuth subnitrate and bismuth subcarbonate. These salts are used in treating conditions characterized by irritation along the alimentary tract. They are insoluble and form a coating over an irritated or ulcerated surface, which protects against mechanical injury from food and chemical injury from digestive juices.

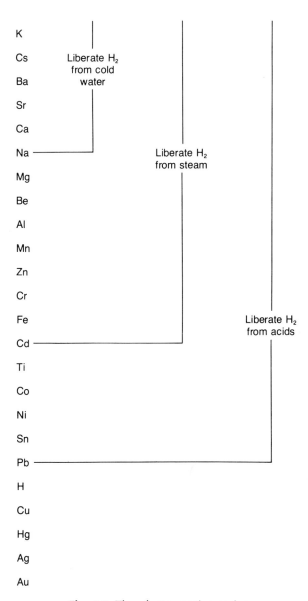

Fig. 4-1. The electromotive series.

muth subnitrate has also been used in taking radiographs of the alimentary tract. The ntgen rays will not penetrate tissues coated with this substance to the same degree at they do the tissues not coated with it; hence after the ingestion of bismuth salts a shadowed outline of the stomach and intestines may be observed with the fluoroscope or may be reproduced on x-ray film. Although this use is now obsolete, it is of interest that bismuth compounds are radiopaque.

Electromotive series

Chemists have prepared a list of metals in the order of their ability to give up electrons. This is called the *electromotive series* (Fig. 4-1). The metals are listed from top to bottom with respect to their decreasing ability to act as electron donors. Fifth from the bottom of the list is found the element hydrogen. All of the metals above hydrogen will replace it from acids. Some of the more active metals will liberate hydrogen from water as well. The metals below hydrogen, that is, copper, mercury, gold and silver, will not react with acids to release hydrogen. The metals at the bottom of the series are usually found free in nature and are relatively unreactive. The metals at the top of the series are active metals and are found in nature united with other elements in the form of compounds. An electrovalently bound metal can be replaced in a compound by any metal appearing above it in the electromotive series.

The ability to accept electrons is called *electronegativity*. In the electromotive series, electronegativity increases from top to bottom. The most electronegative elements are the active nonmetals, with fluorine exhibiting the greatest ability to accept electrons and oxygen running a close second. The least electronegative elements are the active metals (see p. 43).

NONMETALS
Nitrogen

Nitrogen occurs in both the free and the combined states. As the free element, it constitutes about four-fifths of the atmosphere by volume. In the form of nitrogen compounds it exists in the soil in small quantities, and in great abundance in certain natural deposits such as the sodium nitrate (''nitre'') beds of Peru and Chile. Nitrogen is also a constituent of the proteins, a very important class of compounds that make up a large part of the tissues of animals and plants.

Molecular nitrogen is relatively unreactive due to its intermolecular bonding. This chemical inertness is fortunate since nitrogen makes up four-fifths of the atmosphere. A more reactive gas would cause chemical changes in many of the things with which it came into contact.

When breathed into the lungs, some of the nitrogen enters the blood stream, is dissolved, circulates through the body, and returns to the lungs unchanged to be breathed out again. The nitrogen dissolved in its fluids has no effect on the body unless that body changes its altitude suddenly. Gases expand with a lessening of atmospheric pressure. Such expansion of dissolved nitrogen in the blood and tissue fluids causes tissue destruction and, in extreme cases, death. A disorder called ''the bends'' results from unprotected

Some important metals and nonmetals 67

deep-sea divers surfacing too suddenly. Pressurized diving suits prevent such changes. Commercial airplanes have pressure controlled cabins to counteract the effect of sudden changes in altitude. Space explorers wear pressurized suits to protect themselves from airless environments. People suffering from pressure-change illness can be treated by placement in hyperbaric chambers; the pressure within these chambers is allowed to normalize slowly.

Oxides of nonmetals react with water to form acids.

$$N_2O_5 + H_2O \rightarrow 2HNO_3$$
$$\text{Nitrogen pentoxide} \quad \text{Water} \quad \text{Nitric acid}$$

Nitric acid

Preparation. Nitric acid is prepared commercially by distilling a mixture of sodium nitrate and sulfuric acid.

$$2NaNO_3 + H_2SO_4 \rightarrow Na_2SO_4 + 2HNO_3$$
$$\text{Sodium nitrate} \quad \text{Sulfuric acid} \quad \text{Sodium sulfate} \quad \text{Nitric acid}$$

Properties. Pure nitric acid is a colorless liquid. It is an unstable compound and decomposes under the influence of light and when heated mildly. When nitric acid decomposes, it yields a brown gas, nitrogen dioxide (NO_2); therefore a bottle of nitric acid that has been exposed to light is brown because of the presence of this gas in solution. Nitric acid has the typical properties of a strong acid and is also a powerful oxidizing agent. It does not yield hydrogen when mixed with metals because the hydrogen when displaced by a metal is immediately oxidized by the nitric acid, forming water.

Uses. Nitric acid is used in the preparation of explosives. Nitroglycerine (dynamite), nitrocellulose (smokeless powder), picric acid, and trinitrotoluene (TNT) are highly explosive compounds whose explosive nature is due to the presence of nitro (NO_2) groups from nitric acid. Black gunpowder (smoke-producing powder) is a mixture of sulfur, charcoal, and potassium nitrate, a salt of nitric acid. Nitric acid is also used in the manufacture of plastics such as celluloid and in the preparation of dyes and drugs.

Nitrates or the salts of nitric acid must be present in the soil to promote the growth of plants, and they are therefore indispensable as fertilizers. Nitrogen has rightfully been called a "preserver and destroyer of life" by one author[*] because of the role of nitrogen in the preparation of compounds that may be used as explosives to destroy life, as drugs to save life, or as fertilizers for the growth of plants that yield food for man.

Ammonia and ammonium hydroxide

Ammonia. Ammonia (NH_3) is a colorless gas with a sharp, suffocating odor. It is about one-half as heavy as air. It is extremely soluble in water. A saturated solution of ammonia in water at a temperature of 18° C and a pressure of 1 atmosphere contains

---
[*]Slosson: Creative chemistry, The Century Co.

36% ammonia. Ammonia is easily liquefied by applying pressure at ordinary temperatures, and when liquid ammonia is allowed to evaporate, it absorbs a great deal of heat; hence ammonia is used as a refrigerant in the manufacture of ice. Ammonia was widely prepared on a commercial scale in this country as a by-product in the production of illuminating gas from coal. Coal contains a small amount of ammonia combined with other elements, and if heated in the absence of air, the ammonia is evolved and may be collected by passing it into water or sulfuric acid. Ammonia is prepared by direct combination of nitrogen with hydrogen at high temperatures and pressures. Ammonia is one of the products liberated by the action of putrefactive bacteria on nitrogenous organic matter.

Ammonium hydroxide. When ammonia is dissolved in water, two ions are formed

$$NH_3 + H_2O \rightleftarrows NH_4^+ + OH^-$$
$$\text{Ammonia} \quad \text{Water} \quad \text{Ammonium ion} \quad \text{Hydroxide ion}$$

A solution of ammonia water is frequently named ammonium hydroxide. This solution turns red litmus blue and manifests all the properties of a base but in a mild degree. Ammonia gas (NH_3) is easily soluble in water, but the ammonia molecule reacts only slightly with the water to form ammonium ion and hydroxyl ion. If a vessel containing ammonium hydroxide is left open, ammonia gas will pass into the air, and only water will remain after a time. The latter change may be affected more quickly by boiling the solution. "Household ammonia" and "aqua ammoniae" are dilute solutions of ammonia in water. They are used for cleaning and for softening water.

Sulfur

Sulfur is a common element. It was known in biblical times and is probably the "brimstone" of the Bible. It is found as the free element in large deposits in Louisiana and Texas and also in certain volcanic regions in Mexico, Sicily, and Japan. The chief sulfur compounds are the sulfides of iron (FeS), lead (PbS), zinc (ZnS), and copper (CuS) and the sulfates of calcium ($CaSO_4 \cdot 2H_2O$), barium ($BaSO_4$), strontium ($SrSO_4$), and magnesium ($MgSO_4 \cdot 7H_2O$). Sulfur is a constituent of most proteins and is therefore found in the tissues of animals and plants.

Properties. Sulfur is a pale yellow solid, insoluble in water but soluble in carbon disulfide. It is tasteless and odorless. It exists in several allotropic forms. Sulfur is inactive at ordinary temperatures but reacts with many substances when heat is applied. It burns with a blue flame, forming sulfur dioxide.

Uses. Sulfur is used in the manufacture of vulcanized rubber, certain kinds of matches, black gunpowder, and depilatories. Mixed with lime, it is applied as a spray on fruit trees and plants to kill fungi that are harmful to their hosts. Its greatest use is in the production of sulfuric acid.

Important sulfur compounds

Sulfur dioxide. Sulfur dioxide is a colorless gas formed when sulfur is burned in air or oxygen.

$$S + O_2 \rightarrow SO_2$$
$$\text{Sulfur dioxide}$$

When inhaled, it produces a disagreeable, suffocating effect. It is used for bleaching straw, wool, and silks and for fumigating rooms infected with vermin. When dissolved in water, it forms sulfurous acid.

Sulfur trioxide. Sulfur trioxide is a white solid formed by the oxidation of sulfur dioxide.

$$2SO_2 + O_2 \rightarrow 2SO_3$$
$$\text{Sulfur trioxide}$$

This oxide is important in that it forms sulfuric acid when dissolved in water.

Sulfuric acid. Sulfuric acid is prepared by oxidizing sulfur dioxide to sulfur trioxide with the aid of a catalyst and then dissolving the trioxide in water.

$$2SO_2 + O_2 \rightarrow 2SO_3$$
$$\text{Sulfur trioxide}$$

$$SO_3 + H_2O \rightarrow H_2SO_4$$
$$\text{Sulfuric acid}$$

Sulfuric acid is a heavy, oily, white liquid. It is highly soluble in water. When the concentrated acid is mixed with water, there is a marked evolution of heat. It ranks as a strong acid. It has all the characteristic properties of an acid and when hot behaves as an oxidizing agent. It has a strong affinity for water and is a good drying agent. Its affinity for water is so marked that it will extract water from certain organic compounds. For instance, if concentrated sulfuric acid is spilled on wood or is mixed with sugar, water is withdrawn from these materials, and a charred mass of carbon remains. The reactions is as follows:

$$C_{12}H_{22}O_{11} + H_2SO_4 \rightarrow 12C + 11H_2O + H_2SO_4$$
$$\text{Sugar}$$

Sulfuric acid is used in the manufacture of explosives, dyes, and many chemicals and in the purification of petroleum. It is perhaps more indispensable to industry than any other chemical reagent.

Magnesium sulfate (Epsom salts). Magnesium sulfate has the formula $MgSO_4 \cdot 7H_2O$. It is a white crystalline substance whose principal use in medicine is as a laxative.

Sodium sulfate (Glauber's salt). Sodium sulfate ($Na_2SO_4 \cdot 10H_2O$) is used in medicine as a laxative.

Barium sulfate. Barium sulfate ($BaSO_4$) is given internally previous to taking x-ray films of the alimentary tract. It interferes with the passage of roentgen rays. Thus being radiopaque, it permits either the visualization of the alimentary tract (fluoroscopy) or the radiographing of an area through which roentgen rays are passed (radiography). It has replaced bismuth salts for this purpose.

Hydrogen sulfide. Hydrogen sulfide (H_2S) is found in the waters of sulfur springs and in the gases of volcanoes. It is also emitted from decaying organic matter containing combined sulfur, such as bad eggs or meats. Hydrogen sulfide is a colorless gas with a disagreeable odor. It is a poisonous substance, being fatal if inhaled for a considerable time. It is moderately soluble in water and when in solution forms a weak acid. It is an important reducing agent. The sulfide ion of hydrogen sulfide will combine with many metals, producing insoluble sulfides of different colors. It is this substance that blackens silverware by the formation of silver sulfide.

Phosphorus

Phosphorus belongs to the chemical family in Group VA of the periodic table. Other members of this group are nitrogen, arsenic, antimony, and bismuth.

Phosphorus owes its name to the fact that it glows in the dark, the word *phosphorus* meaning "bearer of light." It is too active an element to occur freely in nature and is found chiefly in the form of phosphates, which are salts of phosphoric acid, the principal acid of phosphorus. Phosphates are found in the soil in varying amounts and in large quantities in certain phosphate rock deposits such as in South Carolina and Florida. Phosphorus is a constituent of many important compounds in the human body. About 85% of the solid matter of bones and teeth is composed of calcium and magnesium phosphates. The blood, brain, and nerve and muscle tissues contain phosphorus compounds, and the urine and feces contain phosphates as constant excretory products. Phosphorus compounds are constituents of plant tissues also. A soil, to be fertile, must contain adequate soluble phosphates, which are essential for the promotion of plant growth.

Phosphorus was discovered accidentally in 1669 by the German alchemist Brandt, who prepared this element by heating dried urine and charcoal in a clay retort. It is prepared at the present time by essentially the same procedure, which consists of heating phosphate rock or bone ash with sand and charcoal.

Phosphorus is a solid that exists in three elementary forms—white, black, and red phosphorus. Black phosphorus is difficult to obtain in pure form and has little practical importance. White phosphorus is very inflammable, igniting at 35° C, it is soluble in carbon disulfide and ether, and it is poisonous. Red phosphorus will not burn until a temperature of 240° C is reached, it is not soluble in the solvents of white phosphorus, and it is not poisonous. Red phosphorus is prepared by heating the white form to 250° C in the absence of oxygen. White phosphorus is so inflammable that it must be kept under water. The temperature of the body will ignite it; hence it cannot be held in the hands. Both red and white phosphorus oxidize to phosphorus pentoxide, which forms phosphoric acid when dissolved in water.

$$4P + 5O_2 \rightarrow 2P_2O_5$$
Phosphorus pentoxide

$$P_2O_5 + 3H_2O \rightarrow 2H_3PO_4$$
Phosphoric acid

Uses. The chief commercial use of the element phosphorus is in the preparation of matches. The sodium salts of phosphoric acid are used in medicine. Disodium phosphate (Na_2HPO_4) may be used as a laxative.

Arsenic

Arsenic is a gray solid that resembles metals in appearance and has some other metallic properties. The oxide of arsenic (AsO) reacts with water to form an acid, a reaction of a nonmetal oxide. Arsenic therefore exhibits the chemical characteristics of a metal and a nonmetal. Elements such as germanium, arsenic, and antimony, appearing near the borderline separating metals and nonmetals in the periodic table, exhibit the properties of metals and nonmetals and are called metalloids (see p. 43). Arsenic and its compounds are very poisonous. For this reason they are valuable as insecticides, weed killers, and destroyers of disease-producing microorganisms. Paris green or copper arsenite and lead arsenate are used by agriculturists as sprays for certain vegetable crops and fruit trees to kill destructive insects. Calcium arsenate has been found especially valuable in combating the boll weevil that infests the cotton plant.

Boron

Boron is of importance chiefly as a constituent of boric acid (H_3BO_3) and borax ($Na_2B_4O_7$). Boric acid is an extremely weak acid. It is a mild antiseptic, and its saturated solution makes a very satisfactory eyewash. Borax forms a basic solution when dissolved in water. Its mild alkalinity makes it applicable to laundry work. Boric acid is slightly poisonous. Care should therefore be taken in using this substance since cases of serious poisoning from boric acid have been reported.

The halogen family

Occurrence. Group VIIA of the periodic table is a chemical family of closely related elements. It consists of five elements—fluorine, chlorine, bromine, iodine, and astatine. Astatine is highly radioactive and generally is disregarded in a study of the halogens. They are called "the halogens," from a Greek word meaning "salt-former," because they combine with many metals to form a great number of salts. The halogens are found in nature as compounds, being too active chemically to exist as free elements. Fluorine occurs in the mineral fluorspar, or calcium fluoride. Chlorine is found chiefly in sodium chloride in seawater and in certain local deposits such as the beds of ancient dried-up salt lakes. Bromine occurs along with chlorine in salt-bed deposits in the form of sodium and magnesium bromides. Iodine is found principally as sodium iodide in seawater and in seaweeds and as sodium iodate ($NaIO_3$) in Chile saltpeter. In the human body, fluoride is found in the bones and teeth in small amounts; chlorine in the form of sodium chloride makes up a part of the inorganic salts of the tissues and body fluids and is a constituent of the hydrochloric acid of gastric juice; and iodine is a constituent of thyroxin, the active principle of the thyroid gland. Bromine compounds do not occur naturally in the human body in significant amounts.

Properties and periodic relationships. The properties and certain chemical relationships of the elements in the halogen family are shown in Table 4-1. This table shows gradual changes in the physical and chemical properties of these elements—changes that appear in the same order as would the elements listed in relation to their atomic number. As the atomic number of these elements increases, their melting point and boiling point rise; their physical state changes in the order of gas, liquid, and solid; their color changes

Table 4-1. Gradation of properties of the halogen family

Element	Atomic number	Melting point in °C (rises)	Boiling point in °C (rises)	Physical state	Color	Chemical activity
Fluorine	9	−223	−187	Gas	Pale yellow	Most active
Chlorine	17	−102	−33.6	Gas	Greenish yellow	Second in order of activity
Bromine	35	−7	+59	Liquid	Brownish red	Third in order of activity
Iodine	53	+107	+175	Solid	Dark red	Fourth in order of activity

Table 4-2. Atomic composition of the halogen elements

| Element | Nuclear composition | | Number of electrons in shell | | | | |
	Number of protons	Number of neutrons	1	2	3	4	5
Fluorine	9	10	2	7			
Chlorine	17	18	2	8	7		
Bromine	35	45	2	8	18	7	
Iodine	53	74	2	8	18	18	7

from pale yellow to a dark red; and their chemical activity decreases with increase in atomic number. This gradation of properties according to change in atomic number is one of the fundamental relationships between the elements, which were previously discussed in Chapter 2. It definitely indicates that the properties of the elements are related to their atomic numbers.

Relation to atomic structure. Table 4-2 gives the numbers of protons and neutrons in the nucleus and the number and grouping of the electrons in the shell of each halogen element. The data in Table 4-2 help to explain the physical and chemical properties of these elements. The melting point and boiling point, as shown in Table 4-1, rise with increase in the atomic number (and hence the atomic weight), which is determined by the number of protons and neutrons in the nucleus. Likewise, the variation in the physical state of these elements from gas to liquid to solid is directly related to the increase in the atomic weights.

The chemical properties of the halogen elements are determined by the number of electrons in the outer shell. Each of these elements has an outer electron shell containing 7 electrons (Table 4-2), of which 6 are paired and 1 is unpaired. When the halogens react chemically, the unpaired electron of a halogen atom is brought into chemical union with an unpaired electron belonging to another element. This gives the halogen atom an outer shell containing 8 electrons, a stable configuration. The halogens usually have a valence of −1 because 1 electron is needed to fill out their outer electron shell. In most cases they form negative ions because the halogen atom after a chemical reaction tends to become an ion with more electrons than protons. The halogen elements thus form a natural family with similar chemical properties, these properties being due to the identical outer electron structure of these elements. The halogens also form covalent bonds with certain elements. In covalent bonding they exhibit a variety of combining powers. In sodium iodate, the compound previously mentioned (p. 71), the iodine has a valence of 5.

$$NaIO_3$$

$$Na^+ \left[\begin{array}{c} \ddot{O} \\ \ddot{I}:\ddot{O} \\ \ddot{O} \end{array} \right]^-$$

Uses. Fluorine and hydrofluoric acid (HF) will attack glass. Hydrofluoric acid is therefore used for etching glass. Many synthetic fluorine compounds, such as Teflon and Freon, have become widely used.

Chlorine is a powerful disinfectant. It is used with excellent results in the purification of city water supplies, being introduced in the form of liquid chlorine or chlorinated lime (bleaching powder). Solutions containing chlorine in water have been used for irrigating and disinfecting wounds. One of these preparations is Dakin's solution, which is a mixture of sodium hypochlorite, sodium chloride, and boric acid. Chlorine, used in the form of bleaching powders ($CaOCl_2$), is an efficient bleaching agent but is subject to the limitation that chlorine will injure many materials. Chloroform, chloral, and carbon tetrachloride are chlorine compounds that are used in medicine. In the human body the compounds of chlorine are indispensable. Sodium chloride is essential for the vital activities of the tissues, and hydrochloric acid is necessary for gastric digestion.

One of the salts of bromine, sodium bromide, is used as a sedative.

Iodine is a valuable disinfectant. It is used chiefly in the form of tincture of iodine, which is a 7% solution of iodine in alcohol, to which potassium iodide (5%) is added. Potassium iodide is introduced into tincture of iodine because it makes iodine more soluble in water and will prevent iodine from precipitating if the tincture is diluted with water. Iodine in the form of some of its salts such as potassium iodide is essential in the diet to prevent the development of simple goiter. The introduction of potassium iodide into the diet in regions where iodides are deficient in the drinking water has given excellent results in the prevention of this condition. The human body uses iodine to synthesize the hormone thyroxin.

Questions for study

1. State the properties of sodium and potassium.
2. State the properties of calcium. Name several important calcium compounds in nature.
3. Relate the properties of magnesium. What is the principal use of magnesium?
4. State the properties of pure iron.
5. Describe mercury.
6. What is the difference between mercuric and mercurous chlorides? Explain why mercurous chloride is not poisonous.
7. What is the importance of bismuth in medicine?
8. What is the electromotive series? What does it tell us?

9. Describe the occurrence of sulfur.
10. What are the properties of sulfur? Give its uses.
11. What are the principal uses of sulfur dioxide?
12. How is sulfuric acid prepared?
13. State the properties of sulfuric acid.
14. What is the formula for Epsom salt? Glauber's salt?
15. What is the use of barium sulfate in medicine?
16. What are the properties of hydrogen sulfide?
17. Discuss the occurrence of phosphorus in the human body.
18. Discuss the contrast in the properties of white phosphorus and red phosphorus.
19. What are the uses of phosphorus?
20. What are the uses of arsenic?
21. What are the uses of boric acid?
22. What is a metalloid?
23. What is a chemical family?
24. Name the members of the halogen family. Why are they called the halogens?
25. Referring to the table of halogens, give two important properties of fluorine, of chlorine, of bromine, and of iodine.
26. Consulting Table 4-2, show the relationships of the members of the halogen family.
27. What is the relationship of the properties of these elements to their atomic numbers?
28. Why do the halogen elements behave similarly in chemical reactions?
29. What are the uses of fluorine? Chlorine? Bromine? Iodine?

5 Hydrogen, oxygen, oxidation-reduction, and equilibrium constant

HYDROGEN

Occurrence. Hydrogen occurs widely in nature. It constitutes about 1% of the mass of the earth's crust, but it has a much greater importance than is indicated by this percentage. The name "hydrogen" is derived from a Greek word meaning "producer of water," and the element was so named because it forms water when it combines with oxygen. Hydrogen exists in traces as a free element in the air and in the gases of volcanic eruptions and in larger amounts in the gaseous mantles surrounding the sun and the hotter stars. The chief occurrence of hydrogen, however, is in the combined state. Combined with other elements, it is found in water, acids, alkalies, plant and animal tissues, natural gases, and many other substances.

Preparation. Hydrogen is obtained from some of the compounds in which it occurs. Of these compounds it is most readily secured from acids and from water. The simpler methods of preparation from these two sources will be described.

Reaction of acids with certain metals. The reaction of acids with certain metals, for example, hydrochloric acid with zinc, produces hydrogen.

$$\text{Zn} + 2\text{HCl} \rightarrow \text{ZnCl}_2 + \text{H}_2\uparrow$$
$$\text{Zinc} \quad \text{Hydrochloric} \quad \text{Zinc} \quad \text{Hydrogen}$$
$$\text{acid} \quad \text{chloride}$$

In this equation zinc chloride is formed; zinc combines chemically with chlorine more readily than hydrogen does, and free hydrogen passes off as a gas, which is collected by displacement of water, as shown in Fig. 5-1.

Hydrogen may be prepared from many acids, but hydrochloric and sulfuric acids are the most convenient and the most satisfactory. A number of other metals, such as iron, tin, nickel, calcium, and magnesium, may be used instead of zinc in the preparation of hydrogen by this method, but there are certain metals—copper, mercury, silver, gold, and platinum—that will not react with acids and liberate hydrogen. The action of metals

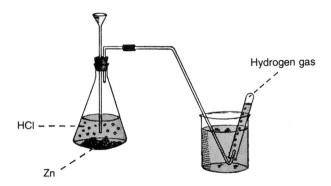

Fig. 5-1. Preparation of hydrogen using zinc and hydrochloric acid.

on acids is a convenient method of preparing hydrogen and is the one used in laboratory work.

Reaction of certain metals with water. If a small piece of potassium is tossed on water, hydrogen is evolved; the heat of the reaction is so great that the hydrogen is ignited and burns as it leaves the surface of the water, frequently in a rather violent explosion.

$$2HOH + 2K \rightarrow 2KOH + H_2\uparrow$$
Water — Potassium — Potassium hydroxide — Hydrogen

In this reaction an atom of hydrogen is removed from each water molecule, and the other hydrogen atom remains attached to the oxygen; the oxygen atom and its attached hydrogen atom form a hydroxyl ion that combines with an atom of potassium, forming potassium hydroxide. The released hydrogen atoms combine in pairs, forming molecules of hydrogen (H_2), and the hydrogen molecules, being very light, pass off as a gas.

Water is also decomposed by the metal sodium, hydrogen being liberated.

$$2HOH + 2Na \rightarrow 2NaOH + H_2\uparrow$$
Water — Sodium — Sodium hydroxide — Hydrogen

This reaction is the same as that of potassium and water except that less heat is evolved. The hydrogen liberated is often ignited by the heat of the reaction. Both reactions, the one with potassium as well as the one with sodium, are quite dangerous. They should never be attempted by students except under careful supervision.

To demonstrate that hydrogen is the gas evolved by these metals, place a small piece of sodium in a gauze spoon and plunge it very quickly under an inverted test tube of water in a vessel of water, as shown in Fig. 5-2. The water in the test tube will be displaced by the gas produced, and this gas can be identified as hydrogen by the tests given in the discussion of the properties of hydrogen (p. 77).

Calcium will decompose water and liberate hydrogen but at a much slower rate than potassium and sodium. Other metals such as magnesium, iron, and aluminum will react with water at high temperatures (superheated steam) and liberate hydrogen (see electromotive series p. 65).

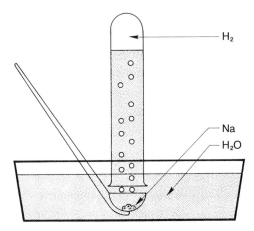

Fig. 5-2. Preparing hydrogen from water.

Physical properties. Pure hydrogen is colorless and odorless. It is the lightest of all the elements, its atomic mass being 1.00797. It is a gas at ordinary temperatures. By the application of considerable pressure and very low temperatures, hydrogen may be liquefied and even solidified. Solid hydrogen passes into the liquid state at $-257°$ C ($16°$ K); liquid hydrogen becomes a gas at temperatures warmer than $-253°$ C ($20°$ K).

Hydrogen is very slightly soluble in water, 100 ml of water dissolving 2 ml of hydrogen gas at standard conditions of temperature and pressure. For practical purposes hydrogen may be considered insoluble in water, and such low solubility can be used to an advantage in its preparation by collecting it by displacement of water in a jar inverted in a vessel of water.

Chemical properties. Hydrogen is not very active at ordinary temperatures, but at higher temperatures it combines with most nonmetallic elements and with some metals. It burns readily, producing a nonluminous flame, but it does not support combustion. A burning splinter will be extinguished if placed in a jar of pure hydrogen. When hydrogen is burned in the presence of sufficient oxygen, the hottest flame known is produced. The hydrogen flame produced with an oxyhydrogen blowpipe yields a temperature as high as $2500°$ C. Such a flame will vaporize gold and silver and will melt platinum and asbestos. Hydrogen combines with oxygen at high temperatures with explosive violence. It is this familiar "pop," heard when igniting a test tube of hydrogen mixed with air, that the student uses as a test for hydrogen in the laboratory.

The sole product obtained by burning hydrogen in oxygen or air is water. This can be demonstrated by collecting hydrogen in a dry test tube and then igniting the mixture. After the ignition, water will be plainly visible on the sides of the test tube.

Hydrogen also has the very important property of removing oxygen from the oxides of certain metals at high temperatures. For example, if hydrogen is passed over copper oxide heated to redness, the oxygen is removed from its combination with copper and unites with the hydrogen. Water will appear on the cool portions of the tube, and the black copper oxide will be changed into reddish metallic copper. Substances that remove oxygen from compounds are called reducing agents.

$$CuO + H_2 \rightarrow Cu + H_2O$$
Copper   Hydrogen   Copper   Water
oxide

Sometimes a hydrogen atom may complete its outer orbit by accepting an electron and becoming the *hydride* ion, H^-. Hydrides are formed when positive ions combine with the hydride ion.

$$H_2 + 2Na \rightarrow 2NaH$$

Uses. Hydrogen, being the lightest gas, was once used extensively for inflating balloons. Since the lifting power of a balloon is proportional to the difference between the weight of the gas used for its inflation and the weight of an equal volume of air, it follows that hydrogen has the greatest lifting power of all gases. But during World War I, and for several years thereafter, serious accidents occurred with hydrogen-filled balloons, which were caused by the ignition of hydrogen and its explosive violence when burning. Because of the danger from combustion, the use of hydrogen has been discontinued, and helium, a gas that will not burn, has been substituted for hydrogen in the construction of balloons, especially of the dirigible type.

Another commercial use of hydrogen is in the hardening of liquid fats. Many vegetable fats are unpalatable liquids and as such have been considered undesirable for human consumption. These "soft" fats or oils can be hardened by treating them with hydrogen in the presence of finely divided nickel. However, the dietary use of polyunsaturated oils has diminished the importance of this process. In addition to the preparation of solid fats for consumption as foods, the hydrogenation of oils has been extended to the hardening of certain oils that are used in the manufacture of soaps.

OXYGEN

Occurrence. Oxygen is the most abundant of all the elements. It exists both in the combined and in the free elementary state. Combined with other elements, oxygen constitutes by weight one-half of the substances composing the earth's crust, eight-ninths of water, two-thirds of the human body and the body of animals, and one-half of the tissues of plants. As the free element, oxygen makes up about one-fifth of the air by volume and is found in small quantities in the soil and dissolved in the natural waters of the earth. The abundance of oxygen as compared with other elements is shown graphically in Fig. 5-3.

Preparation. Oxygen may be prepared from mercuric oxide, from potassium chlorate, or from air.

From mercuric oxide. Perhaps the simplest method of preparing oxygen is by heating mercuric oxide. This is the method that was used by Priestley, who discovered this element in 1774. Priestley's method is represented by the following equation:

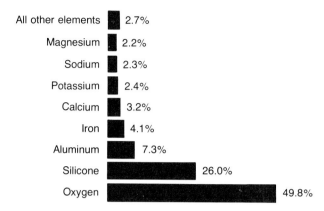

Fig. 5-3. Graph showing relative abundance of the elements in the crust of the earth.

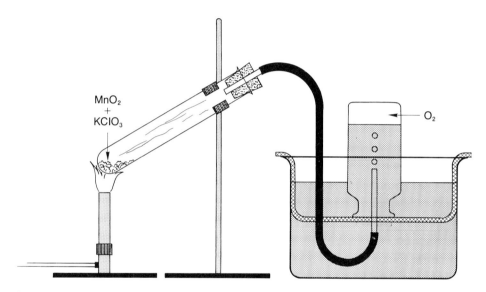

Fig. 5-4. Preparation of oxygen by heating a mixture of potassium chlorate and manganese dioxide.

$$2HgO \;+\; \text{Heat} \;\rightarrow\; 2Hg \;+\; O_2\uparrow$$
Mercuric oxide　　　　　　　Mercury　Oxygen

In this reaction the chemical attraction between mercury and oxygen is overcome by heating, and free mercury and oxygen are produced. The oxygen, being a gas that is only slightly soluble in water, is collected by displacement of water.

From potassium chlorate. If potassium chlorate, a compound containing oxygen, is heated to about 400° C, oxygen will be liberated and can be collected by displacement of water, as shown in Fig. 5-4. If the potassium chlorate is mixed with a small amount of powdered manganese dioxide and the mixture heated, oxygen will be evolved at only one-half as high a temperature. The reactions are as follows:

$$2KClO_3 + \text{Heat } (400°\,C) \rightarrow 2KCl + 3O_2\uparrow \quad (1)$$
Potassium chlorate → Potassium chloride + Oxygen

$$2KClO_3 + MnO_2 + \text{Heat } (200°\,C) \rightarrow 2KCl + 3O_2 + MnO_2 \quad (2)$$
Potassium chlorate + Manganese dioxide → Potassium chloride + Oxygen + Manganese dioxide

In preparing oxygen according to equation 2, manganese dioxide is introduced because it acts as a catalyst and makes it possible to liberate oxygen at a much lower temperature (200° C) than is necessary to prepare oxygen by heating potassium chlorate alone (400° C).

From air. Since the air consists of a mixture of gases, chiefly oxygen and nitrogen, the problem of preparing oxygen from air is the separation of these gases. This separation can be accomplished by converting air into the liquid state. Air is liquefied by pumping it into strong chambers where it is submitted to great pressure at low temperatures. Later it is permitted to evaporate. Advantage is taken of the fact that liquid nitrogen evaporates at a temperature 13° lower than that of liquid oxygen. Therefore the nitrogen escapes first when liquid air is evaporated. Oxygen prepared in this manner is not pure, since it contains a small percentage of nitrogen and other gases of the atmosphere. Nevertheless, it is satisfactory for commercial use, and this is the method used at the present time to obtain oxygen for commercial purposes.

Isotopes of oxygen. Naturally occurring oxygen exists in three isotopic forms. The atomic mass numbers of these forms are 16, 17, and 18; the atomic number of all is the same, 8. The amounts of the isotopes with atomic weights 16, 17, and 18 in the atmosphere are 99.76%, 0.04%, and 0.20%, respectively. The number of protons and electrons in the atoms of these three isotopes is the same, 8. The essential difference in these forms is in their atomic mass (16, 17, or 18), which is due to the different number of neutrons in the nucleus (8, 9, or 10). Since the number of electrons is the same, there is no difference in the chemical properties of the three isotopes.

Physical properties. Oxygen is a colorless, odorless, and tasteless gas. It is slightly heavier than air. A liter of oxygen weighs 1.43 g and a liter of air, under the same conditions of temperature and pressure, weighs 1.27 g. The element is slightly soluble in water, 100 ml of water dissolving 3 ml of oxygen at room temperature and at a pressure of 1 atmosphere. Oxygen may be liquefied by applying great pressure to the gas at very low temperatures, and it may be converted into the solid state by applying pressure and producing still lower temperatures. Liquid oxygen is a bluish fluid that evaporates at $-183°$ C. Solid oxygen is a pale blue snowlike substance with a melting point of $-218°$ C.

Chemical properties. Oxygen reacts slowly with other substances at ordinary temperatures and vigorously at high temperatures. It is capable of combining with nearly all the other elements, and it will react with a great many compounds.

When substances enter into a chemical reaction that produces heat and light, the process is called combustion. This process is commonly spoke of as "burning." If a substance burns, it is said to be combustible. Examples of combustible materials are wood, ether, gasoline, and hydrogen; these substances combine with oxygen at appropriate temperatures, and the chemical change produces heat and light. Since oxygen is essential for the combustion of so many substances, we consider it a *supporter of combustion* rather than a combustible material. Other elements may take the place of oxygen in producing a combustion, but combustion practically always means a chemical combination of some free element or the elements of some compound with oxygen at higher than ordinary temperatures. Oxygen is therefore the great supporter of combustion.

The influence of oxygen as a supporter of combustion can be easily demonstrated in the laboratory. If we thrust a feebly burning splinter of wood into a jar of pure oxygen, it will glow with a flame many times more brilliant than that which is exhibited in the air. Sulfur, which burns freely in the air, burns with a more brilliant flame when ignited and placed in pure oxygen. Iron in the form of steel wool will burn in pure oxygen with a dazzling flame if it is heated to redness before being placed in the oxygen.

The necessity of oxygen for combustion is further demonstrated in the operation of a stove or furnace. If we shut off the "draught" and stop the admission of air, which is one-fifth oxygen, the fire will not burn. If we desire to increase the burning process, we open the air-regulating device and admit more air, which is equivalent to admitting more oxygen, and the fire burns more vigorously.

Oxidation. The union of oxygen with other substances is called oxidation. Oxidation may be slow or rapid. When slow oxidation takes place, heat is evolved but usually this heat is conducted away as rapidly as it is produced, and no increase in temperature is noted; the amount of heat liberated by the slow oxidation of a substance, however, is the same as that produced by the most rapid oxidation of a similar amount of the same substance. We see many instances of slow oxidation at ordinary temperatures in everyday life. The rusting of iron results from the combining of iron with oxygen; the decay of plant tissues and of waste products of animals is an oxidation; and the drying of paint is the union of oxygen with the linseed oil in which the paint is dissolved. At higher temperatures many substances can be oxidized that are not oxidized at ordinary temperatures. Oxidation also takes place more rapidly when the temperature is raised.

Oxidation is a significant chemical change because of the heat that it liberates. It is by the oxidation of foods in the body that the energy stored in foods is made available for work and the body is kept warm. It is the oxidation of wood or coal or oil that produces the heat that warms our dwelling places and runs the engines of commerce and industry. Oxidation is, in short, a chemical reaction without which life would not exist.

Oxides. An oxide is a compound in which oxygen is combined with some other element or elements. Oxides are usually formed by oxidation or direct combination of oxygen with another element, as shown in the following equations:

$$4Al + 3O_2 \rightarrow 2Al_2O_3$$
Aluminum   Oxygen   Aluminum oxide

$$C + O_2 \rightarrow CO_2$$
Carbon Oxygen Carbon dioxide

$$S + O_2 \rightarrow SO_2$$
Sulfur Oxygen Sulfur dioxide

A few oxides are prepared more conveniently by heating compounds containing several elements, of which oxygen is one.

$$CaCO_3 + \text{Heat} \rightarrow CaO + CO_2$$
Calcium carbonate   Calcium oxide   Carbon dioxide

Spontaneous combustion. We have previously seen that oxidation of many substances takes place at ordinary temperatures. Sometimes such substances are packed in places where they oxidize slowly, and the heat produced is not removed but accumulates until the temperature gets high enough to ignite the material. In this manner fires are often started. Fires of such origin are said to be due to spontaneous combustion. *Spontaneous combustion is a self-starting combustion; it originates from the heat that accumulates in a confined region as a result of slow oxidation at ordinary temperatures.* Familiar examples of spontaneous combustion are fires that occur in heaps of oily rags, especially rags containing linseed oil, which oxidizes readily, fires in coal yards when fine-grained soft coal is packed together in large amounts, and fires occurring in factories when organic matter is piled too compactly.

Relation to life. Most living things require free molecular oxygen. This is because oxygen will unite with other substances and produce heat. Oxidation is thus an important process in living things since heat is liberated and energy is made available for activity by the oxidation of foods.

Animals require large quantities of oxygen, and they must obtain it constantly and rapidly; without oxygen, animal life would cease in a few minutes. Hence most animals have specialized organs (lungs, gills) for absorbing oxygen from the air or water and making it available to their tissues. In the lowest forms of animal life no special oxygen-absorbing mechanism is necessary since these forms consist of relatively few cells or of a single cell, and in these instances oxygen can be absorbed directly from the environment.

A few forms of life do not require free oxygen for oxidation reactions and are called *anaerobes,* a name that means "without air." Certain bacteria such as tetanus, botulinus, and anthrax are examples of anaerobic organisms. These bacteria will not grow except in the absence of free oxygen.

When foods are oxidized in the animal body, carbon dioxide, a colorless gas, is produced. This substance is undesirable in the body in excessive amounts and is consequently carried by the blood to the lungs and excreted through the lungs. Respiration than has a twofold function: taking oxygen into the lungs, where it can be absorbed by the blood and carried to the tissues, and removing carbon dioxide, an undesirable product, to the exterior.

Hydrogen, oxygen, oxidation-reduction, and equilibrium constant

From this discussion it is evident that the constant withdrawal of oxygen from the air by animals will soon exhaust the oxygen content of the atmosphere unless there is some process by which oxygen can be returned to the air. Furthermore, since animals are constantly giving off carbon dioxide and a great deal of carbon dioxide is produced from the oxidation of organic matter, which is composed of carbon compounds, some mechanism to prevent the accumulation of carbon dioxide is equally important. Nature has provided well for this situation by the activities of plants. Plants absorb carbon dioxide from the air, water from the soil, and, in the presence of sunlight, manufacture their own food by a process called *photosynthesis*. During photosynthesis, oxygen is given off by the plant, and in this way the oxygen supply of the atmosphere is continuously renewed. It is true that the plant uses some oxygen in its life processes, but by active photosynthesis a great deal more oxygen is produced than is necessary for the plant's needs, and the extra oxygen is given off by the plant into the atmosphere. Plants thus have a purifying influence on the air; they remove carbon dioxide and supply free oxygen. As oxygen passes from the plant to the animal in the form of free oxygen and back to the plant again in the form of carbon dioxide, the process is a complete cycle and is commonly referred to as the *oxygen cycle in nature*.

Uses. Oxygen is a useful element. Its importance in various uses will be summarized.

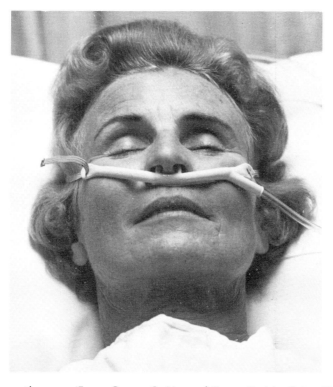

Fig. 5-5. Oxygen therapy. (From Gragg, S. H., and Rees, O. M.: Scientific principles in nursing, ed. 7, St. Louis, 1974, The C. V. Mosby Co. Courtesy Hoag Memorial Hospital, Newport Beach California.)

All plants and animals except a few anaerobes require free molecular oxygen for the living processes that take place in their tissues.

Oxygen therapy is given as a life support technique through a mask or nasal catheter applied directly to the patient's nose and mouth (Fig. 5-5).

Miners who undertake rescue work following a mine accident are subject to the hazards of getting into regions in which the air contains large amounts of poisonous gases. To save themselves when such an atmosphere is encountered, each miner is equipped with a tank of oxygen, carried on his back, from which he can breathe when a region of foul air is entered. Deep sea divers are supplied with pure oxygen mixed with helium from tanks attached to their helmets by long connecting tubes.

The human body is adapted to carry on respiration successfully in an atmosphere of 20% oxygen at altitudes varying from sea level or slightly lower to around 5,000 to 10,000 feet above sea level. As a person ascends from the surface of the earth the atmosphere around him becomes less dense, the oxygen tension is reduced, and breathing becomes more difficult. In rapid ascents to high altitudes it is therefore necessary to provide the needed oxygen. Modern high-altitude aviation and space flight depend on the use of apparatus for supplying oxygen to the occupants of the airplane or space capsule at a concentration and pressure that maintain normal breathing.

Oxygen is an important factor in the removal of organic matter from the surface of the earth by the process called decay. Decay is a slow oxidation of organic matter. It is facilitated by the action of bacteria. If it were not for the process of decay or continuous slow oxidation, the surface of the earth would soon be covered with the tissues of dead plants and animals, and life on the earth would not be possible.

Oxygen is a great disinfecting agent. The bactericidal power of many substances such as hydrogen peroxide and potassium permanganate is due to the oxygen that is liberated from these compounds by chemical action. Oxygen is one of the disinfecting agents used by nature to free natural waters of microorganisms. Natural waters in modern times often become polluted with sewage, which contains myriads of bacteria, and it is through the action of dissolved oxygen that many of these organisms are destroyed.

As previously noted, hydrogen mixed with oxygen in correct proportions and ignited yields the hottest flame attainable by combustion. The oxyhydrogen torch is an important industrial application of this principle. Instead of pure hydrogen, acetylene, a hydrogen-containing compound, is now generally used because acetylene is less expensive and produces a very satisfactory flame, the apparatus in this case being called the oxyacetylene torch. Pure oxygen as the supporter of combustion is necessary for the application of these industrial devices.

Liquid oxygen is used to supply the oxygen necessary for the combustion of fuels that produce the energy for driving rockets into space.

Ozone

The atoms composing oxygen and ozone are identical. The essential difference between these two substances is in the number of atoms in their molecules. The oxygen molecule contains 2 atoms of oxygen; the ozone molecule is composed of 3 atoms of oxygen and is an allotropic form of oxygen.

Hydrogen, oxygen, oxidation-reduction, and equilibrium constant

Preparation. Ozone is prepared by passing electric discharges through oxygen or air. The discharge of electric energy into oxygen gas decomposes the oxygen molecules into atoms, and these atoms recombine and form molecules containing 3 atoms of oxygen.

$$3O_2 \xrightarrow{\text{Electric discharge}} 2O_3$$

Ozone may therefore be detected in the air following electric storms and in the vicinity of electrical apparatus that produce "sparks" or discharges. In the reaction described, a mixture of oxygen and ozone is produced. Ozone is separated by cooling the mixture until it becomes a liquid. The ozone will liquefy at around $-112°$ C, a temperature at which oxygen remains in the gaseous state.

Properties. Ozone is a colorless gas with a penetrating, characteristic odor. It is 1.5 times heavier than oxygen. It is a powerful oxidizing reagent, being more active in this respect than molecular oxygen. It is unstable; when it decomposes, atomic oxygen is formed.

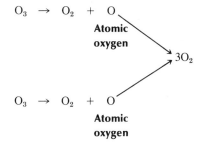

Atomic oxygen is a strong disinfectant and bleaching agent. Ozone has been used for destroying bacteria in air and water and for removing obnoxious odors from the air in restaurants, theaters, and public buildings.

OXIDATION AND REDUCTION

Oxidation. The term *oxidation* can be used to designate the union of oxygen with other elements. This term has been extended, however, to include other, similar chemical reactions. As it is now used, oxidation includes many chemical changes in which oxygen does not take part. Thus the following reactions are oxidations:

$$2Mg + O_2 \rightarrow 2MgO \qquad (3)$$

$$Mg + Cl_2 \rightarrow MgCl_2 \qquad (4)$$

$$Mg + Br_2 \rightarrow MgBr_2 \qquad (5)$$

Reaction 3 is an oxidation in the initial meaning of the term. Reactions 4 and 5 are oxidations in which oxygen does not take part. From these examples it is apparent that oxidation, as is now used, means more than chemical reactions in which oxygen takes a part. An examination of these three reactions reveals one underlying change that is common to all of them. There is a loss of electrons by the magnesium atom. This reaction may be written in either of the following ways:

$$Mg^0 \rightarrow Mg^{++} + 2 \text{ electrons} \tag{6}$$

$$Mg\colon \rightarrow Mg^{++} + 2 \text{ electrons} \tag{7}$$

The neutral, uncombined magnesium atom, which has 2 electrons in its outer shell giving it a valence of 0, loses 2 electrons and becomes a positive ion with a valence of +2. Therefore *oxidation is a loss of electrons, a change that brings about an increase in positive valence, or oxidation number*. Reactions 6 and 7 show one-half the oxidation reaction of the total reaction shown in 3, 4, or 5. The accompanying reduction is shown in reactions 8, 9, and 10. When one-half of a reaction is shown in this way it is called a *half-reaction* or a *half-cell reaction*.

Reduction. Note that the magnesium atoms were oxidized in the oxidation reactions just described. But this is only a part of the changes that occurred. We must next consider what happened to the oxygen, the chlorine, and the bromine. When the atoms of oxygen, chlorine, and bromine reacted with magnesium, an electron change occurred in them also but in an opposite direction. The atoms of these elements gained electrons, a change that decreased their positive valence. The electrons these atoms gained came from the magnesium atoms with which they reacted. The changes in these elements may be indicated as follows:

$$\colon\ddot{\text{O}}\cdot + 2 \text{ electrons} \rightarrow \colon\ddot{\text{O}}\colon^{-2} \tag{8}$$

$$\colon\dot{\text{Cl}}\cdot + 1 \text{ electron} \rightarrow \colon\ddot{\text{Cl}}\colon^{-1} \tag{9}$$

$$\colon\dot{\text{Br}}\cdot + 1 \text{ electron} \rightarrow \colon\ddot{\text{Br}}\colon^{-1} \tag{10}$$

In reaction 8 the oxygen atom shows a gain of 2 electrons (from magnesium), which brings about a decrease in positive valence, the valence of oxygen becoming -2. In reactions 9 and 10 a gain of electrons with a decrease in positive valence also occurs, the chlorine and bromine atoms finally showing a negative valence of 1. The term used to indicate such changes is reduction. *Reduction is thus a chemical change in which there is a gain of electrons with a consequent decrease in positive valence, or oxidation number*.

The assignment of oxidation numbers is a way of showing valence based on the number of valence electrons involved and the electronegativity of the elements reacting. The oxidation number of elements in the free state is 0. When elements react, the more electronegative elements are assigned negative oxidation numbers indicating the number of electrons involved in the reaction. The less electronegative elements are assigned positive oxidation numbers, these numbers again indicating how many electrons are involved. Oxidation numbers also are assigned to covalently bonded elements indicating positive and negative charges when no transfer of electrons actually takes place. For example:

$$H^{+1}N^{+5}O_3^{-2}$$

or more exactly

$$H^{+1}\left[N^{+5}O_3^{-2}\right]^{-1}$$

Hydrogen, oxygen, oxidation-reduction, and equilibrium constant

Oxidation involves reduction. Oxidation and reduction are inseparable chemical changes. One always involves the other. It is impossible to oxidize one element without bringing about a reduction in some other element. Whenever an oxidation occurs, there is always a corresponding reduction.

When an element is oxidized, its positive valence is increased; this means that it has lost electrons.

$$Cu^0 \rightarrow Cu^{++} + 2\,e^- \tag{11}$$

$$Fe^{++} \rightarrow Fe^{+++} + 1\,e^- \tag{12}$$

$$Sn^{++} \rightarrow Sn^{++++} + 2\,e^- \tag{13}$$

In reaction 11, copper has its valence increased to a positive value of 2; to do this, it loses 2 electrons. In reaction 12, iron (II) ion, with a positive valence of 2, is changed to iron (III) ion with a positive valence of 3; this results from the loss of 1 electron. In reaction 13, tin (II) ion, with a positive valence of 2, is changed to tin (IV) ion with a positive valence increased to 4, a change that is brought about by the loss of 2 electrons.

Reduction, which is the opposite of oxidation, is a chemical change in which there is a gain in electrons by the atoms of an element. In reactions 3, 4, and 5, the oxidizing agents (oxygen, chlorine, and bromine) had their positive valence decreased; this was brought about by the addition of electrons to each of these elements. Reduction thus occurs when electrons are added to an element.

$$Cu^{++} + 2\,e^- \rightarrow Cu^0 \tag{14}$$

$$Fe^{+++} + 1\,e^- \rightarrow Fe^{++} \tag{15}$$

$$Sn^{++++} + 2\,e^- \rightarrow Sn^{++} \tag{16}$$

Oxidation is a chemical change in which there is a loss of electrons by the atoms of an element; oxidation is therefore an increase in positive valence, or oxidation number. Reduction is the opposite of oxidation. Reduction is a chemical change in which there is a gain in electrons by the atoms of an element; reduction is therefore a decrease in positive valence, or oxidation number.

A chemical reaction that involves a change in the oxidation state of some of its elements is called an *oxidation-reduction reaction* or a *redox reaction*.

Balancing redox equations. Equations for redox reactions must be balanced to allow for an even exchange of electrons. A brief outline of this procedure follows.

Consider a sample of iron (II) sulfate (ferrous sulfate) reacting with an aqueous acid solution of potassium permanganate. Analysis of the products have revealed that potassium, iron (III) (ferric), and manganous sulfates have been formed.

Step 1. Write the unbalanced equation for the reaction. Water can be assumed to be present in whatever quantity is needed, since the reaction takes place in an aqueous medium.

$$FeSO_4 + H_2SO_4 + KMnO_4 \rightarrow K_2SO_4 + Fe_2(SO_4)_3 + MnSO_4 + H_2O$$

Step 2. Write oxidation numbers for each element involved in the equation.

$$Fe^{+2}(S^{+6}O_4^{-2})^{-2} + H_2^{+1}(S^{+6}O_4^{-2})^{-2} + K^{+1}(Mn^{+7}O_4^{-2})^{-1} \rightarrow$$
$$K_2^{+1}(S^{+6}O_4^{-2})^{-2} + Fe_2^{+3}(S^{+6}O_4^{-2})_3^{-2} + Mn^{+2}(S^{+6}O_4^{-2})^{-2} + H_2^{+1}O^{-2}$$

Step 3. Identify the elements that have changed oxidation numbers.

$$Fe^{+2} \rightarrow Fe^{+3}$$
$$Mn^{+7} \rightarrow Mn^{+2}$$

Step 4. Calculate the electron change for each change in oxidation number of the two half-reactions.

$$Fe^{+2} - 1e^{-1} \rightarrow Fe^{+3} \qquad\qquad Mn^{+7} + 5e^{-1} \rightarrow Mn^{+2}$$
$$\text{Oxidation} \qquad\qquad\qquad\qquad \text{Reduction}$$

Step 5. Multiply by the proper numerical coefficients to allow the electron exchange to be balanced. Add the two half-reactions.

$$5Fe^{+2} - 5e^{-1} \rightarrow 5Fe^{+3}$$
$$\underline{Mn^{+7} + 5e^{-1} \rightarrow Mn^{+2}}$$
$$5Fe^{+2} + Mn^{+7} \rightarrow 5Fe^{+3} + Mn^{+2}$$

Step 6. Rewrite the original equation showing the results of the balancing of the electron exchange in step 5.

$$5FeSO_4 + H_2SO_4 + KMnO_4 \rightarrow 0.5K_2SO_4 + 2.5Fe_2(SO_4)_3 + MnSO_4 + H_2O$$

Step 7. By inspection, balance the rest of the equation. Four molecules of sulfuric acid need to be taken on the left-hand side of the equation, since 8 sulfate ions are used on the right-hand side of the equation and iron sulfate can give only 5 sulfate ions. At this time the coefficients of the molecules containing elements that undergo exchange of electrons must not be increased or decreased. These values have been set by step 5.

$$5FeSO_4 + 4H_2SO_4 + KMnO_4 \rightarrow 0.5K_2SO_4 + 2.5Fe_2(SO_4)_3 + MnSO_4 + H_2O$$

At this stage the equation is balanced, but convention dictates the use of coefficients that are whole numbers. Multiplying the whole equation by two will give whole number coefficients.

Step 8. Multiplying the equation by two gives a conventional, balanced equation.

$$10FeSO_4 + 8H_2SO_4 + 2KMnO_4 \rightarrow K_2SO_4 + 5Fe_2(SO_4)_3 + 2MnSO_4 + 8H_2O$$

EQUILIBRIUM CONSTANT

In a reversible reaction (see p. 14) equilibrium between the forward reaction and the reverse reaction is attained when the speed of the forward reaction equals the speed of the reverse reaction. The speed of a reaction depends on many things, among which are

1. The chemical nature of the reactants—this factor takes into account the chemical reactivity of the substances themselves.
2. The physical state of the reactants—as a rule, finely divided substances react more quickly than do large aggregates. Dissolved substances usually react more quickly than do those that are undissolved.
3. The temperature—most reactions are speeded by the application of heat.
4. The pressure—the reaction of gasses are greatly affected by the pressure over the the reaction vessel. Increased pressure favors a reaction resulting in a reduction of

Hydrogen, oxygen, oxidation-reduction, and equilibrium constant

volume. Decreased pressure favors a reaction resulting in an increase of volume.
5. The presence or absence of catalysts—catalysts only change the rate at which a state of equilibrium is reached.
6. The concentration of the reactants—these concentrations are expressed in moles. The effect of the concentration can be expressed as follows:

$$A + B \rightleftarrows C + D$$

For the forward reaction, the speed varies with respect to the product of the molar concentrations of the reactants A and B, each raised to the power of the coefficients that accompany them.

$$s_1 \propto [A]^1 \times [B]^1$$

For the reverse reaction, the speed varies with respect to the product of the molar concentration of reactants C and D, again each raised to the power of their coefficients.

$$s_2 \propto [C]^1 \times [D]^1$$

For any single reaction, the conditions described in the preceding items 1 through 6 are constant and can be expressed by a numerical constant, k_1 for the forward reaction and k_2 for the reverse reaction. Now the following expressions can be written:

$$s_1 = k_1[A]^1[B]^1$$
Forward reaction

$$s_2 = k_2[C]^1[D]^1$$
Reverse reaction

At equilibrium $s_1 = s_2$ and

$$k_1[A]^1[B]^1 = k_2[C]^1[D]^1$$

$$\frac{k_1}{k_2} = \frac{[C]^1[D]^1}{[A]^1[B]^1}$$

Since k_1 and k_2 are numerical constants, the quotient resulting from the division of k_1 by k_2 results in another numerical constant called the equilibrium constant, or K_{eq}.

$$K_{eq} = \frac{[C]^1[D]^1}{[A]^1[B]^1}$$

Using this equation, one can solve the following problem. At a fixed temperature and pressure, the reaction

$$H_2 + Cl_2 \rightleftarrows 2HCl$$

revealed the following data at equilibrium:

$$[H_2] = 6 \times 10^{-8} \text{ moles}$$
$$[Cl_2] = 6 \times 10^{-8} \text{ moles}$$
$$[HCl] = 10 \text{ moles}$$

What is the equilibrium constant?

$$K_{eq} = \frac{[HCL]}{[H_2]^1[Cl_2]^1}$$

$$K_{eq} = \frac{10^2}{(6 \times 10^{-8})^2} = \frac{100}{36 \times 10^{-16}} = 2.9 \times 10^{16}*$$

The K_{eq} is a large number indicating conditions favoring the production of HCl.

---

*See Appendix for a review of exponential numbers.

Questions for study

1. Where does hydrogen occur?
2. Describe two methods for the preparation of hydrogen.
3. Give three physical properties and three chemical properties of hydrogen.
4. What is formed when hydrogen burns in oxygen?
5. Give three uses of hydrogen.
6. Discuss the use of hydrogen in the preparation of vegetable shortenings.
7. Discuss the occurrence of oxygen.
8. Describe two methods for the preparation of oxygen.
9. Give three physical and two chemical properties of oxygen.
10. What is combustion?
11. What is oxidation? Why is oxidation such an important chemical change?
12. What is an oxide? How are oxides prepared?
13. Explain how spontaneous combustion originates. Give examples.
14. What is the relationship of oxygen to life in general? What are anaerobes?
15. How is oxygen carried to the tissues of the human body?
16. The oxygen of the air is constantly being used by animals. Why does not the supply become exhausted?
17. How is the air freed from excess carbon dioxide?
18. What are the applications of pure oxygen?
19. Describe in detail the medical uses of pure oxygen.
20. Mention several nonmedical uses of oxygen.
21. What is the essential difference between oxygen and ozone?
22. What are the properties of ozone?
23. What is allotropism?
24. What is oxidation? Give illustrations.
25. What is reduction? Give illustrations.
26. Balance the following redox equations:
    a. $K_2CrO_7 + HCl \rightarrow KCl + CrCl_3 + H_2O + Cl_2$
    b. $KMnO_4 + Na_2SO_3 + H_2SO_4 \rightarrow K_2SO_4 + MnSO_4 + Na_2SO_4 + H_2O$
27. What is the equilibrium constant of a reaction? To what reactions is it applicable? What can it tell us?

6 Water, hydrogen peroxide, solutions, standard solutions, and milliequivalents

WATER

Occurrence. Water is the most abundant compound of all on the earth. It exists in all three states of matter: solid, liquid, and gaseous. Three-fourths of the earth's surface is covered by water, and the atmosphere contains water vapor in considerable amounts. Water is also a large and essential constituent of living matter. The human body is about two-thirds water, and the bodies of animals have a somewhat similar composition. Plant tissues have a water content varying from 50% to 90%.

Composition. The composition of water is as follows:

> By volume: 2 parts of hydrogen to 1 part of oxygen
> By weight: 1 part of hydrogen to 8 parts of oxygen

The difference between the volume and weight ratios is explained by the fact that the atomic weight of oxygen is 16 times that of hydrogen. The composition of water may be demonstrated by several methods, two of which are analysis and synthesis.

Analysis. Decomposing a substance and identifying its elements is called analysis. Electrolysis is an analytical method of showing that water contains the elements hydrogen and oxygen and of showing the relationship between these elements by volume. The word "electrolysis" literally means "split by electricity." *Electrolysis is the decomposition or splitting of a substance by passing an electric current through it.*

Pure water is a poor conductor of electricity. It is therefore necessary in performing electrolysis to add some substance that will enable an electric current to pass through it. For this purpose a little acid is added.

The electrolysis of water is performed with apparatus such as is shown in Fig. 6-1. This apparatus is filled with water to which a little sulfuric acid has been added, and an electric current is passed through the solution. As the current passes through the acidified water, bubbles of gas appear near two metal plates and pass to the top of the tubes containing these plates. These plates are called *electrodes*. The gas that collects in the tube

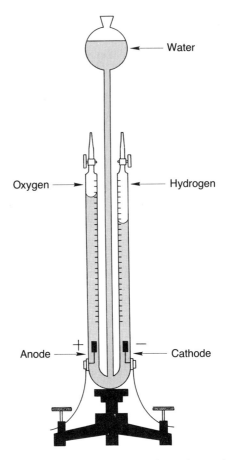

Fig. 6-1. Apparatus used for the electrolysis of water.

containing the anode is oxygen, and that in the tube containing the cathode is hydrogen. After an electric current has been passed through the acidified water for several minutes, it can readily be observed on the milliliter graduations of the two collecting tubes that the volume of hydrogen is twice that of oxygen. For every milliliter of oxygen produced, 2 ml of hydrogen are formed. The gases appearing above the cathode and anode can be drawn off through the stopcocks and proved to be hydrogen and oxygen by the chemical tests previously mentioned.

Synthesis. The composition of water can be demonstrated by *synthesis,* that is, *by putting together the elements of which it is composed.* The synthesis of water may be carried out by burning hydrogen in air or by collecting a mixture of the hydrogen and oxygen produced by electrolysis and igniting it. If a mixture of hydrogen and oxygen in a dry test tube is ignited, the two elements will combine explosively, and water will be formed on the sides of the tube.

Physical properties. Water may be identified by its characteristic physical properties.
Color, odor, and taste. Pure water is colorless, odorless, and tasteless. The pleasing

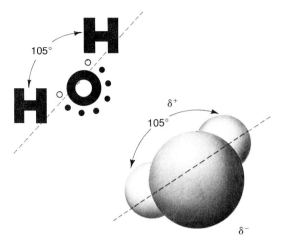

Fig. 6-2. Polar water molecule.

taste of good drinking water is due to the small amounts of other substances that are dissolved in it.

Solvent power. Water is a great solvent; that is, it has great power to dissolve substances or make them go into solution. It is considered one of the best solvents since so many substances readily dissolve in it. The water molecule is a polar molecule with a concentration of electrons on one side. The effect of this concentration of electrons is to make the molecule more negatively charged on one side and less negatively (more positively) charged on the other side (Fig. 6-2). This polarity makes the water molecule a very effective solvent for substances that are made up of ions or that have points of polarity on their molecules. A positive ion is attracted to the negative side, whereas a negative ion is attracted to the positive side of the water molecule. Some nonionic, nonpolar substances will not dissolve in water. Chief among these are fats. It will be shown later that fats need a nonpolar solvent to dissolve them.

The polar molecules of water attract each other forming hydrogen bonds between the molecules. This close association among water molecules is reflected in physical constants, such as the boiling point and freezing point of water and crystal formation of ice. Molecules other than water are held in close association by means of hydrogen bonding. Hydrogen bonds also help to structure complicated organic molecules.

Hydrogen bonding in water

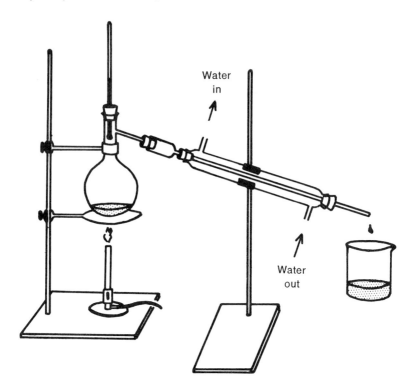

Fig. 6-3. Distillation apparatus.

Chemically pure water in the liquid state does not exist in nature. This is because water is such an excellent solvent. It dissolves some of almost every substance with which it comes in contact; consequently, the waters with which we are familiar contain varying amounts of mineral matter, salts, and gases in solution. These substances do not make water impure from a hygienic viewpoint; some of the mineral substances in natural waters are essential constituents of the diet. We must therefore qualify the term "pure water." From the chemical point of view, pure water means that the compound, whose formula is H_2O, is in a state free from all other substances. This is produced by distilling ordinary water. In the distillation process, steam from boiling water is condensed by cooling, resulting in chemically pure water. Impurities do not vaporize or condense at exactly the same temperature as water and are either distilled off before the water or are left behind in the distillation flask (Fig. 6-3). Pure water, hygienically speaking, is water free from disease-producing organisms or poisonous substances that make it unhealthful for drinking purposes.

Freezing point and boiling point. At 760 mm Hg pressure and 0° C (32° F), water becomes a solid (ice); this temperature is called the freezing point of water. At the same pressure and 100° C (212° F), water becomes a gas (steam); this temperature is referred to as the boiling point of water. Freezing and boiling points are influenced by both temperature and pressure.

Change of volume and density. At 4° C water is at its greatest density. It expands and becomes less dense (lighter in weight) when either heated above or cooled below this temperature. Most liquids expand and become lighter only when heated. The expansion of water as it is cooled below 4° C and becomes a solid at 0° is a property of the utmost importance to marine life. If water did not grow less dense as it freezes, the ice formed in cold weather, being heavier than liquid water, would sink to the bottom of the rivers, lakes, and oceans, and the animal life of the sea would be destroyed. But being lighter than water, ice floats on top of it, and the life of marine animals is not interfered with. The expansion of water in freezing is of further importance in nature in that it causes rocks to crumble and yield to the land their enriching mineral constituents, which are valuable for plant growth.

Chemical behavior. The chemical properties of water that are of special interest are stability, activity, and hydrolysis.

Stability. Water is a stable compound. It is unaffected by heat until high temperatures are reached. At temperatures below 1000° C it is not decomposed, and at 2000° C only 1.8% of its molecules are dissociated into hydrogen and oxygen.

Activity. Water reacts with most of the metallic elements and with some of the nonmetals. It also reacts with many compounds.

Hydrolysis. Another important property of water is its ability to split compounds and introduce its own constituents into the products formed. This process is called hydrolysis, a term literally meaning "split by water." In hydrolysis the hydrogen of the water molecule is incorporated into one of the new compounds formed, and the hydroxyl (OH) of the water molecule becomes a part of the other compound formed. An example is the hydrolysis of ammonium chloride (NH_4Cl) when dissolved in water, as shown by the following equation:

$$NH_4Cl + HOH \rightleftharpoons NH_4OH + HCl$$

| Ammonium chloride | Water | Ammonium hydroxide | Hydrochloric acid |

This reaction takes place only to a slight extent, but it is appreciable and is typical of the hydrolytic action of water on many substances. Another example of hydrolysis is the action of water on sugars when boiled in the presence of a little acid.

$$C_{12}H_{22}O_{11} + H_2O \rightarrow C_6H_{12}O_6 + C_6H_{12}O_6$$

| Sucrose | Water | Glucose | Fructose |

In this reaction sucrose is split by the hydrolytic action of water, and two simpler sugars, glucose and fructose, are formed.

Hydrolysis is one of the most important chemical reactions that take place in the human body. Digestion, for example, is a series of hydrolyses or chemical reactions in which food substances are hydrolyzed through the catalytic influence of digestive enzymes.

Hydrates. Water combines with the molecules of certain substances, forming chemical combinations called hydrates. Hydrates yield well-defined crystals when their solutions are allowed to evaporate slowly, and the water held in such a combination is called *water of crystallization*. This combination is not a stable one since it is easily broken up by heating. For example, copper sulfate will form beautiful blue crystals when a solution of this substance is allowed to evaporate slowly. The formula for crystalline copper sulfate is $CuSO_4 \cdot 5H_2O$. If this hydrate is heated above 200° C, its water of crystallization is driven off and a grayish white powder, whose formula is $CuSO_4$, is obtained. Examples of other hydrates are washing soda, $Na_2CO_3 \cdot 10H_2O$; gypsum, $CaSO_4 \cdot 2H_2O$; crystalline sodium sulfate, $Na_2SO_4 \cdot 10H_2O$; and alum, $K_2Al_2(SO_4)_4 \cdot 24H_2O$. The H_2O in the formula of a hydrate is set off from the rest of the formula by a period to indicate that it is loosely attached to the molecule.

An important hydrate is gypsum ($CaSO_4 \cdot 2H_2O$), the dehydrated form of which is plaster of Paris, a substance used extensively in surgery in making casts for artificially supporting body structures until natural processes have completed their repair. It is also used in making masks, busts, molds, and various decorative objects of art. The chemistry of the "setting" of plaster of Paris is the absorption of water and consequent formation of the crystalline hydrate, gypsum, which is firm and hard. The process is represented by the following equation:

$$(CaSO_4)_2 \cdot H_2O + 3H_2O \rightarrow 2CaSO_4 \cdot 2H_2O$$

Plaster of Paris	Water	Gypsum
(a soft powder)		(a hard crystalline substance)

When plaster of Paris is mixed with about one-third its weight of water, a thick paste is produced that "sets" in a short time, forming hard crystalline gypsum. In setting, the paste expands slightly and forms a tight cast or an exact reproduction of a mold.

Hard and soft waters. Water containing calcium or magnesium salts in solution is called hard water. If soap is added to hard water, it combines with calcium or magnesium and forms an insoluble curd, which is precipitated from solution. As long as water is hard it precipitates soap and prevents the action of soap as a cleansing agent. The hardness of water is therefore appropriately referred to as its "soap-destroying power." Soft waters are those that contain little or none of these salts. Rain water and waters that have never passed through earths containing calcium or magnesium compounds are soft waters.

Hardness of water may be temporary or permanent. Temporary hardness is due to the presence of calcium or magnesium bicarbonates. The temporary hard waters are softened by boiling or by adding chemicals such as lime or washing soda. Boiling removes temporary hardness by decomposing the bicarbonates of magnesium or calcium, liberating carbon dioxide gas, and precipitating the magnesium or calcium from solution as insoluble carbonates. The removal of temporary hardness by boiling is illustrated by the following equation:

$$Ca(HCO_3)_2 + Heat \rightarrow CaCO_3\downarrow + H_2O + CO_2\uparrow$$

Calcium bicarbonate (soluble)	Calcium carbonate (insoluble)	Carbon dioxide

The addition of chemicals also removes temporary hardness of water by forming precipitates with magnesium or calcium. This method is practical and inexpensive and consequently is used in communities where the natural waters require softening.

Permanent hardness of water is due to calcium or magnesium in solution as chlorides or sulfates. This type of hardness cannot be removed by boiling and is dissipated by adding chemicals that will precipitate magnesium or calcium. Washing soda (sodium carbonate) is used to remove permanent hardness. The reaction is represented by the following equation:

$$\underset{\substack{\text{Calcium}\\\text{chloride}\\\text{(soluble)}}}{CaCl_2} + \underset{\substack{\text{Sodium}\\\text{carbonate}}}{Na_2CO_3} \rightarrow \underset{\substack{\text{Calcium}\\\text{carbonate}\\\text{(insoluble)}}}{CaCO_3\downarrow} + \underset{\substack{\text{Sodium}\\\text{chloride}}}{2NaCl}$$

Passing water through a filter containing resins that attract and hold the positive ions (cations) is called *ion exchange,* and the resin is called an *ion-exchange resin.* This process can be used to rid water of the calcium and magnesium ions together with other positive ions. Hard-water laundry problems are also solved by the use of modern detergents, which are not affected by calcium and magnesium ions.

Uses of water in the human body. Water has many important functions in the human body. The chief ones are as follows:
1. It aids in the digestive process.
2. It carries digested food products in solution to the tissues.
3. It removes waste products from the cells and carries them to the proper excretory organs.
4. It distributes internal secretions.
5. It is an essential constituent of body tissues and body fluids.
6. It aids in the dissipation of heat and thus assists in the regulation of body temperature.

The individual cells of the body may be considered as living units largely composed of and surrounded by water. From the surrounding water they receive their food; in it they carry on their vital activities; and into it they discharge their waste products.

HYDROGEN PEROXIDE

Preparation. Hydrogen peroxide (H_2O_2) may be prepared by the action of sulfuric acid on barium peroxide.

$$\underset{\substack{\text{Barium}\\\text{peroxide}}}{BaO_2} + \underset{\substack{\text{Sulfuric}\\\text{acid}}}{H_2SO_4} \rightarrow \underset{\substack{\text{Barium}\\\text{sulfate}}}{BaSO_4} + \underset{\substack{\text{Hydrogen}\\\text{peroxide}}}{H_2O_2}$$

Properties. Hydrogen peroxide is a colorless liquid, with a faint characteristic odor and a bitter, astringent taste. It mixes with water in all proportions. It is an unstable compound, decomposing readily with the liberation of oxygen and the formation of water.

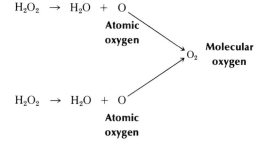

Its decomposition is quite rapid when exposed to light, or to the air; hence it is kept in well-stoppered, dark-colored bottles. It keeps best in dilute solutions to which a small amount of acid and a trace of acetanilide have been added. The preparation given in the *United States Pharmacopeia* is a 3% solution containing a little phosphoric acid and acetanilide as preservatives. Since hydrogen peroxide liberates atomic oxygen when it decomposes, it is an important oxidizing agent.

Uses. Hydrogen peroxide is used for bleaching and as a disinfectant and an antiseptic. All these applications are due to its oxidizing property. It is used for bleaching silks, feathers, hair, oil paintings, and various delicate materials. Its value as a bleaching agent is due not only to its mild oxidizing properties but also to the fact that it leaves a perfectly harmless substance, water, as the end product of its action. When mixed with blood or tissue fluid, hydrogen peroxide liberates atomic oxygen very rapidly; therefore it is applied to wounds as a disinfectant.

SOLUTIONS

A solution is a homogeneous mixture of two or more substances. By homogeneous is meant uniformly mixed or alike in all parts. In a solution the molecules of each substance are uniformly distributed with respect to each other; in other words, any part of a solution has exactly the same composition as any other part, provided that the temperature of all parts is the same. For example, if we place some common salt in the bottom of a beaker and fill the beaker with water, the salt dissolves and a solution is formed. What happens is that the attraction between the salt ions is overcome by a greater attraction between the salt and water molecules, and the result is that the salt ions spread themselves uniformly through the water. If after solution has taken place we remove equal samples of the mixture from the top, bottom, midregion, and any other part of the beaker and evaporate them to dryness, we will obtain the same amount of salt in each of the samples. This shows that the salt was uniformly mixed with the water.

A solution has two parts, the *solute* and the *solvent*. As we ordinarily think of them, the solute is the part present in the smaller amount, and the solvent is the substance present in the greater quantity. However, we can often reverse the relative amounts of the substances present in a solution and thus make the part that was called the solvent become the solute and vice versa.

Solubility

The solubility of a substance means the amount of the substance that will dissolve in a specified amount of solvent at a definite temperature. If the solvent is not mentioned, it is understood to be water. Solubility is generally expressed as the number of parts of solute that will dissolve in 100 parts of solvent, and since temperature affects solubility, the temperature is usually mentioned for the solubility given. If the temperature is not stated in connection with a solubility, room temperature (20° to 25° C) is understood to be the condition.

Factors determining solubility. Whether a substance will dissolve and form a solution is based on three things: the nature of the solvent and solute, the temperature, and the pressure.

Nature of solvent and solute. The nature of the solvent and of the solute has a determining influence on solubility. Some substances are highly soluble in water, whereas others are slightly soluble or highly insoluble in this solvent. Other substances will dissolve best in alcohol, ether, chloroform, or benzene.

Temperature. A marked influence is exerted by temperature. The solubility of most solids in water is increased by heating, but some are little affected by temperature change, and a few are less soluble in hot water than in cold water. Thus 100 ml of water will dissolve 13 g of potassium nitrate at 0° C and will dissolve 246 g of the same substance at 100° C. At 0° C, 36 g of sodium chloride will dissolve in 100 ml of water, but only 39 g of this material will dissolve at 100° C. Calcium hydroxide, on the other hand, is only about one-half as soluble at 100° C as at 20° C. For *gases dissolved in liquids* there is but one rule in regard to temperature: *increasing the temperature decreases their solubility.*

Pressure. Pressure has practically no influence on the solubility of solids or liquids in water, but *an increase in pressure increases the solubility of gases.*

Concentration

Saturated and supersaturated solutions. The concentration of a solution means the amount of solute dissolved in the solvent. A *dilute solution* contains a relatively small amount of solute compared to the solvent. A *concentrated solution* is one that contains a relatively large amount of dissolved substance. When a solvent has dissolved all the solute it can contain at a specific temperature, the solution is said to be saturated. A *saturated solution* contains the greatest amount of solute it can hold permanently under specified conditions. A *supersaturated solution* contains more solute than its solvent holds under usual conditions. It is prepared by dissolving, with the aid of heat, a greater amount of solute than the solvent will hold in solution at lower temperatures and then allowing this solution to cool. If cooled very slowly without agitation, the solvent will hold in solution for a time a greater amount of solute than is held by a saturated solution under the same conditions; hence it is called a supersaturated solution. Supersaturated solutions are not stable; they readily give up their excess solute if allowed to stand, if shaken, or if a few crystals of solute are added.

Table 6-1. Familiar solutions

Type of mixture	Solute	Solvent	Solution
Solids in liquids	Sodium chloride	Water	Aqueous sodium chloride
	Fat	Ether	Ethereal solution of fat
	Iodine	Alcohol	Alcoholic solution of iodine (tincture of iodine)
Liquids in liquids	Alcohol	Water	Aqueous solution of alcohol
	Ether	Alcohol	Alcoholic solution of ether
Gases in liquids	Carbon dioxide	Water	Aqueous solution of carbon dioxide ("carbonated water")
	Ammonia	Alcohol	Alcoholic solution of ammonia (aromatic spirits of ammonia)

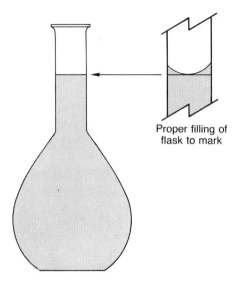

Fig. 6-4. Volumetric flask.

Percent solutions (%). This solution is expressed as the *number of grams of solute present in 100 g of solution.* A 10% aqueous solution of NaCl contains 10 g of NaCl in 100 g of the solution of salt and water. An exact way of preparing this solution is to weigh out 10 g NaCl, dissolve it in a small amount of water (less than 70 ml), and then add water until the total weight of salt and water equals 100 g. When water is the solvent, it is usual to assume that 1 ml of water weighs 1 g. In this case the 10 g of NaCl in the small amount of water is added to a 100-ml volumetric flask and water is added to the 100-ml mark on the neck of the flask.

A volumetric flask is calibrated to hold a specific volume at a certain temperature (usually 20° C) when filled to a mark on the neck of the flask (Fig. 6-4).

The original step of dissolving the solute in a small amount of water allows for the fact that some solutes need to be heated or stirred vigorously, or both, to dissolve in a solvent. If heat is needed, the resulting solution must be allowed to cool to 20° C before the final dilution to the mark on the volumetric flask is made. For accuracy the vessel in which the first solution is made should be rinsed several times with solvent after its contents have been added to the volumetric flask, and these rinse waters should be included in the amount of solvent that is added to the contents of the volumetric flask. This assures that all of the weighed solute is present in the final solution.

Two liquids can be made up in percentage solutions by using volume to volume relationships. A beverage that is 20% alcohol has 20 ml of alcohol in each 100 ml of beverage.

Molar solutions (M). This solution is expressed as *the number of moles of solute present in 1 liter of solution.* A 1M solution contains 1 g-mol of solute per liter of solution. To make a 2M aqueous solution of $CoCl_2$, it would be necessary to weigh out 2 moles of cobalt chloride.

Water, hydrogen peroxide, solutions, standard solutions, milliequivalents **101**

$$\text{Co} = 58.85 \times 1 \text{ (1 atom present in molecule)} = 58.85$$
$$\text{Cl} = 35.5 \ \times 2 \text{ (2 atoms present in molecule)} = \underline{71.00}$$
$$\text{Molecular mass} = 129.85$$

Two moles are needed. If 1 mole equals 129.85 g, 2 moles will be $129.85 \times 2 = 259.7$ g. This amount is weighed out and dissolved in a small amount of water. This solution of cobalt chloride and water is added to a 1-liter volumetric flask and the flask is filled to the 1-liter mark with water.

Normal solutions (N). *The number of gram equivalent masses of solute per liter of solution* is the basis of normal solutions. A gram equivalent mass of a substance contains one combining mass of positive ions and one combining mass of negative ions (see p. 17 for a review of combining masses). A gram equivalent mass can be considered as the amount of any substance that is equivalent to 8 g of oxygen or 1.008 g of hydrogen ions (H^+).

To make a 0.5N aqueous solution of $(NH_4)_2SO_4$, it would be necessary to consider the equivalency of the ammonium sulfate to oxygen or hydrogen. If oxygen ions are substituted for sulfate ions in 1 mole of ammonium sulfate, it would be necessary to use 16 g of oxygen.

$$(NH_4)_2SO_4 + O^{-2} \rightarrow (NH_4)_2O + SO_4^{-2}$$

This is twice the amount of oxygen needed for an equivalent mass of oxygen (8 g). If hydrogen ions are substituted for ammonium ions in 1 mole of ammonium sulfate, it would be necessary to use 2.016 g of them. Again, this is twice the combining mass of hydrogen (1.008 g).

$$(NH_4)_2SO_4 + 2H^+ \rightarrow H_2SO_4 + 2NH_4^+$$

It is obvious from the results of these two substitutions that the molecular mass of ammonium sulfate is equivalent to twice the combining mass. Put another way, the combining mass of ammonium sulfate is one-half the molecular mass. A simple way to calculate equivalent masses of inorganic substances is to divide the molecular mass by the largest electrovalence number occurring in the compound. In the salt, ammonium sulfate, the sulfate radical has the largest electrovalent number, 2.

$$(NH_4)_2^{+1}SO_4^{-2}$$

Therefore an equivalent mass of $(NH_4)_2SO_4$ is the molecular mass divided by two.

$$N = 14.0067 \times 2 \text{ (2 atoms in molecule)} = 28.0134$$
$$H = 1.00797 \times 8 \text{ (8 atoms in molecule)} = 8.06376$$
$$S = 32.064 \ \times 1 \text{ (1 atom in molecule)} = 32.064$$
$$O = 15.9994 \times 4 \text{ (4 atoms in molecule)} = \underline{63.9976}$$
$$\text{Molecular mass} = 132.13876$$

$$132.13876 \div 2 = 66.07 \text{ equivalent mass}$$

A 0.5N solution is needed, so 0.5 of the equivalent mass is $66.07 \times 0.5 = 33.035$. When 33.035 g of $(NH_4)_2SO_4$ is dissolved in a small amount of water, this mixture added to a volumetric flask, and the volume made up to the liter mark with water, the result is a 0.5N aqueous solution of $(NH_4)_2SO_4$.

Molal solution (mole/kg). This solution is expressed as *the number of moles of solute per 1000 g of solvent*. To make a 1-mole/kg aqueous solution of $LiNO_3$, 1 mole of lithium nitrate is dissolved in 1000 g of water.

$$\begin{aligned} Li &= 6.939 \times 1 \text{ (1 atom in the molecule)} = 6.939 \\ N &= 14.0067 \times 1 \text{ (1 atom in the molecule)} = 14.0067 \\ O &= 15.999 \times 3 \text{ (3 atoms in the molecule)} = \underline{47.9982} \\ & \qquad \text{Molecular mass} = 68.9439 \end{aligned}$$

Therefore, 68.94 g of $LiNO_3$ is weighed out and added to 1000 g of water.

Mole fraction (X). This is the fraction formed when the number of moles of a component of a solution is divided by the total moles making up the solution. If 5 g of sugar ($C_{12}H_{22}O_{11}$) are dissolved in 20 g of water, the mole fraction is calculated in the following manner:

$$C_{12}H_{22}O_{11}$$

$$\begin{aligned} C &= 12.01115 \times 12 \text{ (12 atoms in molecule)} = 144.1338 \\ H &= 1.00797 \times 22 \text{ (22 atoms in molecule)} = 22.17534 \\ O &= 15.9994 \times 11 \text{ (11 atoms in molecule)} = \underline{175.9934} \\ & \qquad \text{Molecular mass} = 342.30254 \end{aligned}$$

$$H_2O$$

$$\begin{aligned} H &= 1.00797 \times 2 \text{ (2 atoms in molecule)} = 2.01594 \\ O &= 15.9994 \times 1 \text{ (1 atom in molecule)} = \underline{15.9994} \\ & \qquad \text{Molecular mass} = 18.01534 \end{aligned}$$

$$\text{Number of moles of } C_{12}H_{22}O_{11} = \frac{5}{342.3} = 0.014$$

$$\text{Number of moles of } H_2O = \frac{20}{18} = 1.111$$

$$\text{Total number of moles} = 1.125$$

$$\text{Mole fraction of sugar} = X_{C_{12}H_{22}O_{11}} = \frac{0.014}{1.125} = 0.012$$

$$\text{Mole fraction of water} = X_{H_2O} = \frac{1.111}{1.125} = 0.988$$

The sum of the mole fractions should equal 1.

It is possible to dilute any solution to make solutions of lesser strength from it. A normal solution may be diluted 10 times, 100 times, 1000 times, and so on to make solutions that are 0.1N, 0.01N, and 0.001N, respectively; similarly, solutions that are 0.1M, 0.01M, and 0.001M can be made by diluting a molar solution 10, 100, or 1000 times, respectively. It is conventional to keep stock standard solutions of higher concentrations available for such dilutions, thus making the work of commercial and clinical laboratories less complicated.

In diluting a percent solution to one of lower concentration, an elementary "rule of thumb" may be employed. Stated simply, "Take what you want and bring it up to what you've got." To illustrate, a 20% aqueous solution of ethyl alcohol is on hand and a 7%

Water, hydrogen peroxide, solutions, standard solutions, milliequivalents **103**

solution is needed. Take 7 ml of the 20% solution and dilute it to 20 ml with water. Multiples of the 7 ml and 20 ml can be used if larger quantities are needed; 14 ml can be diluted to 40 ml, 28 ml, to 80 ml, and so on. The result will be a 7% aqueous solution of ethyl alcohol.

If an exact volume, such as 250 ml, is needed, the following calculation can be made (7% means 7 gm in 100 ml):

$$\frac{\text{Grams in 100}}{100 \text{ ml}} = \frac{X}{\text{Milliliters needed}}$$

in this case

$$\frac{7 \text{ g}}{100 \text{ ml}} = \frac{X}{250 \text{ ml}}$$

$$X = 17.5 \text{ g}$$

By definition, 20% alcohol means that in 100 g of a solution of alcohol and water there are 20 g of alcohol; 17.7 g are needed. A simple ratio gives the number of milliliters of 20% alcohol needed to be diluted to 250 ml to make a 7% solution.

$$\frac{100}{20} = \frac{X}{17.5}$$

$$X = 87.5 \text{ ml}$$

MILLIEQUIVALENT

A milliequivalent (mEq) is one-thousandth of an equivalent (eq) and is expressed in milligrams (mg). It is a unit that is used to express the concentration of ions in the tissue fluids of animals and plants. The milliequivalent system was adopted by biologists because it shows the quantitative relationship among those ions in units of convenient size.

Table 6-2. Derivation of the milliequivalent

Ion	Atomic mass	Valence	Equivalent (g)	Milliequivalent (mg)
Na	23	1	23	23
K	39	1	39	39
Ca	40	2	20	20
Mg	24	2	12	12
Cl	35.5	1	35.5	35.5
S	32	2	16	16
P	31	1.8*	17.2	17.2
CO_2 (gas)	—	—	22.4 (liters)†	22.4 (ml)

*The value 1.8 for the valence, or combining power, of phosphorus is used because in extracellular fluid 20% of the phosphate present contains 1 eq of base (BH_2PO_4) and 80% contains 2 eq of base (B_2HPO_4). B represents a positive ion, usually Na^+. This value is derived as follows: $(1 \times 0.2) + (2 \times 0.8) = 1.8$. The value 1.8 is used because it reflects most closely the combining equivalent of phosphate in extracellular fluid.

†The volume of an equivalent (1 mole) of CO_2 under standard conditions is 22.4 liters. A milliequivalent of CO_2, in terms of volume, is therefore 22.4 ml. This value is used in converting volume percent of CO_2 (CO_2-combining power) to milliequivalents per liter.

Derivation. The figures in the fourth column of Table 6-2 are chemical equivalents. In the fifth column the values are milliequivalents. The numerals are the same as those in the fourth column because milliequivalents are expressed in milligrams. This is made clear by the following deductions, using Na as the example:

$$1 \text{ eq of Na} = \frac{\text{Atomic mass}}{\text{Valence}} = \frac{23}{1} = 23 \text{ g}$$

$$1 \text{ mEq of Na} = \frac{23 \text{ g}}{1000} = 0.023 \text{ g} = 23 \text{ mg}$$

Application. Using further the data of Table 6-2, it is clear that for sodium, 1 mEq is 23 mg, 2 mEq total 46 mg, and 3 mEq, 69 mg, and so on; similarly, for chloride, 1 mEq is 35.5 mg, 2 mEq equivalents equal 71 mg, 3 mEq, 106.5 mg, and so on. In considering body fluids, the physician is interested in the relative numbers of the ions present. This relationship is what the milliequivalent system indicates. It is a method by which a direct comparison of the numbers of ions of the different elements or compounds present is made.

In clinical literature, concentrations are sometimes expressed as milligrams per 100 ml. This is called *milligrams percent* (mg%). It is better to express mass per volume ratios as parts per million (ppm), 1 mg per 100 ml being close to 10 ppm. It is preferable to express concentrations as moles per liter (moles/l), millimoles per liter (mmoles/l), and micromoles per liter (μmoles/l). Concentrations of electrolytes can be expressed as equivalents per liter (eq/l) or milliequivalents per liter (mEq/l).

Conversion of data. If the values for a blood or tissue fluid analysis are reported in milligrams percent, the data may be converted to milliequivalents per liter by a simple calculation. The number of milligrams in 100 ml is multiplied by 10 to obtain the number of milligrams in 1 liter. The value obtained is divided by the number of milligrams in 1 mEq. The answer is the number of milliequivalents per liter. For example, the normal content of sodium in blood plasma is 330 mg/100 ml. This value is converted to milliequivalents per liter as follows:

$$\frac{330 \times 10}{23} = \frac{3,300}{23} = 143$$

To convert milliequivalents per liter to milligrams percent, the preceding calculation is reversed.

Summary

1. A milliequivalent is derived by dividing the atomic mass of an element by its valence.

2. Milliequivalents are expressed in milligrams.

3. The milliequivalent is used to designate the amount of electrolyte per liter of solution.

Water, hydrogen peroxide, solutions, standard solutions, milliequivalents

Questions for study

1. Discuss the occurrence of water.
2. What is the composition of water by volume? By weight?
3. Describe the electrolysis of water. What does it show regarding water?
4. What is analysis? What does it show?
5. What is meant by synthesis? How is the synthesis of water carried out?
6. Mention five properties of water. How does water compare with other substances in solvent power? What makes water a good solvent for ions?
7. Distinguish between hygienically pure and chemically pure water.
8. Water becomes less dense when it freezes. What is the importance of this property in nature? Of what value is the expansion of water when freezing?
9. What is hydrolysis?
10. What is a hydrate?
11. What are the uses of plaster of Paris? Explain the chemistry of the setting of plaster of Paris.
12. Distinguish between hard and soft waters. What are the objections to hard water? How is hardness of water removed?
13. What are the uses of water in the human body?
14. What are the properties of hydrogen peroxide? What are its uses?
15. Is hydrogen peroxide a good disinfectant? Why?
16. Define a solution. What are the two parts of a solution?
17. What is meant by solubility?
18. What are the factors that influence solubility?
19. What is a dilute solution? A concentrated solution? A saturated solution? A supersaturated solution?
20. Define percent, molarity, normality, molality, and mole fraction.
21. Describe how to make 1 liter of a 3M solution, a 0.1N solution, a 2 molal solution, and a 0.9% solution of $Cu(NO_3)_2$.
22. Calculate the number of grams of compound that must be dissolved in 1 liter of solution to prepare a molar solution of each of the following: ethyl alcohol (C_2H_6O), magnesium chloride ($MgCl_2$), and sucrose ($C_{12}H_{22}O_{11}$).
23. Calculate the number of grams of compound that must be dissolved in 1 liter of solution to prepare a normal solution of each of the following: hydrobromic acid (HBr), nitric acid (HNO_3), potassium chloride (KCl), barium chloride ($BaCl_2$), potassium hydroxide (KOH), and barium hydroxide ($Ba[OH]_2$).
24. What is a milliequivalent?
25. How is a milliequivalent derived?
26. What are the uses of the milliequivalent system in interpretations in biology?
27. What applications of the milliequivalent system are made by the physician?
28. How are electrolyte data in terms of milligrams percent converted to values expressed in milliequivalents per liter?
29. Derive the milliequivalent value for each of the following elements: Ca, K, S, Al, and iron (II).

7 Electrolytes, ionization, acids, bases, salts, and pH

ELECTROLYTES AND NONELECTROLYTES

Electrolytes are ionic compounds; they are composed of positive and negative ions held together by electrovalent bonds. When an electrolyte is dissolved in water, the electrostatic forces that hold the ions together are overcome and the ions are separated. The ions become closely associated with water molecules because of the attraction between the ions and the polarized water molecule.

The ions of an electrolyte in the solid state exist relatively close together in an orderly arrangement; in aqueous solution they are relatively far apart and are free to move throughout the solution. Thus electrolytes are ionic compounds that in aqueous solution form freely moving ions bearing positive or negative charges, and by virtue of their ions they have the property of conducting an electric current.

Nonelectrolytes are compounds in which the atoms are held together by covalent bonds. They are not ionic in character. When dissolved in water, they do not conduct an electric current because they do not give rise to the formation of ions.

IONIZATION

The separation of ionic compounds in solution into freely moving positive and negative ions is called dissociation. The term ionization is frequently used for the same phenomenon, although ionization means the formation of ions by any method.

In most cases an ion is a single atom that bears one of several electric charges; some ions, however, consist of many atoms that hold their electric charge or charges in common. The name "ion" comes from a Greek word meaning "going." It is applied to these individual particles of electrolytes because they are in a rapid state of motion in a solution and go to oppositely charged electrodes under the influence of an electric current.

When an electric current is passed through a solution of an electrolyte, the positive ions go to the cathode or negative pole, and the negative ions pass to the anode or positive pole. Positive ions are therefore called *cations* because they migrate to the *cathode*

Electrolytes, ionization, acids, bases, salts, and pH 107

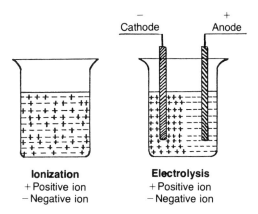

Fig. 7-1. Diagram showing ionization and electrolysis.

when an electric current is passing through a solution, and negative ions are called *anions* since they pass to the *anode* under the influence of an electric current.

The essential features of ionization are outlined in the following summary:

1. An electrolyte (ionic compound) dissociates into ions when in aqueous solution. The process may be called ionization.

2. Ions bear either positive or negative charges of electricity. The total positive charge held by positive ions in a solution of an electrolyte is exactly equal to the total negative charges held by negative ions in the same solution. A solution of an electrolyte is therefore electrically neutral.

3. Ions are the individual carriers of an electric current through a solution. When a current of electricity flows through a solution of an electrolyte, the positive ions pass to the negative pole and the negative ions go to the positive pole. The negative ions serve as carriers of electrons. The behavior of ions in this respect is shown diagrammatically in Fig. 7-1.

4. The number of charges on an ion is equivalent to its oxidation (valence) number.

5. Chemical reactions involving electrolytes in solution are reactions between ions. This is an extremely important concept and must be used in interpreting all chemical reactions of electrolytes. In every reaction between electrolytes the student must visualize the individual ions and not the molecules as the reacting particles. For instance, take the reaction between sodium chloride and silver nitrate; in an aqueous solution these substances dissociate as follows:

$$NaCl \underset{}{\overset{H_2O}{\rightleftarrows}} Na^+ + Cl^-$$
Sodium chloride

$$AgNO_3 \underset{}{\overset{H_2O}{\rightleftarrows}} Ag^+ + NO_3^-$$
Silver nitrate

A chemical reaction between these substances involves the four ions in the formulas just given. Each of the positive ions has an equal opportunity of coming into contact with either of the negative ions and reacting with it. The result will be a combination between the ions and is represented by the following equation:

$$\underset{\text{Sodium chloride}}{NaCl} + \underset{\text{Silver nitrate}}{AgNO_3} \xrightarrow{H_2O} \underset{\text{Sodium nitrate}}{NaNO_3} + \underset{\text{Silver chloride}}{AgCl\downarrow}$$

The silver chloride is precipitated as an insoluble salt as soon as it is formed. This results in all of the sodium and the nitrate ions being left to unite with each other. If the reaction takes place in a water solution, some or all of the sodium nitrate will be ionized, depending on the amount of sodium nitrate produced.

$$NaNO_3 \underset{}{\overset{H_2O}{\rightleftharpoons}} Na^+ + NO_3^-$$

If a solute is partially ionized in solution, an equilibrium exists between the nonionized and the ionized solute.

$$AB \rightleftharpoons A^+ + B^-$$

This equilibrium has a constant called the ionization constant, K_i.

$$K_i = \frac{[A^+]^1[B^-]^1}{[AB]^1}$$

The K_i is an indication of how much ionization takes place when a certain solute is dissolved in a specific solvent. The solvent may be indicated; $K_{i\ aq}$ means the ionization constant in water. The value of K_i is calculated by dividing the product obtained by multiplying the molar concentrations of the ions formed by the molar concentration of the nonionized solute.

Nonaqueous ionizations

The separation of molecules into freely moving ions may occur under other conditions than being dissolved in water. Thus ionization of certain substances occurs when these substances are dissolved in liquid ammonia, formic acid, or concentrated sulfuric acid. Molecules of gases may also be ionized by the passage of electric discharges through fields in which they exist. Fused salts also have freely moving ions.

THE EFFECT OF SOLUTE ON SOLUTION

The vapor pressure of a solution of a nonvolatile solute in a solvent is lower than the vapor pressure of the pure solvent. Lowering of the vapor pressure affects the freezing and boiling points of the solution. A mole contains an Avogadro's number of particles. This amount of particles lowers the freezing point and raises the boiling point of any solvent by a definite number of degrees of temperature. The exact number of degrees that a mole of solute will raise the boiling point and lower the freezing point varies from solvent to solvent. For water, the molal (gram-molecule per 1000 g of water) boiling-point elevation is 0.56° C and the molal freezing-point depression is 1.86° C. This effect is for a mole of *particles*. If molecules coalesce into larger aggregates, then the tem-

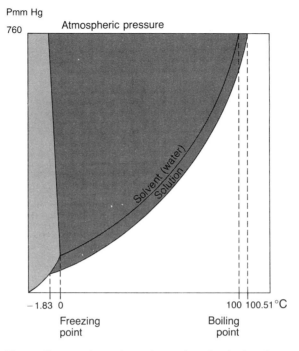

Fig. 7-2. Phase diagram for solvent (water) and solution (one molar).

perature changes will be less than expected. If molecules ionize, each ion acts as a particle and the boiling-point elevation and the freezing-point depression will be greater than expected. Fig. 7-2 shows the boiling-point elevation and the freezing-point depression effected in water by a mole of nonionizing, noncoalescing, nonvaporizing solute. If the solute vaporizes, it will not be as effective in changing the critical temperatures of the solvent as would be expected. In fact it may completely vaporize from the solution. This is what used to happen to early users of alcohol as an antifreeze in automobiles. In cold weather the alcohol acted to lower the freezing point of the water in the radiator, but it would boil away when the temperature of the radiator was raised by long running or a change in the weather. The modern coolants for radiators of cars effectively lower the freezing point of the fluid in the radiator and do not evaporate when the car is heated or the weather turns warm. This molal freezing-point depression and boiling-point elevation of solvents is another method of determining the molecular weight of a solute. The fact that solvents other than water may be used offers a great variety of combinations of solutes and solvents that may be tested in this manner.

Calculations involving critical temperatures

If 40 g of ethyl alcohol (C_2H_6O) are dissolved in 300 g of water, what will be the freezing point of the solution?

Forty grams in 300 is the same as how many grams in 1000?

$$\frac{40 \text{ g}}{300 \text{ g}} = \frac{X}{1000 \text{ g}}$$

$$X = 1000 \times {}^{2}/_{15}$$

$$X = 133.33 \text{ g in } 1000 \text{ g } H_2O$$

Molecular mass of C_2H_5O

C = 12.01115 × 2 (2 atoms in the molecule) = 24.0223
H = 1.00797 × 6 (6 atoms in the molecule) = 6.04782
O = 15.9994 × 1 (1 atom in the molecule) = 15.9994
Molecule mass = 46.06952

A molecular mass in grams is 1 mole. One mole of ethyl alcohol equals 46.07 g.

48.07 g in 1000 g lowers the freezing point 1.89° C (by definition)
133.33 g in 1000 g lowers the freezing point $X°$ C

$$X = \frac{133.33}{48.07} \times 1.89° \text{ C} = 5.24° \text{ below } 0° \text{ C}$$

Twenty grams of a nonvolatile, nonaggregating, nonionizing solute is dissolved in 250 g of water. This solution boils at 101.5° C. What is its molecular mass?

Twenty grams in 250 is the same as how many grams in 1000?

$$\frac{20}{250} = \frac{X}{1000}$$

$$X = 80 \text{ g solute in } 1000 \text{ g of water}$$

$$101.5 - 100.0 = 1.5° \text{ C boiling-point elevation}$$

If 80 g of solvent results in a 1.5° C boiling-point elevation, how much would effect a 0.51° C boiling-point elevation?

$$\frac{80}{1.5} = \frac{X}{0.51}$$

$$X = 27.29 \text{ g (molecular mass)}$$

ACIDS

An acid is a substance that produces hydrogen ions in aqueous solution. Since the hydrogen ion is a proton, an acid is also defined as a substance that yields protons in aqueous solution or is a proton donor. An acid is also defined as an electron pair acceptor. The concept of an acid is well established by all of these definitions. The last two definitions are more extensive and encompass a greater variety of reactions.

When placed in aqueous solution, the hydrogen ion of an acid is attached to one of the free electron pairs of the oxygen atom of a water molecule. This unit is called the hydronium ion. Its structure may be represented as follows:

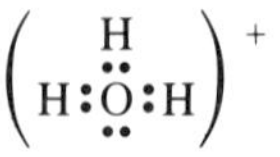

$$\left(\begin{array}{c} H \\ H{:}\ddot{O}{:}H \end{array} \right)^{+}$$

Hydronium ion (H_3O)$^+$

As an example of an acid let us consider hydrochloric acid. When hydrogen chloride is dissolved in water, it undergoes ionization. The hydrogen ion becomes attached to one of the unused electron pairs of the oxygen atom of the water and forms a hydronium ion. Hydrochloric acid in dilute aqueous solution therefore consists of hydronium ions, chloride ions, and mainly water molecules. This behavior is represented as follows:

$$H:\overset{..}{\underset{..}{O}}:\ +\ H:\overset{..}{\underset{..}{Cl}}: \ \rightleftarrows\ \left(H:\overset{..}{\underset{H}{O}}:H \right)^{+} + \left(:\overset{..}{\underset{..}{Cl}}: \right)^{-}$$

Water   Hydrogen   Hydronium   Chloride
**   chloride   ion (H_3O)$^+$ ion (Cl)$^-$**

Water is not always needed as a solvent for acid reactions to take place. An example of this is the reaction between ammonia gas and hydrochloric acid vapor.

$$NH_3 + HCl \rightarrow NH_4Cl$$
** Gas   Gas**

The behavior of all acids in aqueous solution is similar to that shown for hydrochloric acid. For example, when placed in water, sulfuric acid, nitric acid, phosphoric acid, and acetic acid yield hydronium ions and hydrogen sulfate, nitrate, phosphate, or acetate ions, respectively. The first two may be written as follows:

$$H_2SO_4 + H_2O \rightarrow H_3O^+ + HSO_4^-$$
$$HNO_3 + H_2O \rightarrow H_3O^+ + NO_3^-$$

This discussion of the hydronium ion is important. It shows how the hydrogen ion exists in aqueous solution. Solutions containing hydronium ions exhibit all the properties of acids, which are discussed next. However, once this mechanism is understood, it seems more practical to use the briefer term *hydrogen ion*. In referring to acid concentration, it is the almost unanimous choice of writers in biology and medicine to use the term *hydrogen ion concentration* rather than *hydronium ion concentration;* hence the terms hydrogen ion and hydrogen ion concentration will be used primarily throughout this textbook.

Properties. Acids may be either solids, liquids, or gases. They manifest their characteristic properties when ionized. Since all common acids yield hydronium ions in water, there are certain properties that are characteristic of acids in general: (1) have a sour taste, (2) change blue litmus to red, (3) react with many metals, liberating hydrogen, (4) react with carbonates, liberating carbon dioxide, and (5) unless very dilute, will destroy plant and animal cells.

Have a sour taste. Everyone is familiar with the sourness of taste that is associated with acids. The sour taste of certain fruits and vegetables such as lemons, oranges, apples, grapefruit, tomatoes, and rhubarb is caused by the presence of acids in these substances.

Change blue litmus to red. A substance that turns one color in an acid solution and another color in a basic solution is known as an *indicator*. Indicators are useful in showing the hydrogen ion concentration of solutions. Most indicators are weak acids that exhibit one color when they are undissociated and another when they are ionized.

$$\text{H Indicator} \rightleftarrows H^+ + \text{Indicator}^-$$
Color 1   **Color 2**

Litmus is a vegetable dye that is extracted from lichens. It is pink in the presence of acids and blue in the presence of bases. It is therefore used to test for presence of these substances. Litmus is generally used in the form of litmus paper. Blue litmus paper is prepared by soaking white paper in a solution of the dye to which a little alkali has been added, washing it thoroughly, and then drying and cutting it into small strips. Red litmus paper is made in the same way except that the paper is treated with an acid solution of litmus before washing and drying.

React with many metals, liberating hydrogen. Hydrogen ions are displaced from solution by any metal that is above hydrogen in the electromotive series (see p. 65). Therefore any metal that precedes hydrogen in this list will react with acids and liberate hydrogen, and the metals below hydrogen in the series will not displace hydrogen ions from solution and therefore will not liberate hydrogen when treated with an acid.

React with carbonates, liberating carbon dioxide. All carbonates are attacked by acids, with the liberation of carbon dioxide. A typical reaction is as follows:

$$Na_2CO_3 + H_2SO_4 \rightarrow Na_2SO_4 + H_2O + CO_2\uparrow$$

Sodium   Sulfuric   Sodium   Water   Carbon
carbonate   acid   sulfate   dioxide

The behavior of acids with carbonates is the basis of the commercial methods for preparing carbon dioxide, which we will study in Chapter 10.

Unless very dilute, will destroy plant and animal cells. The stomach and the urinary tract are the only parts of the body that will tolerate acids, the tissues of these parts being adapted to withstand acids when very dilute. The urine of carnivorous animals is slightly acid, and the normal gastric juice has an acid concentration of about 0.2% to 0.3% hydrochloric acid. The stomach and urinary tract organs cannot endure higher concentrations of acids, however, and if more than normal amounts of acids get into these parts by accident or disease, their tissues are attacked with serious consequences.

Strong and weak acids. The strength of an acid is dependent on the number of hydronium ions it will produce in aqueous solution. A strong acid will yield a relatively large number of hydronium ions in solution; a weak acid will produce a relatively small number of hydronium ions when dissolved in water. A substance that does not ionize well in a solvent is a poorly ionized substance. It will react weakly in that solvent. A solvent that will ionize it will alter its activity. A strong acid is a highly ionized acid; a weak acid is a poorly ionized acid. To illustrate, take hydrochloric and acetic acids as examples. Relatively, these acids will yield hydronium ions in water as follows:

100 molecules HCl $\rightarrow$ $H^+,H^+,H^+,H^+,H^+,H^+,H^+,H^+,H^+,H^+,$ . . . $92H^+$ ions in all

100 molecules CH_3COOH $\rightarrow$ $H^+,$—————————$1H^+$ ion only

Electrolytes, ionization, acids, bases, salts, and pH 113

For every 100 molecules of 0.36% hydrochloric acid in water, 92 molecules will ionize, producing 92 hydronium ions; for every 100 molecules of acetic acid of the same molecular concentration, only 1 molecule will ionize, forming 1 hydronium ion. Thus hydrochloric acid is a strong acid because it yields such a relatively large number of hydronium ions when placed in water. This acid will dissolve a piece of zinc much more rapidly than acetic acid, it will conduct an electric current much better, and it manifests other acid properties in the same degree of intensity. Other strong acids are nitric and sulfuric acids. Acetic acid is a weak acid in comparison with hydrochloric acid because it yields so few hydronium ions when placed in water. Other weak acids are carbonic and boric acids. Carbonic acid (CO_2 in water) is so weak that we drink it with safety in carbonated drinks, and boric acid is so weak that it can be applied to the eye as an eyewash.

Hypoacidity, hyperacidity, and acidosis. As already stated, the concentration of hydrochloric acid in the gastric juice of a healthy person is approximately 0.2% to 0.3%. The concentration varies in disease, and the conditions known as hypoacidity and hyperacidity often arise. Hypoacidity is a condition in which the gastric juice has less than the normal amount of hydrochloric acid. In hyperacidity the hydrochloric acid concentration of the gastric juice is around or above 0.3%. Acidosis is a condition quite different from those just mentioned and should not be confused with them. Acidosis is usually the result of an overproduction of acids in the tissues of the body or a failure in the excretion of acids. It occurs in certain diseases such as diabetes mellitus or glomerulonephritis and in conditions interfering with the elimination of carbon dioxide from the lungs. Whether hypoacidity or hyperacidity exists is determined by an analysis of the gastric juice removed by means of a stomach tube at certain intervals following the ingestion of a test meal. The existence of acidosis is determined by a chemical analysis of the blood.

Naming of inorganic acids

Binary acids. All acids contain hydrogen. Binary acids contain one other additional element. In naming these acids the prefix *hydro* is used, which is then followed by the name of the other element. The last part of the name is often modified to include the suffix *ic*. Following are several examples:

HCl	*Hydro*chlor*ic* acid
HBr	*Hydro*brom*ic* acid
H_2S	*Hydro*sulfur*ic* acid

Oxyacids. Oxyacids always contain hydrogen and oxygen in addition to a nonmetallic element(s). It is this third element that gives the acid its name. The *hydro* prefix is never used, but the *ic* ending is still retained. A few examples should make this clear:

H_2SO_4	Sulfur*ic* acid (oxyacid containing sulfur)
HNO_3	Nit*ric* acid (oxyacid containing nitrogen)
H_3PO_4	Phosphor*ic* acid (oxyacid containing phosphorus)
H_2CO_3	Carbon*ic* acid (oxyacid containing carbon)

A number of oxyacids exist that are composed of the same elements but vary in their oxygen content. In these cases it becomes necessary to introduce variations in suffixes and prefixes to get distinguishing names. We first apply the suffixes *ic* and *ous*. The acid containing the greater amount of oxygen in its molecule is given a name ending in *ic;* the one containing the lesser amount of oxygen in its molecule is given a name ending in *ous*. If more than two oxyacids of the same elements exist, the prefixes *per* and *hypo* are used. *Per* is prefixed to the name of the acid containing more oxygen than the *ic* acid, and *hypo* is applied to the name of the acid containing less oxygen than the *ous* acid. Examples:

HNO_3	Nit*ric* acid	HClO	*Hypo*chlor*ous* acid
HNO_2	Nitr*ous* acid	$HClO_2$	Chlor*ous* acid
H_2SO_4	Sulfu*ric* acid	$HClO_3$	Chlo*ric* acid
H_2SO_3	Sulfur*ous* acid	$HClO_4$	*Per*chlo*ric* acid

BASES

A base is a substance that produces hydroxyl ions in aqueous solutions. It is also a proton acceptor, since bases combine with protons. Bases can be defined as electron pair donors. The last two definitions are more general and describe a wider range of reactions. Common bases are called hydroxides and dissociate in aqueous solution, giving positively charged metallic ions and negatively charged OH groups, the negatively charged OH groups being called hydroxyl ions. It will be noted that hydroxyl ions are common to each example. A hydroxyl ion consists of 1 atom of oxygen and 1 atom of hydrogen combined so firmly that they remain together in solution and behave as a single ion with a valence number of -1 when a chemical reaction takes place.

$$NaOH \rightleftarrows Na^+ + OH^- \qquad Ca(OH)_2 \rightleftarrows Ca^{+2} + 2OH^-$$
Sodium hydroxide — Sodium ion — Hydroxyl ion — Calcium hydroxide — Calcium ion — Hydroxyl ions

$$KOH \rightleftarrows K^+ + OH^- \qquad NH_4OH \rightleftarrows NH_4^+ + OH^-$$
Potassium hydroxide — Potassium ion — Hydroxyl ion — Ammonium hydroxide — Ammonium ion — Hydroxyl ion

It should be emphasized that a compound containing an OH group is not a base unless it is capable of dissociating in solution and producing hydroxyl "ions." Many substances, notably the alcohols of organic chemistry, for example, ethyl alcohol (C_2H_5OH), contain OH groups but are not bases because they do not dissociate and yield hydroxyl ions when in solution.

The terms *base* and *alkali* are synonyms and are used interchangeably. When a compound is said to have alkaline properties, it is equivalent to stating that it is a basic substance.

Properties. Most bases are solids. They are soluble in water and do not manifest their characteristic properties except in solution. Since all common bases produce hydroxyl ions in solution, there are certain general properties that are true of these bases. These properties are due to the presence of the hydroxyl ion and may be summarized as follows: (1) *bases have a bitter, metallic taste;* (2) *they are slippery or soapy to the touch;* (3) *they turn red litmus blue;* (4) *they react with acids to form salts;* and (5) *they have a corrosive, disintegrating action on organic matter.* Hence some bases are referred to as "caustic," for example, caustic soda, caustic potash.

Electrolytes, ionization, acids, bases, salts, and pH 115

Strong and weak bases. With bases, strength means dissociating or ionizing power just as in the case of acids. A strong base is one that dissociates releasing a relatively large number of hydroxyl ions in solution; a weak base dissociates releasing a relatively small number of hydroxyl ions in solution. Examples of strong bases are sodium hydroxide and potassium hydroxide. In dilute aqueous solutions of these bases about 99% of the molecules are ionized, yielding hydroxyl ions. The terms *dilute* and *concentrated* have the same meaning when used with bases as they do when used with acids and refer to the amount of base present, not to the ability of that base to ionize.

Aqueous solutions of ammonia (ammonium hydroxide) are classified as weak bases. Actually, ammonia gas reacts with water only slightly to form ammonium and hydroxyl ions (see p. 68). Once formed, the ions exist 100% ionized in water. There have been suggestions that solutions of water and ammonia be classified as dilute bases rather than as weak ones. The tendency of such solutions to revert to ammonia and water molecules causes them to give all of the chemical reactions of a poorly ionized base. This characteristic carries over to aqueous solutions of ammonium salts as well.

$$NH_3 + H_2O \rightleftharpoons NH_3 + H_2O \rightleftharpoons NH_4^+ + OH^-$$

Gas   In solution

Naming of bases

A base contains a metallic element combined with an OH or hydroxyl radical. The practice in naming a base is to use the unchanged name of the metallic element of which it is composed, followed by the term *hydroxide*. Examples:

NaOH	Sodium hydroxide
KOH	Potassium hydroxide
Ca(OH)$_2$	Calcium hydroxide
NH$_4$OH	Ammonium hydroxide

Some base-forming elements manifest two valences in entering into combination with the hydroxyl radical. In these instances the suffix *ous* is used where the lower valence is exhibited and *ic* is applied where the higher valence is found. Examples:

		Stock system
CuOH	Cupr*ous* hydroxide	Copper (I) hydroxide
Cu(OH)$_2$	Cupr*ic* hydroxide	Copper (II) hydroxide
Fe(OH)$_2$	Ferr*ous* hydroxide	Iron (II) hydroxide
Fe(OH)$_3$	Ferr*ic* hydroxide	Iron (III) hydroxide

Neutralization

When acids and bases are mixed together in equivalent amounts, they react, and the properties of each disappear. What happens in such a reaction is a combination of the positively charged hydrogen ions of the acid with the negatively charged hydroxyl ions of the base to form neutral unionized molecules of water. A typical reaction is as follows:

$$Na^+ + OH^- + H^+ + Cl^- \rightarrow H_2O + Na^+ + Cl^-$$

Sodium hydroxide   Hydrochloric acid   Water   Sodium chloride

116 Roe's principles of chemistry

In the preceding reaction the hydroxyl and hydrogen ions combine and form water, which is a neutral substance having the same amount of acid properties as it has basic properties. Since the reaction of an acid with a base results in a disappearance of the properties of both, acids and bases are said to neutralize each other, and the process is called neutralization. *Neutralization may be defined as the union of the hydrogen ions of an acid with the hydroxyl ions of a base to form water, a neutral substance.*

An expression to show electron participation in neutralization is as follows:

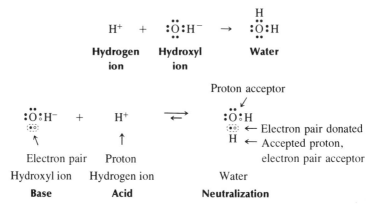

The bond formed when the hydrogen ion, without electrons, combines with a hydroxyl ion that supplies 2 electrons for the chemical union is an example of the formation of a coordinate covalent bond (in which both electrons are supplied by 1 atom).

Titration

Neutralization of an acid of unknown normality by a base of known normality can reveal the concentration of the acid. This process of neutralization is called *titration* (Fig. 7-3). A fixed volume of the acid of unknown normality is placed in an Erlenmeyer flask. An indicator is added to the acid in the flask, and the base of known normality is slowly added to the flask from a calibrated tube called a burette. The contents of the flask are mixed during the addition so that neutralization takes place evenly throughout the flask. A change in the indicator color signals the complete neutralization of the contents of the flask. This point in the titration is called the *end point* or *equivalence point*. It is at this point that the amount of basic ions and the amount of hydrogen ions in the flask are equivalent to each other. The volume of the added base can be read from the burette. At the end point the volume of the acid multiplied by the normality of the acid exactly equals the volume of the added base multiplied by the normality of the base, $V_{acid} \times N_{acid} = V_{base} \times N_{base}$. This process can be used to find the normality of an unknown base by using an acid of known concentration. It is usual to have a second burette containing the acid or base of unknown concentration. This is done in case the titration is not stopped at the exact end point. In this case, a known amount of the unknown solution can be added and the end point reached again.

To illustrate this let us take an example: 10 ml of an acid of unknown normality is placed in a flask, and a drop of phenolphthalein indicator is added. A 0.5N solution of

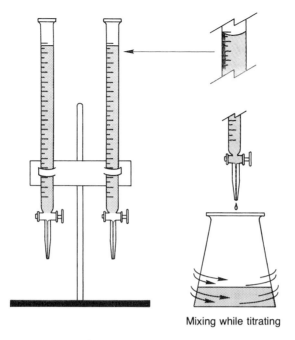

Fig. 7-3. Titration setup.

base is placed in a burette and is added slowly, while mixing, to the flask containing the acid. A change in indicator color (colorless to red) signals that the end point has been reached when 15.65 ml of base have been added. Normality can now be calculated.

$$\text{Volume of the acid} = 10 \text{ ml}$$
$$\text{Volume of the base} = 15.65 \text{ ml}$$
$$\text{Normality of acid} = X$$
$$\text{Normality of base} = 0.5$$

$$V_{acid} \times N_{acid} = V_{base} \times N_{base}$$
$$10 \times X = 15.65 \times 0.5$$
$$X = 0.7825N$$

A similar process can utilize an acid of known normality to determine the normality of an unknown base.

SALTS

A salt is a product of the reaction of an acid with a base. It is a compound containing the negative ion of an acid combined with the positive ion of a base.

Preparation. Salts may be prepared by neutralization and by treatment of certain metals with an acid.

Neutralization. One simple method of preparing salts is by the neutralization of an acid with a base. When an acid is mixed with a base, one of the products formed is water, as previously noted, and a second important product is a salt. The preparation of salts by the interaction of acids with bases is illustrated in Table 7-1.

Table 7-1. Preparation of salts by neutralization

Salt	Method of formation (neutralization)
KCl Potassium chloride	K | OH + H | Cl → H_2O + KCl
NH_4Cl Ammonium chloride	NH_4 | OH + H | Cl → H_2O + NH_4Cl
$MgSO_4$ Magnesium sulfate	Mg | $(OH)_2$ + H_2 | SO_4 → $2H_2O$ + $MgSO_4$
$Ca(NO_3)_2$ Calcium nitrate	Ca | $(OH)_2$ + 2H | NO_3 → $2H_2O$ + $Ca(NO_3)_2$
Na_3PO_4 Sodium phosphate	3Na | OH + H_3 | PO_4 → $3H_2O$ + Na_3PO_4

Treatment of certain metals with an acid

$$Zn + H_2SO_4 \rightarrow ZnSO_4 + H_2\uparrow$$
Zinc   Sulfuric   Zinc
       acid   sulfate

Naming of salts

The name of a salt is derived from the names of the acid and base from which it is prepared. The first part of the name of a salt is the unaltered name of the metallic element of the base from which it is derived; the second part is the name of the nonmetallic element or radical of the acid from which it is made, with a modified suffix. If it is a salt of a hydro acid, the prefix *hydro* is dropped and its name ends with the suffix *ide*. If it is derived from an oxyacid, its name ends in *ate* when the parent acid is an *ic* acid and in *ite* when the parent acid is an *ous* acid. Examples:

KCl	Potassium chlor*ide*—a salt formed from potassium hydroxide (KOH) and hydrochlor*ic* acid (HCl)
Na_2SO_4	Sodium sulf*ate*—a salt formed from sodium hydroxide (NaOH) and sulfur*ic* acid (H_2SO_4)
Na_2SO_3	Sodium sulf*ite*—a salt formed from sodium hydroxide and sulfur*ous* acid (H_2SO_3)
CH_3COONa	Sodium acet*ate*—a salt formed from sodium hydroxide and acet*ic* acid (CH_3COOH)
$NaClO_2$	Sodium chlor*ite*—a salt formed from sodium hydroxide and chlor*ous* acid ($HClO_2$)
$NaClO_4$	Sodium perchlor*ate*—a salt formed from sodium hydroxide and perchlor*ic* acid ($HClO_4$)

Some metallic elements enter the formation of more than one salt because of a variation in their valence. In these instances the same rule is applied as was used previously when two valences were exhibited. For the salt in which the higher valence of the metallic element is manifested, the suffix *ic* is used, and for the salt in which the metal has the lower valence, the suffix *ous* is used. Examples:

		Stock system
$HgNO_3$	Mercur*ous* nitrate	Mercury (I) nitrate
$Hg(NO_3)_2$	Mercur*ic* nitrate	Mercury (II) nitrate
$FeSO_4$	Fer*rous* sulfate	Iron (II) sulfate
$Fe_2(SO_4)_3$	Fer*ric* sulfate	Iron (III) sulfate

A salt having one hydrogen in its molecule is called a *bi* or a hydrogen salt. Two hydrogen atoms in a salt result in a *di*hydrogen salt. These salts result from the partial neutralization of an acid having more than 1 hydrogen atom in its molecule. Examples:

$NaHCO_3$ — Sodium bicarbonate or sodium hydrogen carbonate—a salt formed from sodium hydroxide and carbonic acid reacting in a 1:1 ratio

NaH_2PO_4 — Sodium dihydrogen phosphate or monosodium phosphate—a salt formed from the reaction between 1 mole of sodium hydroxide and 1 mole of phosphoric acid

Na_2HPO_4 — Disodium hydrogen phosphate or disodium phosphate—a salt formed from the reaction between 2 moles of sodium hydroxide and 1 mole of phosphoric acid

$KHSO_4$ — Potassium hydrogen phosphate or potassium bisulfate—a salt formed from potassium hydroxide and sulfuric acid reacting in a 1:1 ratio

A neutral salt is the product of the neutralization of a highly ionized acid with a highly ionized base. An acid salt is the product of the neutralization of a highly ionized acid by a poorly ionized base. A basic salt is formed when a poorly ionized acid is neutralized by a highly ionized base. When a neutral salt is placed in water, neutrality is maintained. When an acid salt is placed in water, the solution is acid, whereas a basic salt gives a basic reaction to its water solution.

Reaction of a highly ionized acid and a highly ionized base

$$HCl + NaOH \rightarrow \underset{\textbf{Neutral salt}}{NaCl} + H_2O$$

Neutral salt placed in water

$$NaCl + H_2O \rightarrow \underset{\substack{\textbf{Equal amounts of OH}^- \\ \textbf{and H}^+ \textbf{ produced}}}{Na^+ + OH^- + H^+ + Cl^-}$$

Any solution having the same amount of H^+ and OH^- is a neutral solution.

Reaction of a highly ionized acid and a poorly ionized base

$$3HCl + Al(OH)_3 \rightarrow \underset{\textbf{Acid salt}}{AlCl_3} + 3H_2O$$

Acid salt placed in water

$$AlCl_3 + 3H_2O \rightarrow Al(OH)_3 + 3H^+ + 3Cl^-$$
Poorly ionized H^+ predominant

Any solution having more H^+ than OH^- is acetic.

Reaction of a highly ionized base and a poorly ionized acid

$$NaOH + H_2CO_3 \rightarrow NaHCO_3 + H_2O$$
Basic salt

Basic salt placed in water

$$NaHCO_3 + H_2O \rightarrow H_2CO_3 + Na^+ + OH^-$$
Poorly ionized OH^- predominant

Any solution having more OH^- than H^+ is basic.

Sometimes a *mono* or *di*hydrogen salt is called an *acid salt* because it can react to neutralize 1 or more hydroxyl ions.

$$NaHCO_3 + NaOH \rightarrow Na_2CO_3 + H_2O$$

However, many of these salts give a basic reaction when placed in water because they yield more hydroxyl ions than hydrogen ions on hydrolysis. The terminology *acid carbonate* or *acid phosphate* is giving away to *bi* or *hydrogen carbonate* and *dihydrogen* and *monohydrogen phosphate,* terms that more clearly describe the chemical nature of the salt.

Ammonium salts ionize well in water solutions, but the reaction of ammonium ions with the hydroxyl ions of the water form ammonia and water molecules. Recall (p. 115) that ammonia reacts poorly with water, a fact that explains why ammonium salts give the reaction of salts of weak bases when dissolved in water.

$$NH_4Cl + H_2O \rightarrow H^+ + Cl^- + NH_4^+ + OH^-$$
H^+ predominant $\uparrow\downarrow$
$NH_3 + H_2O$

Insoluble salts

If a salt is very slightly soluble (less than 1 part in 10,000) it is classified as an insoluble salt. The formation of an insoluble salt in a chemical reaction results in an irreversible reaction. Salts vary as to their solubility from solvent to solvent, an insoluble salt in one solvent may be soluble in a different solvent. Much of the research in modern inorganic chemistry involves a search for solvents in which ionizations can take place and otherwise impossible reactions can be accomplished.

The ionization constant as developed on p. 108 can be used to determine a mathematical constant useful in showing the solubility of solutes in certain solvents. This constant is called the *solubility product,* K_{sp}. The K_{sp} is derived from the K_i.

$$CaSO_4 \rightleftarrows [Ca^{+2}] + [SO_4^{-2}]$$

$$K_i = \frac{[Ca^{+2}][SO_4^{-2}]}{[CaSO_4]}$$

In a case like calcium sulfate, in which the solute is highly insoluble in the solvent, the value of the undissolved salt, the denominator in the K_i value, is much greater than the value for the dissolved solute, the numerator in the K_i value. It is so large that it can be considered to be a constant from reaction to reaction. For comparative purposes only the numerator, the ionized particles, need to be considered, and the denominator can be considered to be unity, or one. This gives a new value, the K_{sp}, which is the value of the numerator of the K_i value. For calcium sulfate the solubility product is

$$K_{sp} = [Ca^{+2}][SO_4^{-2}]$$

As for the K_i and the K_{eq}, the values contained in the brackets are concentrations that are usually expressed in moles per liter.

A small K_{sp} value indicates a solute that is highly insoluble in the solvent in question. Some K_{sp} values in water at 25° C are

Aluminum hydroxide	Al(OH)$_3$	1.0×10^{-33}
Barium sulfate	BaSO$_4$	1.08×10^{-10}
Calcium sulfate	CaSO$_4$	3.24×10^{-4}
Lead carbonate	PbCO$_3$	3.3×10^{-14}
Silver chloride	AgCl	1.56×10^{-10}
Zinc sulfide	ZnS	1.1×10^{-21}

ANTIDOTES FOR WOUNDS PRODUCED BY ACID OR BASE

The chemical principle that is used in treating a wound caused by an acid or a base is neutralization. Acids and bases are natural antidotes for each other because each neutralizes the corrosive constituent of the other. Hence the most logical treatment for a burn produced by an acid is to wash with water and apply a solution of a weak base. The water will dilute and wash away the acid to an extent but is insufficient as a treatment. It is suggested because water can usually be obtained more quickly than any other remedy, and quick action is important when an accident occurs. The treatment with water is followed as quickly as possible by the application of a solution of a weak base, the most satisfactory of which is sodium bicarbonate. In case this substance is not available limewater or dilute ammonia may be used.

The same principles are applicable for the treatment of a burn caused by a basic substance. A wound produced by a base should be washed with water, and a dilute solution of a weak acid such as acetic acid should be applied as quickly as possible.

The following rules for treatment of wounds produced by acids or bases should be memorized:

For an acid burn apply water and a dilute solution of either sodium bicarbonate, ammonia, or limewater. In case none of these remedies is available any weak base may be used. Stronger bases may be used also *if greatly diluted.*

For a burn produced by a base apply water and very dilute acetic acid or saturated

boric acid. In case these acids are not available any weak acid may be used and also strong acids, *if greatly diluted*.

The same chemical principles enter into the treatment of hyperacidity and hypoacidity. For hyperacidity, sodium bicarbonate (baking soda) or magnesium hydroxide (milk of magnesia) solutions are ingested because these substances are bases and will neutralize gastric acidity. Furthermore, they are weak bases and will not harm the tissue as strong bases would. Since hypoacidity is a deficiency of hydrochloric acid in the gastric juice, one remedy for this condition is to give very dilute hydrochloric acid by mouth, but this should only be done under the direction of a physician.

pH

The designation, pH, means the negative logarithm to the base 10 of the hydrogen ion concentration (moles per liter).* pH is an abbreviated expression for denoting the hydrogen ion concentration of a solution. This term was proposed by the chemist, Sörensen. On the Sörensen scale, pH values range from 0 to 14. A pH of 7 means a neutral solution. At any pH value below 7 the solution is acid, and the lower the value, the greater the hydrogen ion concentration. At any pH value above 7 the solution is basic, and the higher the value, the greater the hydroxyl ion concentration.

Suppose a solution has 0.00001 g of hydrogen ions in 1 liter of solution. In the case of hydrogen ions the number of grams equals the number of moles, since 1 hydrogen ion has a mass of approximately 1 amu. This 0.00001 g of hydrogen ions is equal to 10^{-5} moles of H^+ per liter of solution or the H^+ concentration per liter, $[H^+] = 10^{-5}$. The pH is the exponent of 10 without the minus sign or the negative log of the hydrogen ion concentration. In this case

$$[H^+] = 10^{-5}; \text{ so the pH} = 5$$

In doing this we must remember that pH always means a negative logarithm (a numerical value less than one), it always applies to the base 10, and it specifically indicates the moles (grams) of hydrogen ions in 1 liter of solution. Since the moles of hydrogen ions per liter of solution are always less than one, it is clear that pH has significance only in dilute solutions of hydrogen ions.

We may illustrate further the meaning of pH by considering its application to water. In 1 liter of water there is 1/10,000,000 g of H^+ ions.

$$\frac{1}{10,000,000} = \frac{1}{10^7} = 10^{-7}$$

Therefore the pH of water is 7. But water is a neutral substance; it is as basic as it is acid. When a molecule of water ionizes, there are as many OH^- ions formed as H^+ ions.

$$H_2O \rightleftarrows H^+ + OH^-$$

Hence the amount of OH^- ions in 1 liter of water is the OH^- equivalent of 10^{-7} gram of H^+ ions. Therefore the pOH of water is 7. In other words, at pH 7 the pOH also is 7; hence it becomes clear why pH 7 represents a neutral point on the Sörensen scale.

---

*It is suggested that the reader review exponential numbers as outlined in the Appendix.

In the beginning of this discussion it was stated that at any pH below 7 the solution is acid, and at any pH above 7 the solution is basic. This may be proved by selecting points on the pH scale and calculating the relative concentrations of H^+ and OH^- ions. Let us take pH 4 as an example. A pH of 4 means the following:

$$10^{-4} \text{ g of } H^+ \text{ ions in 1 liter of solution}$$

$$\frac{1}{10^4} \text{ g of } H^+ \text{ ions in 1 liter of solution}$$

$$\frac{1}{10,000} \text{ g of } H^+ \text{ ions in 1 liter of solution}$$

The entire pH scale goes up to 14. The pOH concentration on this scale is obtained by subtracting the pH value from 14. At pH 4 the pOH is 10 (since $14 - 4 = 10$), and

$$10^{-10} = \frac{1}{10^{10}} = \frac{1}{10,000,000,000} \text{ g of } OH^- \text{ ions in 1 liter of solution}$$

We know that 1/10,000 g of H^+ ions is much greater than the OH equivalent (1/10,000,000,000 g) because the smaller the denominator of a fraction, the greater the value of that fraction. Actually the solution contains 1 million times more H^+ than OH^- ions and therefore is acid.

In a similar way, it is possible to calculate the pH of any point on the scale and show that at any point below pH 7 there are more H^+ than OH^- ions and at any point above pH 7 there are more OH^- than H^+ ions. In other words, calculation will show that all values on the Sörensen scale below pH 7 are acid and all values above pH 7 are basic.

The importance of pH becomes apparent when we begin to think of physiological substances. Thus the pH of normal human urine ranges from 5 to 7 and that of normal gastric juice is 1 to 2.5; these are examples of physiological fluids that are acid. The pH of saliva ranges from 6 to 7.9. Saliva is a substance whose reaction is close to the neutral point. Normal blood has a pH around 7.4; hence it is very slightly basic. A more basic body fluid is the bile, which has a pH of 7.8 to 8.6 (Fig. 7-4).

Buffers

Buffers are substances that resist a change in the hydrogen ion concentration of a solution when acid or base is added. Buffers are usually not very acidic or basic in themselves, but they have marked ability to keep the pH of a solution fairly constant.

To illustrate the action of buffers, let us take a salt of phosphoric acid, K_2HPO_4. If we place some K_2HPO_4 in an aqueous solution in a beaker and add, in drops, a solution of the strong acid, hydrochloric acid, there is essentially no change in the hydrogen ion concentration of the solution until a considerable number of drops of the HCl solution have been added. What happens is shown in the following reaction:

$$HPO_4^= + HCl \rightleftarrows H_2PO_4^- + Cl^- \quad (1)$$

Hydrogen phosphate ion   **Dihydrogen phosphate ion**

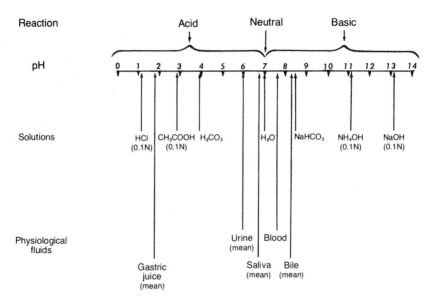

Fig. 7-4. Diagrammatic representation of the pH system.

In reaction 1 the HCl is used up as it reacts with the $HPO_4^=$ ion to form the dihydrogen phosphate, $H_2PO_4^-$ ion. As long as there is some $HPO_4^=$ ion in solution the HCl is consumed as it is added. The H^+ ion concentration, and therefore the pH, does not change appreciably until considerable HCl is added. This resistance to change in pH of the solution is a buffer action.

Now let us carry out a similar experiment, placing the acid salt of phosphoric acid, KH_2PO_4, in the solution in the beaker and adding, in drops, a solution of the strong base, potassium hydroxide. The reaction is as follows:

$$H_2PO_4^- + OH^- \rightleftarrows HPO_4^= + H_2O \qquad (2)$$

Dihydrogen phosphate ion   Hydrogen phosphate ion

In reaction 2 the OH^- ion is used up as it reacts with the $H_2PO_4^-$ ion. As long as there is some $H_2PO_4^-$ ion in solution, the OH^- ion is consumed as it is added. Since the OH^- ion concentration is virtually unchanged until appreciable quantities of base are added, the pH of the solution remains just about constant. This is a buffering action in the direction opposite to that in reaction 1.

Now, if we place an equimolar mixture of these two salts of phosphoric acid (KH_2PO_4 and K_2HPO_4) or the sodium salts in water, we have an ideal buffer solution. Such a solution will resist changes in pH toward either acidity or alkalinity if acid or base is added.

Buffers can be defined as a weak acid or a weak base in the presence of its salt. If two salts are used for a buffer, the one with more hydrogen atoms in it is considered to be the acid, whereas the salt with less hydrogen is considered the salt.

Buffer

$$\frac{KH_2PO_4 \text{ (acid)}}{K_2HPO_4 \text{ (salt)}}$$

Reaction with an acid

$$\frac{KH_2PO_4}{K_2HPO_4} + HCl \rightarrow 2KH_2PO_4 + KCl$$

Reaction with a base

$$\frac{KH_2PO_4}{K_2HPO_4} + KOH \rightarrow 2K_2HPO_4 + H_2O$$

The strong acid and base are neutralized, giving a dihydrogen salt and a neutral salt in one case and a monohydrogen salt and water in the other.

The phosphates are efficient buffers in solutions with a pH range of 6 to 8. The phosphates function very effectively as buffers in assisting in the maintenance of a constant pH in living tissues. Their action in the animal body and also the behavior of other buffers are discussed in greater detail on p. 346.

Buffer solutions of known pH are used a great deal in the laboratory. With a knowledge of buffers, one may make up buffer solutions of fixed and known pH, effective in ranges of 2 to 3 pH units from pH 1 to pH 12. In preparing a buffer solution for use in the laboratory, two substances are usually selected. Pairs of compounds used in the preparation of buffer solutions can be chosen from the following: KCl and HCl, potassium hydrogen phthalate and HCl, sodium acetate and acetic acid, KH_2PO_4 and K_2HPO_4, $NaHCO_3$ and H_2CO_3.

Questions for study

1. What are electrolytes? Nonelectrolytes? Give examples of each.
2. Why is a solution of an electrolyte electrically neutral?
3. What are the carriers of an electric current through a solution?
4. Does ionization occur in water only? Give examples of other instances.
5. Give two definitions of an acid.
6. What is a hydronium ion?
7. Give five general properties of acids.
8. Distinguish between strong and weak acids. Give examples of each.
9. What is hypoacidity? Hyperacidity? Acidosis?
10. What is the general rule for naming hydro acids? Oxyacids?
11. Write the chemical names for the following acids:
    a. An acid composed of hydrogen and fluorine
    b. An acid of hydrogen, oxygen, and boron
12. What is a base?
13. Give five general properties of bases.
14. Distinguish between strong and weak bases. Give an example of each.
15. What is neutralization?
16. What is the general rule for naming bases?
17. When two bases of the same metallic element occur, what is the practice in naming them?
18. Write the chemical names for the bases containing the elements indicated below:
    a. Lithium   b. Barium   c. Magnesium

19. What is a salt?
20. How are salts prepared?
21. What should be done in case of an acid burn?
22. How should one proceed to treat a wound produced by a basic substance?
23. What remedies are used for hyperacidity? For hypoacidity? Explain.
24. What is the general rule for naming salts?
25. A salt of an *ic* acid ends in what suffix? A salt of an *ous* acid ends in what suffix?
26. Write the chemical names for the following salts:
    a. Sodium salt of hydrofluoric acid
    b. Magnesium salt of carbonic acid
    c. Potassium salt of sulfuric acid
    d. Calcium salt of phosphoric acid
    e. Sodium salt of nitrous acid
27. Define pH.
28. At what pH values is a solution acid? Basic?
29. At what pH is a solution neutral?
30. A solution of HCl is on hand, but the normality is unknown. A 0.5N solution of NaOH is available. Tell how one would discover the normality of the acid. If it took 12.4 ml of the base to neutralize 10 ml of the acid, what is the normality of the acid?
31. The hydrogen ion concentration of a solution is 1/10000M. What is the pH?
32. What is a buffer? What are the uses of buffers in the human body?
33. Fifteen grams of a nonvolatile, nonionizable solute are dissolved in 250 ml of water. The solution freezes at $-5.58°$ C. What is the molecular weight of the solute?
34. Twenty grams of a nonvolatile, nonionizable solute are dissolved in 100 ml of water. The solution boils at $100.67°$ C. What is the molecular weight of the solute?

8 Crystalloids, colloids, diffusion, dialysis, and osmosis

CRYSTALLOIDS AND COLLOIDS

If we dissolve gelatin in water, we obtain a solution that has special properties. This solution appears thick and viscous; it is somewhat opalescent; and, if sufficiently concentrated, it will set to a jellylike mass. We can pour it through an ordinary filter paper, but if we attempt to pass it through cellophane, the water will pass through but the gelatin will remain on the filter. The cellophane has a texture too fine for gelatin particles to filter through. The gelatin particles must therefore be much larger than the molecules of such substances as glucose and sodium chloride, which will readily pass through cellophane. *Substances that pass through membranes such as cellophane when in solution are called crystalloids. Substances that will not pass through cellophanelike membranes are known as colloids.* Glucose, sodium chloride, and hydrochloric acid are examples of crystalloids; gelatin, glue, egg albumin, and caramel are examples of colloids. Crystalloids and colloids behave with respect to a number of other membranes in the same manner as described for cellophane. Kidney, intestinal, and egg membrane, parchment paper, and collodion are examples of materials through which crystalloids will pass but colloids will not.

A liquid containing colloidal particles in suspension is known as a colloidal solution. A true solution differs from a colloidal solution in the size of the dissolved or suspended particles. In a true solution the dissolved molecules are small and will readily pass through cellophane. In a colloidal solution the suspended particles are either very large molecules or are particles composed of many molecules, neither of which will pass through membranes with a texture like that of cellophane.

The term *colloid then indicates a state of matter. It refers to matter subdivided into particles ranging in size from 1 to 100 mμ.* The average colloidal particle is about 1000 times smaller than a red blood cell or an anthrax bacterium. Such particles cannot be seen with a high-power microscope but can be made apparent to the eye with an instrument called the ultramicroscope, which does not show actual colloidal particles but reveals their presence by the zigzag reflections the particles produce while moving rapidly in a highly illuminated field.

128 Roe's principles of chemistry

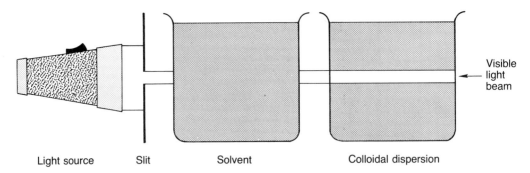

Fig. 8-1. Tyndall effect.

If a beam of light enters a darkened room, its pathway is revealed by the light rays that are reflected to the eye of the observer by dust particles in the air. This is known as the *Tyndall phenomenon* (Fig. 8-1). The same effect is observed when a beam of light is passed through a colloidal solution. The pathway of the light appears as a more highly illuminated region because the light rays are reflected by the colloidal particles. This is another way to show the presence of colloidal particles in a colloidal solution.

One of the most interesting and important properties of colloids is their power to absorb large quantities of water. This is called *imbibition*. It occurs in living cells and is the process that enables the tissues to hold such relatively large quantities of water. The human body is about two-thirds water; yet the tissues are quite firm and solid. Such quantities of water could never be maintained if the cells did not contain a large amount of substances in the colloidal state.

Solvated colloidal particle

Increased surface area. The surface area of a material is greatly increased when its mass is distributed among very small particles. All surfaces have surface charges, but these are very weak in comparison to their mass. The surface area of colloidal particles is great in comparison to the mass, and surface charges are important to matter in the colloidal state. These charges are not to be confused with the charges on an ion. The same colloidal particle can be charged positively or negatively depending on circumstances. In a constant environment, colloidal particles of the same substance have similar surface charges. These charges repel each other and prevent the colloid from forming larger aggregates, which would precipitate from solution because of their mass. Because of its charges, the colloidal particle has great adsorptive powers. It can attract and hold, for example, molecules of solvent and ions.

Lyophilic and lyophobic colloids and emulsions. When a colloid particle has adsorbed a layer of its solvent, it is considered as a *solvated* particle. A colloid that has this attraction to its solvent is called a *lyophilic* (solvent-loving) colloid. If the solvent is water, the colloid may be called a *hydrophilic* (water-loving) colloid. A lyophilic colloid usually forms a stable colloidal suspension. A colloid that does not have this attraction to its solvent is called a *lyophobic* (solvent-fearing) colloid. Lyophobic colloids form unstable solutions with their solvents and are very likely to form large particles and precipitate from solution. If the colloid is lighter than the solvent, the lyophobic colloid will have a tendency to rise to the surface and form a layer on top of the solvent. A lyophobic colloid particle may be made more stable in the solvent by attracting a layer of a lyophilic colloid to its surface. The solvent is attracted to the lyophilic layer, and the colloid is stabilized. The adsorbed lyophilic colloid is called a *protective colloid* or an *emulsifying agent*. The solution of an emulsified lyophobic colloid is called an *emulsion*.

Oil in water is an example of a lyophobic colloid. The oil can be held in solution only if the oil droplets are made very small—a process called *homogenization*. Adding soap, a lyophilic colloid, to a mixture of oil and water stabilizes the solution. The soap acts as a protective colloid to the oil.

A colloidal solution has two phases, the *dispersed* phase and the *continuous* phase. The dispersed phase is the particular matter held in solution. The dispersed phase is held in solution by the dispersion medium (solvent), which is the continuous phase. Any of the states of matter may be either the dispersed or continuous phase, but a gas in gas combination is more accurately classed as a mixture. Molecules of gases are too small to be correctly classified as colloidal particles.

The physical appearance of a colloidal solution is dictated by the continuous phase. If the continuous phase is a liquid, then the colloidal solution is liquid—a solution, or a *sol*. If the continuous phase is a solid, then the colloidal dispersion is a solid or a semisolid—a *gel*. Some colloids, especially the lyophilic colloids, have the ability to exchange phases, at one time appearing as a liquid solution, a sol, and at other times appearing as a semisolid, a gel. Gelatin is an example of this. At first a solution of gelatin in water is a liquid and can be poured into molds. The gelatin is the dispersed phase, the water is the continuous phase, and the solution is a sol. On standing, the gelatin swells and becomes the continuous phase, the water being dispersed in the solvated gelatin. The gelatin is said to have "set"; it has become a semisolid, a gel. Certain colloidal solutions have the ability to repeatedly change from a sol to a gel and back again. The protoplasm of the cell is such a colloid. Protoplasm exists in a sol-gel equilibrium state. The "flowing" motion of the amoeba is the result of sol-gel state interchange.

Isoelectric point. For every type of colloidal particle there exists a pH at which its surface charges are at a minimum. This pH is called the *isoelectric point* and it differs from colloid to colloid. At pH values greater or less than the isoelectric point, the colloid particles exist with a surface charge. On the acidic side of the isoelectric point the charge is positive; on the basic side of the isoelectric point the charge is negative.

When the surface charges are reduced, the colloid particles are less solvated and also repel each other less. At the isoelectric point colloidal solutions are the least stable and the particles are more likely to coalesce and precipitate from solution.

Electrophoresis. Charged particles of a colloid can be attracted to electrodes as ions can. This is the basis of the separation of colloidal particles in solution by placing them in an electrical field. The particles move toward the electrodes at different rates, depending on their charge and mass. This process is called *electrophoresis*. At the isoelectric point the colloid particles move but slightly in an electric field. It is necessary to have colloids above or below their isoelectric points for electrophoresis. Mixtures of colloids can be separated by the use of electrophoresis.

DIFFUSION

Diffusion is a process in which molecules, ions, or colloidal particles move from a region of higher to a region of lower concentration. It is a physical process in which there is a net change in concentration, the direction of passage being from the higher to the lower concentration. The kinetic energy of the molecules or diffusing particles keeps the diffusing particles in continuous motion.

Diffusion is a process of great importance in biology. It is by diffusion that nutritive substances are distributed to all the cells of a living organism and waste products of metabolism are removed. Diffusion is important also as the process that brings about a distribution of substances of physiological importance in plants and animals, such as water, oxygen, inorganic ions, enzymes, hormones, and vitamins.

DIALYSIS

In studying the chemistry of living things, it is necessary to get clearly in mind the principles involved in the passage of materials through membranes. In the human body the absorption of such substances as water, salts, and food substances takes place through the mucous membranes of the intestinal walls, which are living membranes; the blood corpuscles carry on their functions through processes involving their membranous walls; respiration takes place through the thin membranous sacs of the lung alveoli; the excretion of urine takes place through membranes that line the kidney glomeruli and tubules; and many other processes occur that involve the activity of living membranes. The cells of plants have walls that are typical living membranes, and the plant is dependent for its foods on the passage of materials through its membranous cell walls. The phenomena associated with the passage of materials through membranes are therefore of great significance in biological chemistry.

In studying the diffusion of materials through membranes, no difficulty will be experienced if the student will keep in mind the two points brought out in the discussion of colloids: (1) the size of the molecules of the diffusing substance and (2) the texture of the membrane through which substances pass. As was noted previously, the essential difference between crystalloids and colloids is that crystalloids have a much smaller mole-

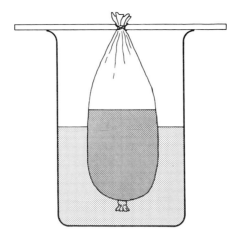

Fig. 8-2. Dialyzing apparatus.

cule. If a solution containing both crystalloids and colloids is placed in a cellophane bag and the bag is tied securely and suspended in a beaker of distilled water, as shown in Fig. 8-2, the crystalloids will pass through the cellophane into the distilled water and in this manner can be separated from the colloids. *The separation of crystalloids from colloids by the passage of the crystalloids through a membrane that holds back colloids is called dialysis.* A small amount of colloidal particles may pass through a dialyzing membrane in time, but they pass so slowly that dialysis is considered a practical process for the separation of crystalloids from colloids. Membranes other than cellophane may be used to demonstrate dialysis, such as a collodion bag or a pig's bladder.

Practical applications of dialysis are in purification processes in which crystalloids are removed from a solution containing crystalloids mixed with colloids. Thus the chemist in preparing a pure protein may have salt as a contaminant in his preparation. He can remove the salt by placing his mixture of protein and salt in a dialyzing bag and immersing it in slowly running water. Salt may be removed from milk by placing it in a dialyzing apparatus for a time. This application of dialysis has been used in the preparation of a low-salt food for patients suffering with hypertension or cardiac decompensation.

OSMOSIS

Vapor pressure is related to osmotic pressure. A pure solvent exists in equilibrium with its gas phase. The particles of this gas exert a pressure on the surface of the solvent. When a nonvolatile solute is added to the solvent, the particles of the solute dilute the particles of the solvent. This results in fewer solvent molecules appearing at the surface of the solution and less solvent molecules that are available to become vaporized into gas particles. There is less vapor pressure over the surface of this solution than existed over the surface of the pure solvent. If the solution of solvent and nonvolatile solute is separated from pure solvent by a membrane that allows molecules of solvent to pass but is not penetrated by the solute molecules (semipermeable), a pressure, *osmotic pressure*, forces molecules of the solvent to pass through the membrane into the solution. The volume of

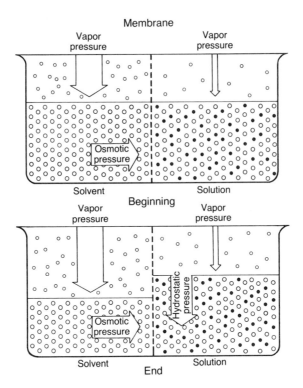

Fig. 8-3. Osmotic pressure.

the solution will increase until its hydrostatic pressure (mass of fluid pulled by gravity) equals the osmotic pressure of the pure solvent (Fig. 8-3). Osmotic pressure is related to the difference in the vapor pressure of a pure solvent and a solution of that solvent and a nonvolatile solute.

It must be noted that although the osmotic pressure is related to the difference between the vapor pressure found over the pure solvent and that found over a solution of the same solvent plus a nonvolatile solute, the osmotic pressure is greater than the numerical difference between the two vapor pressures. The added pressure (such as hydrostatic pressure) needed to offset the osmotic pressure is a measurement of that pressure.

Osmotic pressure also exists between two solutions of different concentrations separated by a semipermeable membrane. The solution having the greater amount of solvent (the more dilute solution with respect to solute) exerts an osmotic pressure across the membrane.

The membranes of living organisms are more or less permeable to solutes. Sugars, for example, readily diffuse through living membranes, but they do not diffuse as rapidly as water. If a more concentrated sugar solution is placed on one side of a living membrane than on the other side, both the sugar and the water will diffuse through the membrane, but the water will diffuse faster and will pass in a direction opposite to that of the sugar; a greater volume and hydrostatic pressure will develop on the side of the membrane having the solution of greater concentration. This situation will continue until the concentration of sugar becomes the same on each side of the membrane. When this occurs, the pressure and volume differences with respect to the membrane will disappear.

It should be stressed that the passage of crystalloids through a membrane is simple diffusion and is not osmosis. Specifically, osmosis is a process in which only the solvent passes through the membrane. In biology the solvent is water; therefore, osmosis in living things refers to the passage of water through a differentially permeable membrane.

Osmosis is of great importance in living organisms. It is the process by which water is absorbed by the living cells and distributed to all parts of the organism. The walls of plant and animal cells behave as differentially permeable membranes, allowing water and small particles such as ions and small molecules to pass through but holding back particles of colloidal size. The differentially permeable membrane has the effect of concentrating the water molecules and small particles on the side of the membrane having the highest concentration of colloidal particles. This creates an osmotic pressure across the membrane.

It is through osmosis that plants absorb and distribute their water. The water enters the plant through the root cells in contact with the soil by operation of the principle of osmosis. From the cells the water diffuses into special vessels in the plant, through which it passes until reaching the leaves. In the leaves there is a high concentration of solutes; this condition is favorable for the passage of water by osmosis from the special vessels of the plant into the individual cells of its leaves. Hence, water passes into the cells of the leaves as a result of the osmotic pressure found in these cells. Thus osmosis plays an important role in plant life.

In the human body or in animals the water that is ingested forms a solution in the alimentary tract that is usually more dilute than the blood or the solution in the cells of the intestinal walls. Under these conditions water readily passes by osmosis into the blood or lymph. After water is absorbed, its distribution throughout the body is influenced very prominently by the principles governing osmosis. Thus in animals osmosis has a prominent role in bringing about the absorption of water and its distribution to the tissues.

Osmolarity is a measurement of the osmotic activity of a solution. *One gram-molecular weight of a dissolved nondiffusible, nonionizable substance is equal to 1 osmol.* The *milliosmole,* equaling *0.001 osmol,* is the measurement of osmotic pressure used to express the osmotic activity of body fluids. For nonelectrolytes, an osmole is the same as the mole. For ionizable substances, osmolarity is determined by the number of particles formed by the ionization process.

Physiological salt solution and Ringer's solution. The normal concentration of salts in the blood and lymph of animals is approximately equivalent to 0.9% sodium chloride. The living tissues of animals undergo no change due to osmosis in a solution of this strength. *A 0.9% solution of sodium chloride is therefore called a physiological salt solution.* A solution of the same concentration of sodium chloride but containing in addition 0.03% potassium chloride, 0.02% sodium bicarbonate, and 0.02% calcium chloride is known as Ringer's solution. Besides being physiologically balanced with respect to osmolarity, Ringer's solution contains the elements calcium and potassium and a trace

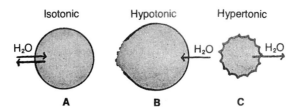

Fig. 8-4. Diagram showing changes in blood cells due to osmosis. **A,** Blood corpuscle in isotonic solution (0.85% NaCl) shows no change in volume. **B,** Blood corpuscle in hypotonic solution (distilled water) swells and bursts. **C,** Blood corpuscle in hypertonic solution (2% NaCl) shrinks and becomes crenated.

of alkali, additional factors under the influence of which a living tissue such as muscle will survive much longer.

Isotonic, hypotonic, and hypertonic solutions. Solutions having the same osmolarity are known as isosmotic or *isotonic* solutions. (*Isos* is a prefix meaning equal.) *Physiological salt solution is isotonic to the tissues of the body.*

A solution is said to be *hypotonic* to another solution when it has a lower osmolarity (*hypo* means below), and a solution is *hypertonic* to another solution when it has a higher osmolarity (*hyper* means above). A solution with a lower osmolarity than blood plasma (lower than physiological salt solution) is hypotonic to the blood; a solution with a higher osmolarity than blood plasma is hypertonic to the blood.

Blood changes due to osmosis. The behavior of blood corpuscles when placed in solutions of different concentrations is an interesting demonstration of osmosis. If blood corpuscles are placed in a physiological salt solution (approximately 300 milliosmoles), they will maintain their normal size and shape. This is illustrated in Fig. 8-4. There is no change in the corpuscles in this solution because the concentration of salt is the same within the cells as on the outside, and consequently water will pass through the walls of the corpuscles with the same rapidity in each direction.

If blood corpuscles are placed in distilled water (zero osmolarity), which represents a hypotonic solution, they will swell and burst. The explanation is that there is a greater concentration of salts within the blood cells. Osmotic pressure causes water to pass into the cells more rapidly than it passes out, since in osmosis the greater flow of solvent is toward the more concentrated solution; consequently the cells swell from increased fluid until they rupture. This is illustrated in Fig. 8-4. The rupture of red blood corpuscles by a hypotonic salt solution (or other factors) is called *hemolysis.* The blood of normal human subjects does not undergo hemolysis in sodium chloride concentrations between approximately 0.44% and 0.9%. The failure of red cells to hemolyze at these concentrations is due to the resistance to rupture exhibited by the walls of these cells. However, red blood cells will swell from osmosis in salt concentrations between 0.9% and 0.44%, a change that may be demonstrated by use of the hematocrit.

If blood corpuscles are placed in a hypertonic solution, for example, a 2% solution of sodium chloride (approximately 700 milliosmoles), they will shrink in volume and have

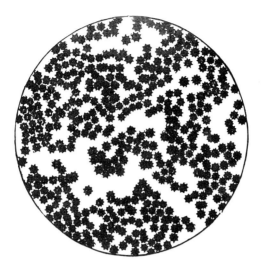

Fig. 8-5. Crenated erythrocytes. A microscopical view showing the shrinking of red blood cells when placed in a hypertonic salt solution. (From Hawk and Bergeim: Practical physiological chemistry, P. Blakiston's Son & Co.)

the appearance represented in Fig. 8-4, C. This change is explained again by the principle of osmosis. The greater concentration of salt this time is on the outside of the cells. Water will therefore pass out of the cells faster than it can pass into them, and the result is a withdrawal of water from the cells with consequent shrinking of blood cells. The shrinking of blood cells in a hypertonic salt solution is called *crenation* (Fig. 8-5).

From the preceding discussion it is apparent that *fluids cannot be safely introduced into the bloodstream unless they are practically isotonic to the blood.* Sometimes it is necessary to give water, salt, glucose, or amino acids to a patient intravenously. In such instances sterile isotonic solutions are used. When drugs are administered intravenously, care must be exercised to use a solution that is of a concentration not far from isotonic, or else the substances must be placed in physiological salt solution. Sterile physiological salt solution is used also in surgery for cleaning wounds, washing out the peritoneal cavity, and so on. In all these applications the use of physiological salt solution is based on the fact that the tissues undergo no changes from osmosis in this solution.

The phenomena concerning the behavior of blood cells are typical of any animal or plant cell possessing a cell wall that acts as a semipermeable membrane.

Questions for study

1. What is a crystalloid? A colloid? Give examples of each. Distinguish between a colloidal solution and a true solution.
2. How may the presence of colloidal particles in a solution be shown?
3. Define diffusion. To what is it due? What is the importance of diffusion in biology?
4. Define dialysis. Give an example of a practical application of dialysis.
5. What is osmosis? Give an explanation of osmosis.
6. Define osmotic pressure.

7. What is the importance of osmosis in relation to the human body? To plants?
8. What is a physiological salt solution?
9. Define isotonic, hypotonic, and hypertonic solutions.
10. What changes will take place in blood cells when placed in a hypotonic solution? When placed in a hypertonic solution? Explain each case.
11. What are the uses of physiological salt solution? Explain the necessity for its use in each case.

9 Radiochemistry

RADIOACTIVITY

Radioactivity is a phenomenon that involves changes in the nucleus of the atom. In striking contrast to ordinary chemical changes, which involve the electrons in the outer orbit of the atom, the changes of radioactivity have their origin in the nucleus and bring about an enormously greater output of energy. Radioactivity was first observed as a property of uranium. In experiments with this element in 1896, Becquerel noticed that it gave off emanations that affected a photographic plate and discharged a charged electroscope. Later the Curies isolated polonium and radium and found that these elements are a million times more radioactive than uranium.

RADIUM

Radium was discovered by Pierre and Marie Curie in 1898. It was first isolated in the form of its salt, radium chloride. Later the pure element was prepared by Madame Curie. Pure radium is a white, lustrous metal. The element is difficult to prepare and to keep in the pure form and is therefore used in the form of radium chloride or radium bromide.

Radium was given its name because it constantly radiates or emits energy. One gram of radium gives off more than enough heat to raise its own weight of water from the freezing point to the boiling point every hour, or—in more exact terms—the energy liberated by 1 g of radium is equivalent to 120 calories per hour. It has been estimated that the amount of heat liberated by 1 g of radium during its period activity is equivalent to the heat that would be obtained from burning about 1 ton of coal. This remarkable radiation of energy is the result of changes that really are a disintegration of the radium atom. These changes will be discussed.

Radium is used in medicine for the treatment of cancer. The radiations from radium will destroy both normal and abnormal tissues. These radiations are applied directly to a malignant growth, care being taken to screen the healthy tissues by means of materials through which the rays from radium do not pass readily. Such treatment has been found valuable in cases of abnormal growths to which radium emanations can be applied satisfactorily. Radium is kept in lead containers to prevent exposure of personnel to its rays, since lead is a substance through which radium emanations pass very poorly or not at all if a thick layer is used.

Investigation of the properties of radioactive elements have shown that three emanations are given off by these substances: α-(alpha), β-(beta), and γ(gamma) rays.

The α-rays are atoms of helium that have been stripped of 2 electrons and therefore bear two positive charges. These particles are shot off from radioactive material at a speed of 18,000 miles per second. They do not have the penetrative power of β- or γ-rays, being stopped by a sheet of paper or a few centimeters of air. This is due to its relatively high mass.

The β-rays are fast-moving electrons. They are ejected from radioactive substances at a velocity of 160,000 miles per second. The β-rays have about 100 times the penetrative power of α-rays; they will pass through several millimeters of aluminum or as much as 10 feet of air before being stopped. The β-rays are the same fundamental particles as the cathode rays emitted from an x-ray tube.

The γ-rays are not particles; they are electromagnetic radiations and have great penetrative power. They will pass through thick layers of metals. In passing through gases they ionize them, a property possessed also by α- and β-rays. There is some parallelism between γ-rays and x-rays. The γ-rays were first observed as radiations emitted by radium. More recently it has been found that γ-rays are formed when positrons are annihilated by impact with electrons. In the latter event, matter is converted into energy, a relatively great amount of energy, which explains the powerful penetrative action of γ-rays.

The relative penetrating powers of alpha, beta, and gamma rays are illustrated by Fig. 9-1. If radium is placed in an electrostatic field, the α-rays, being positive, will pass to the negative pole, and the β-rays, being negative, will pass to the positive pole. The γ-rays are not deflected by an electrostatic field (Fig. 9-2).

Radioactivity is a disintegration of matter. Atoms of such substances as uranium, actinium, thorium, polonium, and radium undergo continuous disintegration, giving off some or all of the three types of rays, and are thus changed into atoms of other elements. To illustrate, let us consider the behavior of uranium, as shown in Fig. 9-3.

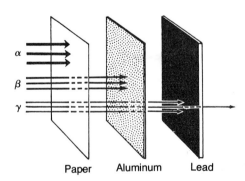

Fig. 9-1. Diagram showing relative penetrating power of alpha, beta, and gamma rays. (From Research Reporter, Chemistry **38**:20, 1965; reprinted with permission of American Chemical Society.)

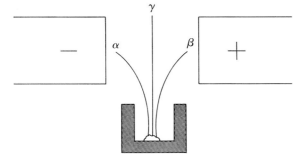

Fig. 9-2. Radioactive particles in an electrostatic field.

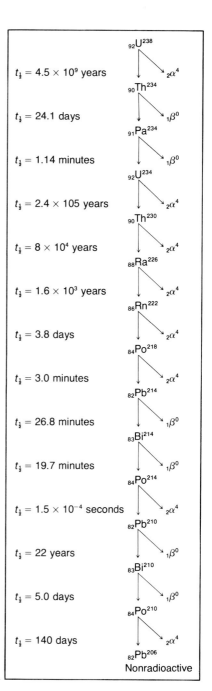

Fig. 9-3. Uranium disintegration series.

Uranium and its disintegration products are listed in the second column of Fig. 9-3. Each element in this column disintegrates, giving rise to the element beneath it. The type of particle emitted by each element is shown in the third column. In the first column the half-life period $(t_{\frac{1}{2}})$ of each element is given. By half-life period is meant the time that elapses while the element loses one-half of its mass by radioactive disintegration.

Half-life

$$_{91}Pa^{234} \xrightarrow[1.14 \text{ minutes}]{t_{\frac{1}{2}}} {}_{92}U^{234} + {}_{1}\beta^{0}$$

The end product resulting from the disintegration of radium is lead. Lead-206 does not undergo transmutation (see Fig. 9-4).

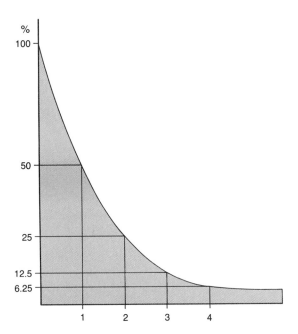

Fig. 9-4. Half-life of radioactive materials.

Artificial radioactivity. Recently, methods have been discovered for imparting radioactivity to elements that do not possess this property naturally. These methods consist essentially of procedures for bombarding an element with very small particles possessing high speed and great energy. The first success along this line was achieved by Frédéric Joliot-Curie and his wife Irène, the daughter of Marie and Pierre Curie. They directed α-particles emitted by polonium on sheets of aluminum and found that the aluminum remained radioactive for several minutes after removal of the bombarding source.

Later, machines (cyclotron and betatron) were developed for projecting bombarding particles with high speed and great energy on targets of substances to be made radioactive. These machines are capable of imparting energies of 2 to 10 million electron volts to particles in their fields of activity. The bombarding particles or "bullets" used in such shooting are protons (hydrogen nuclei), deuterons (nuclei of heavy hydrogen), neutrons, and α-particles. When these tiny projectiles, moving at a high velocity, strike the surface of a substance, a small number of them collide with the nuclei of the atoms of the bombarded material.

Bombardment reaction

$$_7N^{14} + {_2\alpha^4} \rightarrow {_8O^{17}} + {_1H^1}$$

$$_4Be^9 + {_2\alpha^4} \rightarrow {_6C^{12}} + \underset{\textbf{Neutron}}{_0n^1}$$

The "head-on" collisions with atomic nuclei cause some disintegration in the atomic nuclei. From such atomic collisions there are sometimes emitted negative electrons or comparable masses having a positive charge (called "positive electrons," or positrons) and sometimes protons or neutrons. As a result of these collisions new elements are formed that may be radioactive for a time.

By bombardment techniques radioactivity has been imparted to many elements not having this property. In most instances the period of radioactivity is short, lasting only a few seconds or minutes, but a longer-lasting activity has been imparted to some of the elements. Thus radioactive sodium, produced by bombardment of metallic sodium or sodium salts with fast-moving deuterons, has a half-life period of 15 hours; radiophosphorus and radiosulfur have half-life periods of 15 and 87 days, respectively; and radioactive carbon ($^{14}_6C$) has a half-life period of 5,730 years. Great success has been achieved in generating artificial radioactivity in elements, mainly through the use of the cyclotron and the nuclear reactor or atomic pile (described later).

Artificial radioactivity

$$_7N^{14} + \underset{\textbf{Neutron}}{_0n^1} \rightarrow {_6C^{12}} + {_1H^1}$$

$$_{17}Cl^{35} + \underset{\textbf{Neutron}}{_0n^1} \rightarrow {_{16}S^{35}} + {_1H^1}$$

The bombardment of platinum with fast-moving deuterons results in the formation of gold. Thus the dream of the alchemists, the transformation of another element into gold, has been achieved. But the amount of gold obtained in this way is small, the procedure is costly, and the source material, platinum, is more valuable than gold.

Radiocarbon dating. Most of the carbon of the earth and the earth's atmosphere is in the form of the $^{12}_6C$ isotope—the element whose atom has a nucleus containing 6 protons and 6 neutrons. Traces of carbon, however, exist in the form of the $^{13}_6C$ (6 protons and 7 neutrons) and the $^{14}_6C$ (6 protons and 8 neutrons) isotopes. The $^{14}_6C$ isotope is radioactive and is known as radiocarbon.

The origin of radiocarbon is in the upper atmosphere. The process begins with the bombardment of gas molecules by cosmic rays, which produce high-energy neutrons. These energy-charged neutrons react with nitrogen atoms $^{14}_7N$ to produce $^{14}_6C$ atoms (radiocarbon) and hydrogen. The $^{14}_6C$ is oxidized to $^{14}CO_2$. During photosynthesis, plants utilize $^{14}CO_2$ along with the normally occurring CO_2. Thus a small and constant amount of radiocarbon is stored in the tissues of plants. Animals feed on plants, and in this way radiocarbon gets into animal tissues. The amount of radiocarbon in the atmosphere became constant many millions of years ago; hence the amount of radiocarbon incorporated into the tissues of plants and animals has been constant for a very long time.

Radiocarbon undergoes decay into nitrogen at a constant rate. The rate of decay is known; the half-life of radiocarbon is 5730 years. When animals and plants die, the incorporation of radiocarbon into their tissues ceases. Furthermore the tissue radiocarbon decreases at a constant rate after their death. Measurement of the residual content of radiocarbon (by its radioactivity) in the tissues of a dead plant or animal provides data with which there can be calculated the time since the plant or animal lived. For example, if the radiocarbon activity of an old tissue, such as one discovered in a cave, is one-half that of a fresh living tissue, the old tissue is 5730 years old.

Radiocarbon dating is of great interest to the archaeologists who seek to establish the age of materials (bones, dried plant, or animal tissues) found in caves or excavated from the ruins of ancient abandoned human habitations. It is claimed that with these techniques a reliable chronology can be established for carbon-containing materials for the past 50,000 years. Thus radiocarbon dating, used as a tool in archaeological research, has brought about a remarkable advance in the understanding of our ancient heritage.

X-RAYS

Discovery. X-rays were discovered by Roentgen, a German physicist, in 1895. Roentgen noticed a fluorescence was produced on a barium platinocyanide screen when it was placed near a cathode-ray tube in operation. He also observed that covered, fully protected photographic plates left near a cathode-ray tube were acted on the same as if exposed to visible light. Roentgen recognized that these effects were due to a new kind of invisible rays that were capable of penetrating matter. He called these rays x-rays, which are often referred to as roentgen rays in honor of their discoverer.

X-ray tube. The x-ray tube is a highly evacuated glass bulb into which has been sealed two electrodes, the anode or positive electrode and the cathode or negative electrode. The cathode consists of a spiral filament of tungsten wire, which emits electrons when heated by the passage of a current of electricity, and a focusing cup surrounding the filament, which serves to focus the stream of electrons emitted by the spiral filament on the focal spot of the anode. The electrons emitted from the hot cathode are propelled across the gap between the anode and the cathode by a high-voltage current. When the electrons strike on the focal spot of the anode, they give rise to the formation of x-rays. As shown in the diagram (Fig. 9-5), the anode is set at an angle that causes the x-rays to be deflected to the outside of the tube over a hemispherical space, where they may be used in radiographic work or in the treatment of cancer or other lesions.

Nature of x-rays and their application. X-rays are a form of radiant energy. They have a very short wavelength, about 1 ten-thousandth the wavelength of visible light rays. They are invisible. They travel at the same speed as light, and they are similar to light in some properties. They have a close similarity to the γ-rays emitted by radioactive matter.

X-rays are generated when cathode rays, which are fast-moving electrons, strike on

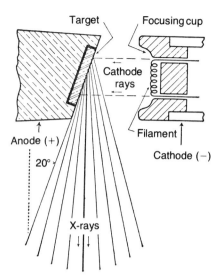

Fig. 9-5. Schematic diagram of the generation and bombardment of electrons from a heated filament in the cathode of an x-ray tube and subsequent production of x-rays from the focal spot of the anode. (From Fundamentals of radiography, courtesy Eastman Kodak Co.)

solid matter. In the x-ray tube they are formed at the anode by the impingement of the cathode rays on the anode. The cathode rays possess enormous energy. Most of this energy is transformed into heat and dissipated through the anode; however, a small amount of this energy is used in the production of x-rays.

The outstanding property of x-rays is their ability to penetrate matter. This is because of their very short wavelength. The degree of penetration of matter depends on the density of the matter. Advantage is taken of this property in making radiographs. In radiography a chemically treated film is used that is sensitive to x-rays just as ordinary photographic film is sensitive to visible light. If an x-ray film is placed on one side of an object and x-rays are passed through this object, the x-rays will affect the film in direct proportion to the amount that gets through. The less dense regions of the object will permit more x-rays to go through than the more dense regions; hence a greater exposure of the film will occur beneath the less dense regions of the object. After the film has been developed there will appear on it an outline of the internal structure of the object, delineated according to the density of the structures present.

In a radiograph of the human body, bone absorbs more x-rays than soft tissues do; hence the parts of the film exposed to x-rays passing through bone will appear lighter than parts exposed through the soft tissues of the body. Similarly, organs or other tissues of the body are outlined on the radiograph in direct relation to their density. Thus a radiograph of the human body presents a shadow outline of the internal structures, based on the differences in penetrability of the body structures to x-rays.

Materials such as barium sulfate may be introduced into the alimentary tract, and because the barium has a greater density than the soft tissues of the body, the stomach and intestinal tract will be clearly outlined on a radiograph. By x-ray techniques the presence of stones in the gallbladder may be revealed, and the gallbladder may also be visualized after injection of a radiopaque dye, which is then excreted in the bile and concentrated in the gallbladder. Similarly, x-ray apparatus may be used to show the presence of kidney stones or to visualize the urinary tract after intravenous injection of a radiopaque dye that is excreted in the urine. The radiopaque dye may also be introduced into the urinary tract by direct administration into the ureters.

X-ray film. X-ray film is a thin transparent plate of cellulose acetate coated with an emulsion consisting of fine silver bromide crystals in gelatin. The emulsion is placed on both sides of the film in layers about 0.001 inch thick. When this film is exposed to x-rays, an ionization of some of the silver bromide crystals into silver and bromine ions occurs. Obviously the greater the exposure, the greater will be the amount of ionization. Later the film is subjected to a chemical development in which the silver ions are reduced to black metallic silver, and finally the film is treated with another chemical solution that removes the unaffected silver bromide crystals. The blackening of the film is due to the silver deposited on it. The degree of blackening is proportional to the intensity of exposure to x-rays. Therefore the black regions of an x-ray film of the body are the result of a large amount of x-rays getting through the soft tissues, and the lighter regions of the film are due to a lesser quantity of x-rays getting through the tissues of greater density.

Fluoroscopy. Another application of x-rays in medicine is fluoroscopy. In fluoroscopy a sensitized screen coated with materials that fluoresce when struck by x-rays (zinc sulfide, barium platinocyanide, and calcium tungstate) is placed behind the patient, and x-rays are sent through the body. A visualization of the inner structure of the body is produced on the fluorescent screen by virtue of the fact that the x-rays will penetrate the less dense tissues more readily than the more dense structures. Thus a means of direct observation of the deep structures of the body is offered the operator who observes on the fluorescent screen the image made by the differential penetration of the tissues by x-rays.

ATOMIC FISSION

Atomic fission is a splitting of the atoms of matter. This phenomenon is taking place spontaneously in the radioactive elements, as previously described. It has been made to occur in the laboratory on a significant scale only since 1939. Much of the work of atomic fission has been centered around two of the isotopes of uranium having atomic weights of 235 and 238. These isotopes are designated by the symbols ^{235}U and ^{238}U. The ^{235}U isotope is fissionable. The ^{238}U isotope is the more abundantly occurring, the ratio of occurrence of ^{238}U to ^{235}U being about 140:1. An outstanding difficulty in atomic fission work has been the separation of ^{235}U from ^{238}U.

The fission of an atom is brought about by techniques that split its nucleus. Nuclear disintegration occurs through the bombardment of a fissionable element by very small, fast-moving particles. The neutron is the most effective particle for splitting the nucleus because it has no charge and hence does not encounter the resistance to its entrance into the nucleus that is offered to protons that bear a charge (+) like that of the nucleus (Fig. 9-6).

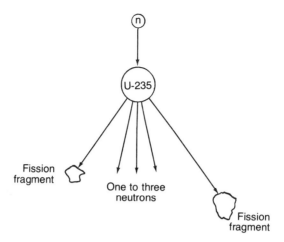

Fig. 9-6. Diagram showing how a neutron may strike an atom of ^{235}U, splitting the atom into fission fragments and neutrons. One of the neutrons released may strike another ^{235}U atom, repeating the process and thus continuing the chain reaction.

When ^{235}U is bombarded with neutrons, it is split into elements of smaller masses. One pair of elements that may be formed from the splitting of ^{235}U is barium and krypton. If we count the masses of the elements formed and the neutrons that are emitted by the fission, we get a quantity that is less than the mass of the original ^{235}U. This means that in the fission of ^{235}U a loss of mass has occurred. This loss of mass brings about the release of a great amount of energy. It is known that matter and energy are interconvertible. The results obtained in the splitting of the atom appear to prove this principle in one direction, the conversion of mass into energy.

The amount of energy obtained by the splitting of the atom is very great. It has been estimated that the fission of 1 pound of ^{235}U would release energy equivalent to the amount produced by the combustion of 10,000 tons of coal. The great amount of energy released by atomic fission accounts for the destructiveness of the atomic bombs used at the end of World War II.

The energy that we are familiar with results from chemical changes that involve an interchange of the electrons of the outer shells of reacting atoms. This is chemical energy. It has been estimated that the chemical energy resulting from ordinary chemical changes is less than 1 billionth of the total conceivable energy locked in the atom. The term *atomic energy* refers to the total energy within the atom. About 99.99% of the total energy of matter is locked in the nucleus of the atom. Hence the work that is being done to obtain control of and develop uses for atomic energy is exceedingly important.

Nuclear fission is a means by which a self-sustaining or chain reaction is produced. Directing a stream of neutrons on ^{235}U sets off a chain reaction. It takes a single moving neutron to split the nucleus of a ^{235}U atom, but this splitting releases 1 to 3 neutrons, as shown in Fig. 19-6, which may split other nuclei, and thus a chain reaction may be continued.

When a mixture containing ^{235}U and ^{238}U is bombarded with neutrons, one of the neutrons emitted by the fission of ^{235}U may enter the atomic nucleus of a ^{238}U atom, producing elements that undergo radioactive decay and finally become the element plutonium, which has an atomic weight of 239. Plutonium (^{239}Pu) is of great interest as a new synthetically prepared element, and also because it is fissionable like ^{235}U.

Atomic pile. In a chain reaction the neutrons liberated by fission of an atom strike other atoms, which in turn are split and release more neutrons that split more atoms; thus the process builds up rapidly, liberating increasing numbers of neutrons and a large amount of energy. When the process is uncontrolled, an explosion such as that produced by the atomic bomb occurs. Chain reactions are controlled in the laboratory by placing graphite around the tubes containing the reacting materials, to absorb or slow down the neutrons. The apparatus in which fission reactions are carried out is called a *nuclear reactor* or an *atomic pile*. The heat liberated in a reactor may be carried away by water piped through the system, and this energy may be used to run a turbine, or it may be transformed into electrical energy that can be stored. Thus atomic energy may be used to generate power for industry and transportation. This source of energy is now a practical reality.

In the fission of ^{235}U more neutrons are released than are used to make the reaction take place, the average gain being about 2½ neutrons per atom of split ^{235}U. The excess neutrons may be reacted with ^{238}U to produce plutonium or with other heavy elements capable of absorbing neutrons and becoming radioactive, and thus the store of fissionable material on the earth may be increased. The process of increasing the amount of fissionable material is known as *breeding*. This process offers possibilities of accumulating nuclear fuel, which otherwise is limited by the relative scarcity of radioactive materials on the earth.

ATOMIC FUSION

The intense energy released from the nuclei of atoms by atomic fission is believed to arise from a conversion of matter into energy. As previously stated, when ^{235}U is split by neutrons, a small amount of mass disappears. The mass that is lost is converted into energy, and the amount of energy released in this process is very great.

There is yet another way to obtain energy from the nucleus of the atom. This is by fusion of the atoms of light elements to form heavier elements. The principle of energy release by atomic fusion is similar to that occurring in atomic fission, the conversion of matter into energy.

Fusion reactions are simple in principle. They are reactions in which the nuclei of atoms of low atomic mass elements are fused at very high temperatures to form elements with higher atomic mass. The classic example of this type of reaction is the fusion of hydrogen to form helium in the sun. This is associated with the production of a great amount of energy and very high temperatures.

Fast-moving neutrons are not required to produce fusion reactions, as is the case for fission reactions. The only requirement is a very high temperature to activate or start the process. Atomic fusion is the process that creates the explosive force of the hydrogen bomb. Fusion reactions thus offer deadly possibilities for military use and remarkable opportunities for the production of energy for peacetime applications.

Loss of mass. That mass is lost when hydrogen nuclei or protons are fused to form helium may be shown by the following equation for the reaction in which 4 protons are fused to form a helium nucleus.

$$4H^+ \text{ (protons)} \rightarrow 1He^{++} \text{ (2 protons + 2 neutrons)} + \text{Energy}$$

Counting the masses on both sides of the equation, we have the following:

$$4 \times 1.0073 \text{ (Mass of a proton)} = 4.0292$$
$$\text{Mass of helium nucleus} = 4.0017$$
$$\text{Loss of mass} = 0.0275$$

Thus for each gram-atomic mass of helium formed by the fusion of hydrogen, 0.0275 g of mass is lost.

Calculation of energy from nuclear fusion. The amount of energy liberated when matter is converted into energy may be calculated by an equation published by Einstein in 1905. This is as follows:

$$E = mc^2$$

In this equation, E is the amount of energy expressed in ergs, m is the mass of matter converted into energy expressed in grams, and c is the velocity of light per second expressed in centimeters.

Substituting in the Einstein equation the mass (m) lost when 1 gram-atomic mass of helium is formed by the fusion of protons, we have the following:

$$E = 0.0275 \times c^2$$

Since $c = 3 \times 10^{10}$, the equation becomes

$$E = 0.0275 \times (3 \times 10^{10})^2$$

Since E is in ergs and 1 cal = 4.18×10^7 ergs, the energy expressed in calories may be calculated as follows:

$$E \text{ (in calories)} = \frac{0.0275 \times (3 \times 10^{10})^2}{4.18 \times 10^7}$$

Simplifying, we have the following:

$$E \text{ (in calories)} = \frac{0.0275 \times 9 \times 10^{20}}{4.18 \times 10^7}$$

$$= \frac{0.2475 \times 10^{20}}{4.18 \times 10^7}$$

$$= \frac{0.2475 \times 10^{13}}{4.18}$$

$$= 592,000,000,000 \text{ calories}$$

The student in a course of this kind need not be disturbed by the mathematics used. However, by making use of the equation and data available we can make calculations and get a good idea of the tremendous energy that may be released from the atomic nucleus by thermonuclear reactions.

It is interesting to compare the energy obtained by burning 4 g of hydrogen in the presence of oxygen to form water, an ordinary chemical reaction in which the valence electrons and not the nuclei of the atoms are concerned, with the energy obtained from the fusion of 4 g of hydrogen to form helium. When 4 g of hydrogen are burned, about 120,000 calories are released; but when 4 g of hydrogen undergo a thermonuclear fusion, the amazingly large amount of energy just calculated, 592 billion calories, is liberated. These figures explain the enormous explosive energy of the hydrogen bomb. They also suggest the great possibilities for the application of nuclear energy to peacetime use.

USES OF RADIOACTIVE ISOTOPES IN BIOLOGY AND MEDICINE

Research. Radioactivity is readily detected and measured by the Geiger-Müller counter. When radioactive material is brought within range of this instrument, radiations penetrate the thin window of the tube and ionize the gas in the tube. The ions are attracted to electrically charged plates in the tube causing a flow of electricity that can be recorded. The flow of electricity is proportional to the amount of radiation. If the radioactive material is of the proper dilution, the number of radiations striking the recording device per minute may be counted. Thus with the Geiger-Müller apparatus the presence of radioactive material may be demonstrated and the amount may be measured (Fig. 9-7).

Radioactive isotopes are of great value in research because they serve as labels. When elements such as iodine, phosphorus, calcium, sodium, iron, or cobalt are made radioactive and are fed or injected into an animal, their pathway can be traced through the animal's body by determining the amount present in the tissues and in the excretions with a radioactivity counter. In this way the metabolism of these elements can be studied, since the radioactive element behaves the same in the living organism as the naturally occurring, nonradioactive element does.

A further extension of the principle of using atoms labeled by radioactivity is the use of radioactive carbon (^{14}C). Techniques are known by which radioactive carbon can be introduced into practically any organic compound or foodstuff. The labeled compound can then be fed or injected into an animal, and its metabolic pathway can be followed

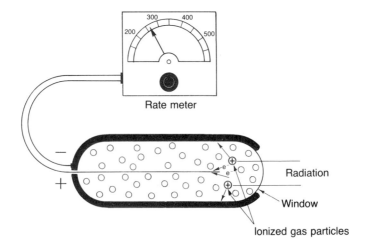

Fig. 9-7. A Geiger-Müller tube.

through the animal's body by analysis of the tissues and excretions for compounds exhibiting radioactivity. Similarly, other radioactive elements incorporated into organic compounds are exceedingly valuable in carrying out crucial experiments in metabolism and also in making critical diagnostic tests in medicine.

Diagnosis and treatment. The radiations given off by radioactive matter may be used in the diagnosis and treatment of certain disease conditions. Two examples of this application of radioactive isotopes (the use of radioiodine and radiophosphorus) will be discussed. Other isotopes receiving some use in the practice of medicine are radioactive sodium, iron, chromium, calcium, gold, cobalt, and gallium.

Radioiodine. Radioiodine (^{131}I) emits β-rays and the highly penetrating γ-rays. Since the site of metabolic transformation and storage of iodine in the body is the thyroid gland, this isotope, which behaves metabolically the same as nonradioactive iodine does, has been adapted to the diagnosis and treatment of thyroid disease.

The rate of uptake of iodine by the thyroid gland is a measure of the functional capacity of the gland. This may be determined by giving the patient a solution of radioiodine, an "atomic cocktail," to drink and later measuring the number of radiations that are given off from the thyroid region of the neck. In this test the radioiodine, following its administration by mouth, is absorbed from the alimentary tract and passes by way of the bloodstream to the thyroid gland where it is stored. At a definite time after administration the patient lies on a bed and a radioactivity-counting apparatus is placed at a measured distance above the thyroid region of the neck. The γ-radiations from the radioiodine penetrate the tissues anterior to the thyroid gland and strike the counter. The number of counts during a standard period of time is recorded. In this way the rate of uptake of iodine by the thyroid gland is determined. This test shows whether the patient has a normally functioning thyroid gland or has hyperthyroidism or hypothyroidism.

After the administration of radioiodine to a patient, the amount of the isotope in the blood, urine, sputum, and feces may be determined with counting apparatus, and thus information of diagnostic importance may be obtained.

Radioiodine is used in the treatment of hyperthyroidism and of cancer of the thyroid gland. For these conditions radioactive iodine, in doses larger than used in the diagnostic test, is administered to the patient. The radioactive iodine collects in the thyroid gland or in the abnormal thyroid tissue in relatively large amounts. The radiations given off from this isotope attack the hyperfunctioning or the malignant cells and thus serve to combat the disease present.

Radiophosphorus. Radiophosphorus (^{32}P) emits β-rays. These rays are much less penetrative than the γ-rays emitted by radioiodine, but they have some application in medicine. Radioactive phosphorus has been used for the treatment of polycythemia vera. Radiophosphorus has an inhibitory effect on the production of red cells and thus aids in the control of polycythemia vera.

UNITS OF MEASUREMENT OF RADIOACTIVITY

Radioactivity is measured by the use of a unit called the curie and by subdivisions of this unit, the millicurie and the microcurie. A microcurie is the amount of radioactive substance that decays at the rate of 37,000 atoms per second. A microcurie is 1 millionth of a curie or 1 thousandth of a millicurie; a millicurie is 1 thousandth of a curie. A roentgen is that amount of radiation that will produce 1 electrostatic unit of ions per cubic centimeter of volume.

Questions for study

1. What are the properties of radium?
2. To what property does radium owe its name? Describe this unusual property.
3. What are the medical uses of radium? Why is radium so expensive?
4. What is radioactivity?
5. Describe the three types of emanations ejected from radioactive substances.
6. Tell how radioactivity is imparted to substances that do not have this property naturally.
7. How is transmutation of the elements accomplished?
8. Who discovered x-rays?
9. Describe the x-ray tube.
10. How are x-rays produced?
11. Mention several properties of x-rays.
12. In making a radiograph of the body why do dark areas appear beneath the region of soft tissue and lighter areas beneath the more dense tissues?
13. What is atomic fission? How is it brought about?
14. How much of the total energy of the atom is locked in the nucleus?
15. How much energy would be released by the fission of 1 pound of ^{235}U?
16. What is a nuclear chain reaction? How is the chain reaction continued?
17. What is meant by the equivalence of mass and energy? What becomes of the mass that is lost in the splitting of the atom?
18. What is a nuclear reactor? How does it serve to create atomic energy?
19. What is the significance of the process known as "breeding" of fissionable material?
20. What is meant by atomic fusion?
21. What is the main source of the heat of the sun?
22. What is the reason that explains why atomic fusion is a means of obtaining such great amounts of energy?
23. How much energy is obtained from the burning of 4 g of hydrogen? From the fusion of 4 g of hydrogen to form helium?
24. How is radioactivity detected and measured?
25. Explain how radioactive isotopes are used in biological investigations.
26. Explain how radioiodine is used to determine thyroid function.
27. Explain why radioiodine is effective in the treatment of hyperthyroidism and cancer of the thyroid gland.
28. How is radiophosphorus used for diagnosis? For treatment?

10 The chemistry of carbon compounds

CARBON

Occurrence and importance. Carbon is an exceedingly important element. As a free element in nature it exists in the form of coal, graphite, and diamonds. It is also the characteristic element of the great class of compounds known as organic compounds. Over 2 million organic compounds are known at the present time, and the number is increasing rapidly because of the advances of synthetic chemistry. The carbon compounds are so numerous and so important that a separate branch of chemistry known as organic chemistry is devoted to a special consideration of them.

As the central element in organic compounds, carbon is closely related to life. The tissues of living things are composed essentially of carbon compounds, and the energy exchanges that maintain life result from the oxidation of carbon compounds. Carbon is also important as the principal element in materials that act as storehouses of energy from sunlight. This energy is used by plants in the formation of certain carbon compounds and is later released when these compounds are utilized as foods or are burned as fuels.

Properties. Carbon is an allotropic element; it exists in several elementary forms. All elementary forms of carbon are odorless, crystalline solids. The crystalline nature of diamond and graphite is apparent to the eye, but the crystals of the other forms of carbon such as coal, coke, lampblack, and charcoal are too minute to be seen even with a microscope, and their existence has been revealed only by x-ray methods of examination. The different forms of carbon are highly insoluble and are infusible and nonvolatile. Chemically, carbon is inactive at ordinary temperatures, but it will combine with oxygen or burn at elevated temperatures. A great deal of energy is liberated when carbon burns.

Diamond. Pure diamond is a colorless, transparent, crystalline substance that is very hard and has a high refractive index. It is inactive chemically, but it will burn when heated in oxygen to a temperature of about 800° C.

Natural diamonds are found chiefly in South Africa, India, and Brazil. The naturally occurring diamonds are of different grades of purity. Some are practically pure and are used for jewelry since pure diamond has the ability to refract light and hence sparkles in the light. Other grades of diamonds have small amounts of impurities in them, which gives them poor optical qualities, and for this reason are of no value as jewelry. The impure forms of diamond are very valuable, however, because of their hardness and are used for grinding the perfect diamonds into shape, for cutting glass, and for pointing rock drills.

Graphite. Graphite is found as natural deposits in Ceylon, Siberia, and Madagascar. It differs widely from the diamond in its physical properties even though it is the same element. It is lighter in weight than the diamond, is not hard, and is black in color. It is inactive chemically. It is used in the manufacture of lead pencils, stove polish, shoe polish, black paints, and crucibles for heating to high temperatures.

Charcoal. Charcoal is prepared by heating organic matter containing carbon in the absence of oxygen or air. If organic matter is heated in the presence of oxygen, the carbon is oxidized to carbon dioxide, which passes off into space; if, on the other hand, oxygen is excluded when carbon compounds are heated, the volatile matter is removed but carbon, being nonvolatile, remains as a residue. Heating organic matter to high temperatures with the exclusion of oxygen is therefore the principle involved in the preparation of charcoal.

Wood charcoal. When wood is heated to high temperatures in kilns that do not permit the access of oxygen, the hydrogen and oxygen of the wood are removed as water, and wood alcohol, acetone, acetic acid, and other volatile substances are driven off. The residue is wood charcoal, an elementary form of carbon. Wood charcoal is used in metallurgy for the extraction of metals from their ores, in the manufacture of black gunpowder, in the manufacture of steel, and as a filtering agent to remove undesirable substances from a solution.

Animal charcoal. Animal charcoal is made by heating animal tissues (blood, bones, horns, hides, and hoofs) in the absence of oxygen. The residue obtained in this process is a charcoal, mixed with calcium phosphate and other inorganic salts that are peculiar to animal tissues. Animal charcoal is used in the refining of sugar by a process called adsorption.

Adsorption. Charcoal has the property of collecting gases to which it is exposed and of removing particles of matter from liquids that are poured through it. This behavior is called *adsorption*. In adsorption, charcoal does not form a compound with the substances it collects but condenses and binds them on its surfaces. Charcoal is a highly porous substance and has a relatively large amount of surface area in comparison to the size of the sample. In addition to an enormous surface area, charcoal has a high degree of surface attraction for certain substances. Charcoal is therefore an excellent adsorbing agent.

One of the greatest applications of adsorption by charcoal in industry is in the refining of sugar. In sugar refineries a raw "brown" sugar solution is passed through large tanks, 20 to 30 feet deep, of animal charcoal. In passing through charcoal, brown organic matter and other substances (but not sugar) are removed by adsorption from the crude syrup, and by the time the syrup reaches the bottom of the filtering bed, it has been freed of all objectionable matter. The colorless solution obtained is drained into appropriate containers, where it is concentrated and the sugar is crystallized. In this manner crude brown sugar is converted into the pure refined sugar of commerce.

Another application of adsorption by charcoal is in the preparation of gas masks. Coconut charcoal is especially efficient in removing poison gases from the air. It has application in cigarette filters as well.

Formation of coal. Coal consists chiefly of carbon and compounds of carbon. It originated from plant material that was produced on the surface of the earth in past ages. To understand its formation we must apply the principles noted in the preparation of charcoal. When heated in the presence of oxygen, plant matter is oxidized, forming chiefly carbon dioxide and water, but when the supply of oxygen is removed, carbon does not burn; it remains at the end of the heating process as a residue called charcoal. A similar process took place on the earth's surface millions of years ago. Great forests of trees, ferns, and other plant material were inundated first with water and later with solid matter by the sudden changes in the crust of the earth. In this manner layers of plant matter were stored beneath the earth's surface and out of contact with air. The internal heat of the earth drove off the volatile products from this organic matter, and the pressure of deep columns of earth completed the process by squeezing the carbon residue into the various forms of coal, graphite, and diamonds. The greater the pressure exerted on the carbon deposit, the more ordered the crystalline structure of the carbon form produced. The diamond is the product of the greatest pressure, which accounts for the fact that deposits of diamonds are rarely found. The scarcity of diamond deposits results in the monetary value of diamonds.

SOME INORGANIC CARBON COMPOUNDS
Carbon monoxide

Carbon monoxide (CO) is a colorless, odorless gas. It is formed when carbon is burned in an insufficient supply of oxygen.

$$:C:::O:$$

Carbon monoxide

$$2C + O_2 \rightarrow 2CO$$

Carbon monoxide

In this reaction there is not enough oxygen to oxidize the carbon to carbon dioxide. In the presence of sufficient oxygen, carbon and carbon monoxide will form carbon dioxide when they burn, as shown in the following examples:

$$:\overset{..}{O}::C::\overset{..}{O}:$$

Carbon dioxide

$$C + O_2 \rightarrow CO_2$$

Carbon dioxide

and

$$2CO + O_2 \rightarrow 2CO_2$$

Carbon monoxide   **Carbon dioxide**

Carbon monoxide burns with a blue flame. The blue flame that flickers over the top of a heap of burning coal is the flame of carbon monoxide. Since the valence of carbon in carbon monoxide is not satisfied, this compound combines readily with many substances. It bonds firmly to the hemoglobin of the blood and is therefore a dangerous poison. One volume of carbon monoxide in 800 volumes of air will cause death if breathed for about 30 minutes. Carbon monoxide combines with hemoglobin to form carboxyhemoglobin, a stable compound, the hemoglobin part of which is no longer able to combine with oxygen. Carbon monoxide poisoning is thus a lowering in the amount of effective hemoglobin that can be used for carrying oxygen, which results in a lowering of the oxygen-carrying power of the blood. The victim of such poisoning dies from asphyxiation or inability to get sufficient oxygen because his active hemoglobin has been lowered to an amount below that which is necessary for the proper oxygenation of his tissues.

Carbon dioxide

Occurrence. Carbon dioxide (CO_2) occurs in the air to the extent of 3 or 4 parts per 10,000 parts of air and is found in many natural waters. The carbon dioxide of the air originates principally from the combustion of carbon. This arises from the fires of dwelling places, the furnaces of industry, the respiration of animals, and the decay of dead animal and plant tissues.

Preparation. Carbon dioxide may be prepared as follows:
1. The simplest method of preparing carbon dioxide is by burning carbon in an abundance of oxygen.

$$C + O_2 \rightarrow CO_2$$

2. Another easy method, the one used in the laboratory, is to mix an acid with a carbonate.

$$Na_2CO_3 + 2HCl \rightarrow 2NaCl + H_2O + CO_2$$

Properties and uses. Carbon dioxide is a colorless, odorless gas and is slightly soluble in water. It will not burn and therefore is a valuable fire extinguisher. The explanation of the nonburning of carbon dioxide is that the carbon atom in the molecule is already oxidized completely. Its valence is completely satisfied with the 2 atoms of oxygen it holds, and therefore it cannot combine with more oxygen. Carbon dioxide is heavier than air; hence it will settle on a fire and smother the fire by preventing access of air. Carbon dioxide is not poisonous and is, in fact, the normal stimulus of the respiratory center of the brain. If the amount in the blood and tissues is increased, however, either by breathing air that contains excess carbon dioxide or by faulty elimination of this substance from the body, respiration is stimulated unduly, and the individual develops a respiratory acidosis. Plants use carbon dioxide in the synthesis of foods.

Principle of the carbon dioxide fire extinguisher. The carbon dioxide fire extinguisher contains a solution of sodium bicarbonate and a small bottle of sulfuric acid placed above the solution near the top of the tank. The sulfuric acid bottle is stoppered with a heavy lead stopper, which remains in place when the apparatus is in an upright position but falls out when the tank is inverted. In case of fire the apparatus is inverted, the stopper

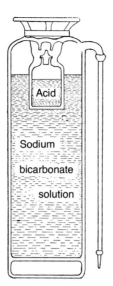

Fig. 10-1. Diagram of a carbon dioxide fire extinguisher.

falls from the sulfuric acid bottle, and the acid and bicarbonate become mixed and react to form carbon dioxide. The carbon dioxide accumulates in the tank, creating pressure, and finally water saturated with carbon dioxide is forced from the tank and sprayed on the fire through a small nozzle attached to the tank. Fig. 10-1 illustrates the construction of a carbon dioxide fire extinguisher.

Another type of fire extinguisher consists of a metal container that holds liquid carbon dioxide under pressure, with a valve to release the carbon dioxide when needed, and a discharge horn that can be manipulated to apply the carbon dioxide effectively to the fire. This type has advantages over the sodium bicarbonate and acid extinguisher in fires involving organic solvents.

Carbonic acid. When carbon dioxide is passed into water, carbonic acid (H_2CO_3) is formed.

$$CO_2 + H_2O \rightleftarrows H_2CO_3$$

This is a very weak acid. Carbon dioxide dissolved in water under pressure is known as "carbonated water." When the pressure is released, the gas comes out of solution giving the effervescense common to carbonated beverages.

Sodium carbonate. Sodium carbonate (Na_2CO_3) is a salt that produces an alkaline solution when dissolved in water. Its alkalinity in solution is explained by the fact that it is a salt of a very weak acid (carbonic acid) and a strong base (sodium hydroxide). When dissolved in water, it undergoes hydrolysis, yielding a base that is more highly ionized than the acid formed.

$$Na_2CO_3 + 2H_2O \rightleftarrows 2NaOH + H_2CO_3$$

In the preceding reaction, sodium hydroxide, being more highly ionized, has a greater influence on the solution than carbonic acid; consequently the solution is alkaline. Sodium carbonate is called "washing soda" because it can be used to precipitate calcium and magnesium salts from hard water allowing soap to lather more effectively (see p. 96).

Sodium bicarbonate. Sodium bicarbonate ($NaHCO_3$) is another salt of carbonic acid that produces a mildly alkaline solution when dissolved in water. It is a constituent of baking powders and has medicinal applications as a neutralizing agent for hyperacidity and as an antidote for acid burns.

Calcium carbonate. One of the common forms of calcium carbonate ($CaCO_3$) is limestone, the substance of which much rocky material on the earth's surface is composed. Other familiar forms of calcium carbonate that are more or less impure are chalk, marble, seashells, coral, pearls, and eggshells. Pure calcium carbonate is a white solid that crystallizes in two forms. Crystalline calcium carbonate is a very hard substance; hence it is valuable for building and construction purposes. It is insoluble in water but decomposes rapidly in acids. When heated to high temperatures, it forms lime or calcium oxide and gives off carbon dioxide.

$$CaCO_3 + \text{Heat} \rightarrow CaO + CO_2$$
$$\text{Calcium carbonate} \qquad \text{Calcium oxide} \quad \text{Carbon dioxide}$$

ORGANIC CHEMISTRY

The term *organic* arose from the relationship of this branch of chemistry to organic or living matter. In the early history of chemistry it was believed that there were many substances that could not be made by the chemist working in his laboratory and could result only from the vital activity of living things. But in 1828 the German chemist Wöhler prepared urea, a compound found in the urine of animals, from ammonium cyanate, an inorganic substance, and this synthesis in the laboratory showed that the vital influence of the living organism is not necessary to produce organic compounds.

$$NH_4CNO \xrightarrow{\text{Heat}} NH_2-\underset{\underset{\|}{O}}{C}-NH_2$$

Since Wöhler's epoch-making discovery chemists have prepared many other compounds that it was once believed could be produced only by plants and animals. Sugars, fats, proteinlike substances, hormones, vitamins, and many other products of living matter have been made in the chemical laboratory. Thus the old idea that vital or living influences create a different class of compounds and a different kind of chemistry has been shown to be untrue. The early division of chemistry into inorganic and organic chemistry is still retained, however, because of its convenience; but organic chemistry is now defined as the chemistry of carbon compounds, since carbon is the characteristic element in all organic compounds.

When beginning the study of organic chemistry, the student may be amazed at the size of organic molecules and may be discouraged with the apparent difficulty of understanding organic reactions. But organic reactions are really less difficult to understand than they at first seem. Although the molecule of organic substances may be quite large, the student will soon realize that organic reactions involve only a small part of the molecule, and it should be noted that the many reactions which take place in organic chemistry can be grouped under a few general types. It is important in studying organic reactions to fix one's attention on that part of the molecule that is responsible for the reaction and to realize that certain atoms or groups of atoms react the same way at all times and that the rest of the molecule remains inactive. The reactive part of the molecule is the *functional group,* and it is on the basis of functional groups, which have characteristic properties, that organic compounds are classified. In a course of this kind it is neither possible nor necessary to enter deeply into the chemical reactions of organic compounds. This text will attempt to present only a general idea of organic chemistry by first giving a classification of organic compounds, emphasizing the characteristic group of each class, and later pointing out a few important physical and chemical properties of the compounds selected as examples.

Some fundamental principles

1. The valence of carbon is 4. When uncombined, the atom of carbon has 4 unpaired electrons in its outer shell. When this atom reacts chemically, changes occur that bring 1 electron into chemical union with each unpaired electron, forming 4 stable electron pairs; this gives the carbon atom an outer shell containing 8 electrons. The carbon atom thus contains 4 electrons in its outer shell when uncombined and shares 4 additional electrons when combined chemically with itself or with atoms of other elements. The chemically combined carbon atom is designated graphically as follows:

$$:\!\ddot{\underset{\cdot\cdot}{C}}\!: \quad \text{or} \quad -\overset{|}{\underset{|}{C}}-$$

2. Carbon always reacts chemically by a *sharing* of valence electrons. It therefore exhibits the type of valence known as *covalence.* For example, the combination of carbon with hydrogen may be represented as follows:

$$\cdot\!\overset{\cdot}{\underset{\cdot}{C}}\!\cdot + 4H\cdot \rightarrow H\!:\!\overset{H}{\underset{H}{\overset{\cdot\cdot}{\underset{\cdot\cdot}{C}}}}\!:\!H \quad \text{or} \quad H-\overset{H}{\underset{|}{\overset{|}{C}}}-H$$

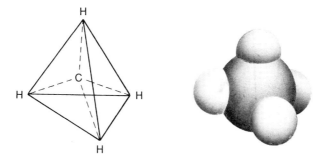

Fig. 10-2. The methane molecule.

A three-dimensional illustration of this combination (Fig. 10-2) shows the molecule to be in the shape of a regular tetrahedron. In this compound carbon *shares* 4 pairs of electrons with hydrogen, or forms four covalent bonds. In organic chemistry the valence bond of carbon is considered as the equivalent of 2 shared electrons.

3. Carbon may combine with (a) 4 univalent atoms, (b) 2 divalent atoms, (c) 2 univalent atoms and 1 divalent atom, or (d) 1 univalent atom and 1 trivalent atom. The following examples illustrate each combination:

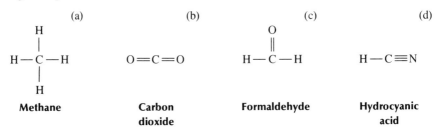

4. Carbon atoms have a great tendency to combine with themselves and form molecules containing a large number of atoms. Because of this property, organic compounds as a class contain far more atoms in their molecules than do inorganic compounds, and this property explains why there are so many organic compounds (over 2 million), either naturally occurring or prepared by chemists. The proteins, for example, are organic compounds whose molecules contain hundreds of atoms; and carbohydrates, fats, dyes, and certain drugs are organic compounds whose molecules are very large when compared with those of inorganic substances.

5. In most carbon compounds only one valence bond is held between adjacent carbon atoms, the other valences being satisfied by atoms of other elements, in which instances the compounds are said to be *saturated;* however, carbon atoms may hold two or three valence bonds in common, and in these instances the compounds are said to be *unsaturated*. The following examples illustrate the three linkages that may exist between carbon atoms:

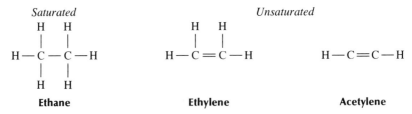

6. Different degrees of affinity for the carbon valences are exhibited. For this reason some elements will readily replace other elements attached to carbon, and this explains further why so many compounds of carbon may be prepared. Thus the compound methane

$$H-\underset{\underset{H}{|}}{\overset{\overset{H}{|}}{C}}-H$$

when treated with chlorine under varying conditions will give the following:

$$\begin{array}{cccc}
\text{H} & \text{Cl} & \text{Cl} & \text{Cl} \\
| & | & | & | \\
\text{H}-\text{C}-\text{Cl} \quad \text{or} \quad & \text{H}-\text{C}-\text{Cl} \quad \text{or} \quad & \text{H}-\text{C}-\text{Cl} \quad \text{or} \quad & \text{Cl}-\text{C}-\text{Cl} \\
| & | & | & | \\
\text{H} & \text{H} & \text{Cl} & \text{Cl} \\
\text{Methyl} & \text{Methylene} & \text{Chloroform} & \text{Carbon} \\
\text{chloride} & \text{dichloride} & & \text{tetrachloride}
\end{array}$$

The chlorine in these compounds may be replaced by other groups of atoms:

$$\begin{array}{cc}
\text{H} & \text{H} \\
| & | \\
\text{H}-\text{C}-\text{Cl} + \text{KOH} \quad \rightarrow \quad & \text{H}-\text{C}-\text{OH} + \text{KCl} \\
| & | \\
\text{H} & \text{H} \\
\text{Methyl} & \text{Methyl} \\
\text{chloride} & \text{alcohol}
\end{array}$$

Thus the process of replacement of one group by another may be continued indefinitely, and a great number of compounds may be made.

Classification of organic compounds

Organic compounds have been placed in three broad groups, a classification based on similarities in structure: (1) *Acyclic*—compounds in which the carbon atoms are attached to each other in an open-chain structure; (2) *carbocyclic*—compounds in which the carbon atoms are attached to each other in a closed-chain structure; and (3) *heterocyclic*—compounds containing atoms of carbon and of some other element or elements arranged in a closed chain.

The acyclic compounds are more commonly called the *aliphatic* compounds because of a resemblance in structure to fats (from the Greek word *aleiphatos,* meaning fat). Because of the widespread use of the term aliphatic in chemical literature, this designation will be used in our discussion in preference to the term acyclic. Aliphatic compounds can also be found as carbocyclic structures.

A most important subdivision of the carbocyclic group is the class of substances called the *aromatic* compounds. These compounds consist of the six-carbon, closed-chain compound benzene and its derivatives. The term aromatic came into use because many of these compounds have a pleasant odor or aroma and hence may be described as aromatic. This very important subdivision will be discussed at some length.

Heterocyclic compounds include many important substances that occur in nature and that have been synthesized by organic chemists. Examples of important naturally occurring heterocyclic compounds are heme in hemoglobin, chlorophyll (the green pigment in plant leaves), several amino acids, some of the vitamins, alkaloids, and some drugs.

The three major divisions of organic compounds are subdivided into the following nine classes:

Hydrocarbons	Ketones	Ethers
Alcohols	Acids	Amines
Aldehydes	Esters	Halides

160 Roe's principles of chemistry

A brief discussion of each of these subdivisions of organic compounds is given in the next three chapters.

Naming organic compounds

Some of the common names of organic compounds have been known for a long time. To put order into organic nomenclature, a system was developed using the endings of each name to indicate the type of compound. Thus, the ending *ane* denotes a saturated hydrocarbon, a member of the alkanes (see p. 162). The endings *ene* and *yne* indicate varying degrees of unsaturation in a hydrocarbon chain and show membership in the alkenes (see p. 166) and the alkynes (see p. 164), respectively; whereas the ending *ol* indicates an alcohol (see p. 164).

In this system the number of carbon atoms in the compound is indicated by prefixes, *meth* showing one carbon; *eth,* two; *prop,* three; and *but,* four carbons. Larger numbers of carbon atoms are indicated by the proper Greek number, for example, *pent,* five; *hex,* six; *hept,* seven; and *oct,* eight.

Differences in structure are indicated by the prefix, normal or *n* being used for straight chained carbon compounds, whereas *iso* or *neo* indicate branched, chained carbon compounds.

As the number of known organic compounds grew, this system of nomenclature became inadequate to show differences in structure. The International Union of Pure and Applied Chemistry suggested a revision of the nomenclature for organic compounds known as the IUPAC system of nomenclature. The following rules offer a brief and simplified summary of the IUPAC system.

1. The longest continuous chain is the parent chain.
2. Hydrocarbon additions or substitutions to the parent chain are given prefix names and are also indicated by a number. Points of unsaturation are also indicated by a number. This numbering of the parent hydrocarbon starts from the end of the chain nearest an addition, substitution, or point of unsaturation.
3. If a group appears more than once, a second prefix (such as *di, tri,* or *tetra*) is included in the name to indicate the number of times the group occurs.
4. Substitutions or additions to the parent chain are listed in alphabetic rather than numeric order.

$$\overset{1}{CH_3}-\overset{2}{CH_2}-\overset{3}{CH}-\overset{4}{CH}-\overset{5}{CH_2}-\overset{6}{CH_2}-\overset{7}{CH_3}$$
$$\underset{CH_3}{|}\underset{CH_2-CH_3}{|}$$

4-Ethyl-3-methyl heptane

not

$$\overset{1}{CH_3}-\overset{2}{CH_2}-\overset{3}{CH}-\overset{4}{CH}-\overset{5}{CH_2}-\overset{6}{CH_3}$$
$$\underset{CH_3}{|}\underset{CH_2-CH_2-CH_3}{|}$$

3-Methyl-4-propyl hexane

In this configuration the longest continuous carbon chain is not used as the parent compound. This does not follow the IUPAC rules.

Questions for study

1. Discuss the occurrence of carbon. What is its relation to life?
2. Give several properties of carbon.
3. What are the properties of diamond?
4. What are the uses of graphite?
5. Explain the principle of charcoal preparation.
6. What is adsorption?
7. Discuss two uses of animal charcoal.
8. Describe the formation of coal.
9. Describe carbon monoxide.
10. Explain why carbon monoxide is poisonous.
11. What is the origin of the carbon dioxide of the atmosphere? How is carbon dioxide prepared? Give the properties of carbon dioxide.
12. What are the uses of carbon dioxide?
13. Explain the principles of the two carbon dioxide fire extinguishers.
14. What is carbonic acid?
15. Explain the alkalinity of sodium carbonate.
16. What are the uses of sodium bicarbonate?
17. What is the importance of calcium carbonate?
18. What is organic chemistry?
19. What was the early concept regarding organic chemistry?
20. What is the valence of carbon?
21. What is the type of valence exhibited by carbon? Explain.
22. Why do so many organic compounds exist?
23. What are the three major divisions of organic compounds? Give examples to distinguish between these three divisions.
24. Name the nine subdivisions of organic compounds.

11 Organic chemistry—hydrocarbons, halogen compounds, alcohols, aldehydes, ketones

HYDROCARBONS, R-H
Saturated hydrocarbons—alkanes

Examples of saturated hydrocarbons include methane, ethane, and propane (Table 11-1). These substances merely introduce the subject of saturated hydrocarbons or alkanes, which are compounds containing the elements carbon and hydrogen. The student will see at once that each compound in the series differs from the one preceding it by CH_2 and will suspect from these examples that many more compounds can be made by adding CH_2 groups, as is the case. This series has been found to extend up to a hydrocarbon with the formula $C_{100}H_{202}$, and it is not known how far it may be continued. This series of hydrocarbons in which each member after methane differs from the one preceding or following it by a CH_2 group is known as a *homologous series*, and the individual members are called *homologues*. The student will notice that the general formula for saturated hydrocarbons is C_nH_{2n+2}; n is the number of carbon atoms, and the number of hydrogen atoms is $2 \times n + 2$.

Radicals. Whenever a saturated hydrocarbon such as methane, CH_4, has a hydrogen atom removed, the resulting substance is called a radical (CH_3-, the methyl radical). The word *radical* is abbreviated to R— to indicate any hydrocarbon radical. Thus R—H is a general formula for any saturated hydrocarbon. As shown in Table 11-1, this R— radical or group derives its name from the parent hydrocarbon by replacing the ending *ane* with *yl*.

Occurrence and importance. The first member of this series, methane, called marsh gas and firedamp, arises from the decomposition of animal and vegetable matter. It is an important constituent of natural gas and artificially prepared coal gas. It is quite flammable and has been a frequent cause of explosions in mines.

Commercially, the hydrocarbons are very important. Gasoline is a mixture of the lower members of this series; petroleum ether, benzine, kerosene, petrolatum, and various lubricating oils are hydrocarbons; the mineral oils used in medicine, such as liquid petrola-

Table 11-1. Examples of saturated hydrocarbons

Name	Formula	Structural formula	Radical
Methane	CH_4	$H-\underset{\underset{H}{\|}}{\overset{\overset{H}{\|}}{C}}-H$	CH_3-methyl
Ethane	C_2H_6	$H-\underset{\underset{H}{\|}}{\overset{\overset{H}{\|}}{C}}-\underset{\underset{H}{\|}}{\overset{\overset{H}{\|}}{C}}-H$	C_2H_5-ethyl
Propane	C_3H_8	$H-\underset{\underset{H}{\|}}{\overset{\overset{H}{\|}}{C}}-\underset{\underset{H}{\|}}{\overset{\overset{H}{\|}}{C}}-\underset{\underset{H}{\|}}{\overset{\overset{H}{\|}}{C}}-H$	C_3H_7-propyl

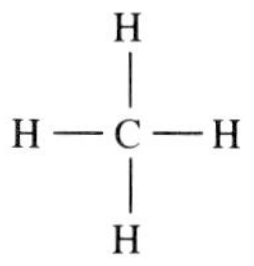

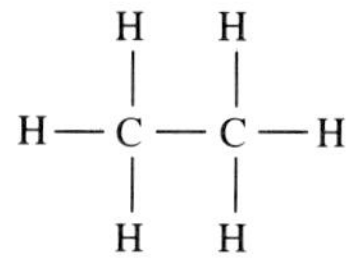

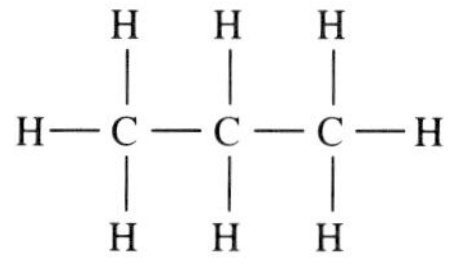

tum, are mixtures of hydrocarbons. Paraffin and petrolatum (solid) are mixtures of hydrocarbons of high molecular weights. The great sources of hydrocarbons are petroleum and coal.

Preparation of methane. The methods of preparing saturated hydrocarbons for commercial purposes are fractional distillation of petroleum and distillation of coal in the absence of oxygen. In this course we will have time to consider only the preparation of methane.

Two laboratory methods for the preparation of methane are as follows:
1. Methane is prepared synthetically by passing hydrogen over elemental carbon at 475° C in the presence of a catalyst (nickel).

$$C + 2H_2 \rightarrow \underset{\text{Methane}}{CH_4}$$

2. It may also be prepared by heating sodium acetate with soda lime (a mixture of NaOH and CaO).

$$CH_3-COONa + NaOH \rightarrow \underset{\text{Methane}}{CH_4} + Na_2CO_3$$

These methods are of importance because they show how methane is prepared in the laboratory. Methane serves as a starting material for the preparation of other hydrocarbons and a great many organic compounds. Natural gas is composed principally of methane.

General properties. The members of the saturated hydrocarbon series containing less than 5 carbon atoms are gases; those having 5 to 16 carbon atoms are liquids; and the ones containing more than 16 carbon atoms are solids. The saturated hydrocarbons are not active chemically. They are called paraffins because the term "paraffin" means little affinity or little activity. They are insoluble in water. They do not react with strong acids

Table 11-2. Examples of unsaturated hydrocarbons

Name	Formula	Structural formula
Ethylene (ethene)	C_2H_4	$H-\underset{\|}{\overset{\|}{C}}=\underset{\|}{\overset{\|}{C}}-H$ (with H's above)
Acetylene (ethyne)	C_2H_2	$H-C\equiv C-H$

and alkalies. They burn readily, however, and react with chlorine and bromine, particularly in the sunlight.

Unsaturated hydrocarbons—alkenes and alkynes

Examples of unsaturated hydrocarbons are ethylene and acetylene (Table 11-2). The saturated hydrocarbons may be treated by chemical procedures that will remove some of the hydrogen atoms. The resulting compounds are known as unsaturated hydrocarbons. In such compounds there are adjacent carbon atoms, which hold two or three valence bonds in common.

When there are two valence bonds (double bond) between adjacent carbon atoms, the compound belongs to the *alkene series*. Ethylene is the first member of this series. When there are three valence bonds (triple bond) between 2 adjacent carbon atoms, the compound belongs to the *alkyne series*. Acetylene is the first member of this series. The functional group in the alkene series is a double bond. In the alkyne series the functional group is a triple bond.

Ethylene. Ethylene is prepared by treating ethyl bromide with alcoholic potassium hydroxide.

$$H-\underset{H}{\overset{H}{\underset{|}{\overset{|}{C}}}}-\underset{H}{\overset{H}{\underset{|}{\overset{|}{C}}}}-Br + KOH \rightarrow H-\overset{H}{\overset{|}{C}}=\overset{H}{\overset{|}{C}}-H + KBR + H_2O$$

Ethyl bromide   Alcoholic   Ethylene
(bromoethane)   (ethene)

Ethylene is a colorless, flammable gas with a pleasant odor. It is a typical representative of the alkene series. The double bond in ethylene gives to this compound the property of being unusually reactive, a characteristic in sharp contrast with the properties of the saturated hydrocarbons.

Acetylene. Acetylene is prepared by treating ethylene dibromide with alcoholic potassium hydroxide, a reaction similar to that used in the preparation of ethylene.

$$CH_2BrCH_2Br + 2KOH \rightarrow H-C\equiv C-H + 2KBr + 2H_2O$$

Ethylene dibromide   Alcoholic   **Acetylene**
(1,2-dibromoethane)   **(ethyne)**

Commercially, acetylene is manufactured by the cracking of methane or by the addition of water to calcium carbide.

Hydrocarbons, halogen compounds, alcohols, aldehydes, ketones 165

$$H_2O + CaC_2 \rightarrow HC\equiv CH + CaO$$

Acetylene is a colorless gas. It is highly flammable and in burning it produces a very hot flame, which is used in the oxyacetylene cutting torch. Acetylene is a typical representative of the alkyne series of unsaturated compounds. The presence of a triple bond in this compound makes it even more reactive than the compounds of the ethylene series.

HALOGEN COMPOUNDS—HALIDES, R-X

Alkyl halogen compounds, or alkyl halides, are formed by replacement of 1 or more hydrogen atoms in the aliphatic hydrocarbons. In the general formula, R-X, X can be any one of the halogens, -Cl, -Br, -F, or -I. The following are some typical examples of halogen compounds.

Chloroform (trichloromethane). Chloroform ($CHCl_3$) may be prepared by the chlorination of methane. It is a heavy, colorless liquid and is insoluble in water. It has a characteristic odor and a sweet, burning taste. It is a good solvent for fats, oils, and certain organic compounds.

Chloroform has the interesting pharmacological effect of producing anesthesia. It has been used considerably as a general anesthetic in surgery and obstetrics, but this usage has been largely discontinued because prolonged application of this compound may have a toxic effect, and safer anesthetics are now available.

When exposed to light and air, chloroform undergoes decomposition and forms phosgene, a poisonous gas. It must therefore be kept in a dark bottle and stored in the dark.

Carbon tetrachloride (tetrachloromethane). Carbon tetrachloride (CCl_4) is prepared by the complete chlorination of methane or by replacing the sulfur in carbon disulfide (CS_2) with chlorine. It is a colorless, heavy liquid and is insoluble in water. It is a good solvent for fats and oils.

Carbon tetrachloride is nonflammable, and because of this property it is used as a fire extinguisher. Some danger is associated with this usage since carbon tetrachloride, when sprayed on material containing water, may react with steam to produce the poisonous gas, phosgene.

Since carbon tetrachloride is a good solvent for grease and is nonflammable, it was used extensively as a cleaning agent; however, the vapors are too toxic, and its use should be avoided.

Carbon tetrachloride is used in medicine as an anthelmintic drug. Enclosed in gelatin capsules, it is administered in the treatment of intestinal worm infestation, for example, infestation with hookworms. It is an irritant to the skin and intestinal mucosa and has a toxic effect on the liver. Its use as an anthelmintic therefore requires caution.

Iodoform (triiodomethane). Iodoform (CHI_3) is prepared by warming ethyl alcohol or acetone with alkali and iodine. It is a yellowish, crystalline solid, insoluble in water and slightly soluble in alcohol. It has a penetrating, characteristic odor. Applied as a dusting powder or impregnated into gauze dressings, it is used as an antiseptic. It has an objectionable odor, and for this reason its use as an antiseptic has declined.

Ethyl chloride (chloroethane). Ethyl chloride (C_2H_5Cl) is an important organic halogen compound. It is prepared by the reaction of ethyl alcohol with hydrogen chloride. The material boils at 13° C. Ethyl chloride is extremely useful as a local anesthetic, by being sprayed on the skin, and as a general anesthetic.

Fluorocarbons. Fluorine-containing materials are quite useful. For example, tetrafluoroethylene, $CF_2 = CF_2$, can be polymerized to polytetrafluoroethylene (Teflon). A widely used fluorocarbon is Freon 12, CF_2Cl_2, found in air conditioners and refrigerators as the refrigerant. The pressurizing medium in many aerosols, for example, hair spray, is composed of Freon 12. Other names for these compounds are Genetron 12, Isotron 12, or Ucon.

Some new general anesthetics, such as halothane (Fluothane), $CHClBrCF_3$, and methoxyflurane (Penthrane), $CHCl_2CF_2OCH_3$, are finding applications. Note the combination of different halogens in these interesting organic halogen compounds.

Many of the fluorocarbon compounds do not become "wet" when placed in water; thus fluorocarbons are used to repel water and stains. The product Scotch Guard is a fluorocarbon coating on fabrics.

ALCOHOLS, R-OH

Examples of alcohols are methyl, ethyl, and propyl alcohols (Table 11-3).

The alcohols are derivatives of hydrocarbons in which a hydrogen is replaced by an OH group. By substituting an OH group for one of the hydrogens in methane, methyl alcohol is derived. Similarly, ethyl alcohol may be derived from ethane, propyl alcohol from propane, and so on; thus a large series of alcohols may be prepared. The functional group in all alcohols is the hydroxyl (OH) group.

General method of preparation. Alcohols may be prepared by a number of procedures. One general method will be given as a type procedure that may be applied to the preparation of all alcohols. *This method consists of treating a halogen derivative of a hydrocarbon with AgOH, NaOH, or KOH.* Examples:

$$CH_3I + AgOH \rightarrow CH_3OH + AgI \quad (1)$$
Methyl iodide (iodomethane) → Methyl alcohol (methanol)

$$C_2H_5Br + NaOH \rightarrow C_2H_5OH + NaBr \quad (2)$$
Ethyl bromide (bromoethane) → Ethyl alcohol (ethanol)

General properties. The simpler alcohols, such as methyl, ethyl, and propyl, are liquids and are readily soluble in water; the more complex ones are solids and are only slightly soluble in water. The alcohols in general have a sweet taste. They are neutral in reaction. They will burn readily, and then enter into many chemical reactions.

Primary, secondary, and tertiary alcohols. It is of further interest that there are three classes of alcohols—primary, secondary, and tertiary. A primary alcohol has an OH group attached to the carbon atom at the end of a carbon chain. In a secondary alcohol there are two organic radicals attached to the carbon atom holding the OH group. A ter-

Table 11-3. Examples of alcohols

Name	Formula	Structural formula
Methyl alcohol (methanol)	CH_3OH	H–C(H)(H)–OH
Ethyl alcohol (ethanol)	C_2H_5OH	H–C(H)(H)–C(H)(H)–OH
Propyl alcohol (propanol)	C_3H_7OH	H–C(H)(H)–C(H)(H)–C(H)(H)–OH

tiary alcohol contains three organic radicals attached to the carbon atom that holds the OH group. The following examples will make this clear:

Propyl alcohol (primary) (propanol): H–C(H)(H)–C(H)(H)–C(H)(H)–OH

Isopropyl alcohol (secondary) (2-propanol): (CH$_3$)(CH$_3$)CH–OH with H

Tert-butyl alcohol (tertiary) (2-methyl-2-propanol): (CH$_3$)(CH$_3$)(CH$_3$)C–OH

Letting R stand for any organic radical (for example, CH_3—, C_2H_5—, or C_3H_7—), the distinction between the three classes of alcohols may be indicated in a more general way as follows:

Primary alcohol: R—OH

Secondary alcohol: R(R)CH—OH with H

Tertiary alcohol: R(R)(R)C—OH

Reactions of alcohols. Alcohols react with oxygen and behave as weak acids or bases under appropriate conditions.

Oxidation. Alcohols readily undergo oxidation.

PRIMARY ALCOHOLS. When a primary alcohol is treated with an oxidizing agent, an aldehyde is formed. Aldehydes will be discussed later. An example is the oxidation of methyl alcohol to yield formaldehyde:

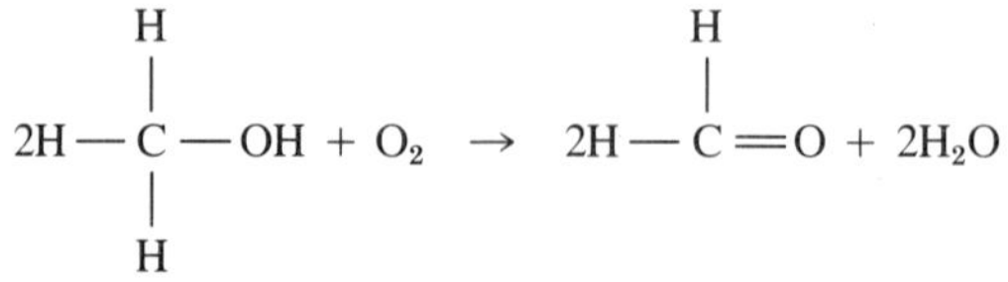

$$2H-\underset{H}{\overset{H}{\underset{|}{C}}}-OH + O_2 \rightarrow 2H-\overset{H}{\underset{|}{C}}=O + 2H_2O$$

Methyl alcohol   **Formaldehyde**
(methanol)   **(methanal)**

SECONDARY ALCOHOLS. When a secondary alcohol is oxidized, a ketone is formed. Ketones will be considered later. An example is the oxidation of isopropyl alcohol, which yields acetone:

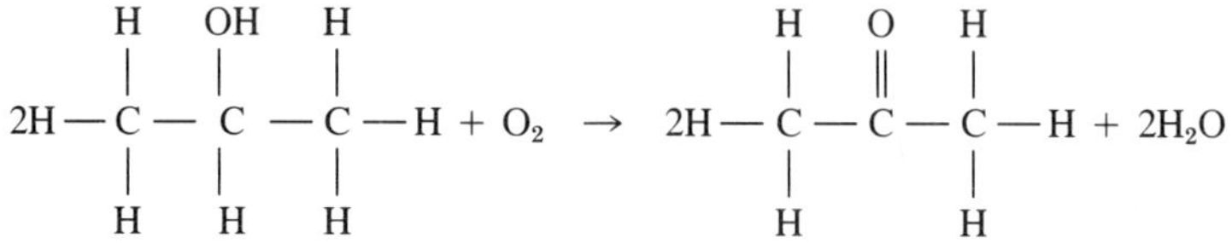

$$2H-\underset{H}{\overset{H}{\underset{|}{C}}}-\underset{H}{\overset{OH}{\underset{|}{C}}}-\underset{H}{\overset{H}{\underset{|}{C}}}-H + O_2 \rightarrow 2H-\underset{H}{\overset{H}{\underset{|}{C}}}-\overset{O}{\underset{}{\overset{\|}{C}}}-\underset{H}{\overset{H}{\underset{|}{C}}}-H + 2H_2O$$

Isopropyl alcohol   **Acetone**
(2-propanol)   **(2-propanone)**

COMPLETE OXIDATION. When an alcohol is ignited, it is brought to a high temperature in the presence of an abundance of oxygen. Under these conditions the alcohol burns; that is, it is completely oxidized, yielding carbon dioxide and water. An example is the burning of ethyl alcohol:

$$C_2H_5OH + 3O_2 \rightarrow 2CO_2 + 3H_2O$$

This reaction yields 7 Calories of heat per gram of alcohol.

Behave amphoterically. Alcohols react with certain basic elements, liberating hydrogen and forming salts, and they therefore behave chemically as weak acids. Alcohols react slowly with sodium, potassium, and calcium.

$$2R-OH + 2Na \rightarrow 2R-ONa + H_2$$

Alcohol   **Sodium alcoholate**

Alcohols can react as bases with strong acids by accepting a proton and forming an *oxonium ion*.

$$R-OH + H^+ \rightarrow ROH_2^+$$

Oxonium ion
(a protonated alcohol)

Some important alcohols

Methyl alcohol (methanol). Methyl alcohol (CH_3OH) is a derivative of methane (CH_4), in which a hydrogen is replaced by an OH group. One method for preparing methyl alcohol was shown in example 1 on p. 166. Methyl alcohol can be produced by the dry distillation of wood; hence its common name is wood alcohol. For many years wood distillation was the principal source of this alcohol, but now the commercial method of preparation is a synthetic process. Synthetic methyl alcohol is produced by the reduction of carbon monoxide with hydrogen in the presence of a catalyst. High temperatures, 300° to 400° C, and high pressures are required to make this reaction take place and give a good yield of the alcohol. The equation for the reaction is as follows:

$$CO + 2H_2 \xrightarrow{Catalyst} CH_3OH$$

Methyl alcohol is used commercially as a solvent and as a reagent or starting material in the preparation of many important organic compounds. It is severely poisonous. Its ingestion or inhalation may result in blindness or cause death. Many alcoholics assume methyl alcohol can be drunk, but the end result is alcohol poisoning.

Ethyl alcohol (ethanol). Ethyl alcohol (C_2H_5OH) is a derivative of ethane (C_2H_6), in which a hydrogen is replaced by an OH group. One method of preparation was given in example 2 on p. 166. An important commercial method for preparing ethyl alcohol is the fermentation of sugars by the enzymes in yeast.

$$C_6H_{12}O_6 \xrightarrow{Yeast} 2C_2H_5OH + 2CO_2$$
Glucose or fructose   **Ethyl alcohol**

Ethyl alcohol, also called grain alcohol, is a colorless, pleasant-smelling liquid. It has been known from the earliest times, since it is so readily made from sugars or from starch that is broken down to sugar by an enzyme in yeast called amylase. Yeast contains many enzymes and is therefore well equipped for promoting the chemical breakdown of starch and different sugars. Fermentation by yeast will produce solutions containing 12% to 18% ethyl alcohol, commonly called wine when the sugar is derived from fruit juices. Purification and concentration of alcohol is accomplished by distillation of the fermented mixture. The boiling point of ethyl alcohol is 78° C.

Distillation will yield an alcoholic solution in which about 95% is ethyl alcohol and 5% is water. The water may be removed by adding quicklime or anhydrous copper sulfate to the mixture and distilling. Pure alcohol, free from water, is known as *absolute alcohol*. The concentration of alcohol in liquor is designated for commercial purposes by the term ''proof.'' The proof is converted to percentage by dividing by 2. Thus ''100 proof'' is 50% alcohol; ''200 proof'' is 100% or absolute alcohol.

The alcohol used for sterilizing the skin prior to giving an injection is ethyl alcohol diluted with water to make a mixture containing approximately 70% alcohol. Coagulation of cellular protein results in destruction of bacteria. Coagulation of protein occurs more completely using 70% alcohol instead of higher percentages because the more dilute alcohol penetrates the cells of the bacteria to a higher degree.

Ethyl alcohol is used commercially as a solvent for gums, resins, varnishes, and perfumes and as a reagent in the manufacture of many important articles. In medicine, ethyl alcohol is used as a vehicle for dispensing drugs, the tinctures, spirits, and extracts being alcoholic solutions. Ethyl alcohol is an antidote for poisoning by phenol or carbolic acid.

Denatured alcohol is alcohol that has been treated so that it is unfit for human consumption. Small quantities of various chemicals such as methyl alcohol, gasoline, benzene, and so on are added to alcohol to render it useless as a beverage but useful as a solvent in industry. Grain alcohol is much more expensive than denatured alcohol because of the taxes levied on beverage alcohol. Tax-free absolute alcohol is used in hospitals and educational institutions and for research purposes.

Complex alcohols. Many alcohols in nature are of a much more complex character than those suggested by the previous discussion. Some alcohols contain more than one OH group in the molecule. Examples are as follows:

```
                                                        H
                                                        |
                                                   H — C — OH
                                                        |
                                                   H — C — OH
                                                        |
                             H                     HO — C — H
                             |                          |
      H                 H — C — OH                 H — C — OH
      |                      |                          |
 H — C — OH             H — C — OH                 H — C — OH
      |                      |                          |
 H — C — OH             H — C — OH                 H — C — OH
      |                      |                          |
      H                      H                          H
    Glycol                Glycerol                   Sorbitol
(1,2-ethanediol)     (1,2,3-propanetriol)     (1,2,3,4,5,6-hexanehexol)
```

Glycol or ethylene glycol. Glycol is a viscous, sweet-tasting liquid with a high boiling point, 197.5° C. When dissolved in water, it lowers the freezing point of the water; hence it is used as a permanent "antifreeze" in automobile radiators. It is useful as a starting material for the synthesis of important organic compounds, such as Dacron polyester fiber.

Glycerol. Glycerol is also called glycerin. It is a viscous, oily liquid highly soluble in water. Its biological importance is that it is a part of the fat molecule. It is prepared by boiling fats with alkali, a process that also forms soaps; it is thus obtained as a by-product in the manufacture of soaps. It is also made from propylene. It is a good solvent and preservative. It absorbs moisture from the air; thus is used in hand lotion.

A compound, *glyceryl trinitrate,* is formed by treating glycerol with nitric acid in the presence of sulfuric acid at a temperature of 10° to 25° C. This compound, also called nitroglycerin, is a violent explosive. Dynamite is a mixture of nitroglycerin and siliceous earth. Glyceryl trinitrate is a potent vasodilator. In medicine it is used in the treatment of angina pectoris. The common name for this material is "nitro pills."

Sorbitol. Sorbitol is important as a type of the polyhydroxy alcohols of which sugars are derivatives. The sugars, for example, glucose and fructose, when reduced with hydrogen are converted to sorbitol. Sorbitol is used for the synthesis of vitamin C.

Hydrocarbons, halogen compounds, alcohols, aldehydes, ketones   171

Table 11-4. Examples of aldehydes

Name	Formula	Structural formula
Formaldehyde (methanal)	CH_2O	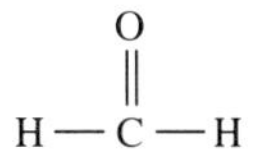 $H-\underset{}{\overset{O}{\overset{\|\|}{C}}}-H$
Acetaldehyde (ethanal)	CH_3CHO	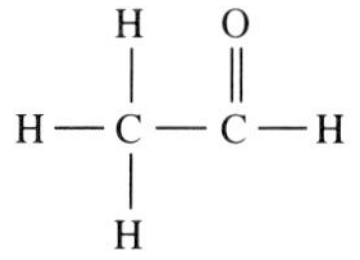 $H-\underset{H}{\overset{H}{\overset{\|}{C}}}-\overset{O}{\overset{\|\|}{C}}-H$

ALDEHYDES, RCHO

Examples of aldehydes are formaldehyde and acetaldehyde (Table 11-4). *The aldehydes may be regarded as derivatives of hydrocarbons in which a hydrogen atom is replaced by a*

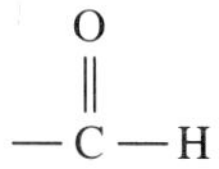

$$-\overset{O}{\overset{\|\|}{C}}-H$$

group. It is the latter group that gives aldehydes their characteristic properties, since the remainder of the molecule is hydrocarbon and will naturally behave like the hydrocarbons. All aldehydes contain the functional group, —CHO.

Preparation. *When a primary alcohol is oxidized, an aldehyde is formed*. This is of interest not only as a general method for preparing aldehydes but also because it recalls an important chemical reaction of alcohols. The reaction follows.

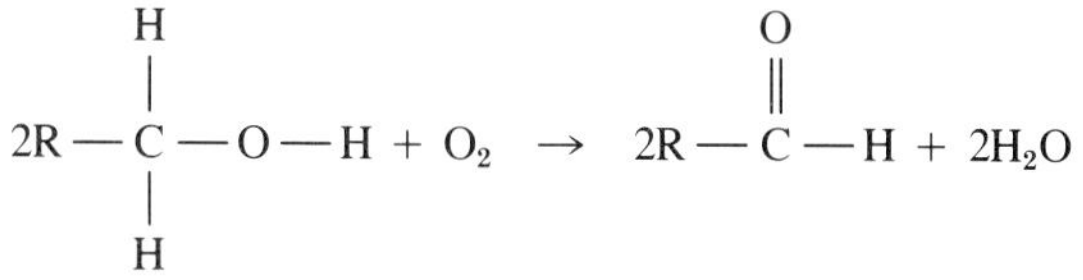

$$2R-\underset{H}{\overset{H}{\overset{\|}{C}}}-O-H + O_2 \rightarrow 2R-\overset{O}{\overset{\|\|}{C}}-H + 2H_2O$$

An alcohol   An aldehyde

General properties. Formaldehyde and acetaldehyde are gases at room temperatures; the other aldehydes are liquids or solids. In general, the aldehydes have penetrating, suffocating odors, and as a class they are strong reducing agents.

The aldehydes have a special importance in that their functional group is a significant constituent of the molecule of certain carbohydrates, substances to be studied later.

Aldehydes as reducing agents. An important property of all aldehydes is their ability to reduce other compounds. For example, aldehydes readily reduce oxides of metals such as silver or copper oxides. When a faintly alkaline solution of silver oxide is mixed with an aldehyde, for example, formaldehyde, in a test tube, a metallic silver mirror forms on the sides of the tube. This is known as Tollens' test.

$$Ag_2O + CH_2O \rightarrow 2Ag + HCOOH$$

Silver oxide   Formaldehyde (methanal)   Silver metal   Formic acid (methanoic acid)

This is a typical oxidation-reduction reaction. The formaldehyde removes oxygen from the silver oxide. It therefore reduces this compound and thus is acting as a reducing agent. The silver, which had a valence of +1 before the reaction, has its valence reduced to 0. The formaldehyde is oxidized to formic acid by the oxygen removed from the silver oxide.

Another example is the reduction of cupric (copper II) hydroxide. When an alkaline solution of this compound is mixed with an aldehyde and the mixture is heated gently, a reddish brown precipitate, cuprous (copper I) oxide, is formed. This is known as Benedict's or Fehling's test for an aldehyde.

$$2Cu(OH)_2 + R-\underset{\underset{O}{\|}}{C}-H \rightarrow Cu_2O + 2H_2O + R-\underset{\underset{O}{\|}}{C}-OH$$

Cupric hydroxide (copper II hydroxide)   An aldehyde   Cuprous oxide (copper I oxide)   An organic acid

This reaction is important in biochemistry because it is the basis of tests for sugars (which contain aldehyde groups) in the urine or other body fluids. The reddish brown precipitate, cuprous (copper I) oxide, is readily recognized, and this test is therefore a practical procedure for detecting the presence of an aldehyde or a compound containing an aldehyde group.

Some important aldehydes

Formaldehyde. Formaldehyde is prepared by the catalytic oxidation of methyl alcohol.

$$2CH_3OH + O_2 \xrightarrow[Cu]{Heat} 2H-\underset{\underset{O}{\|}}{C}-H + 2H_2O$$

Methyl alcohol (methanol)   Formaldehyde (methanal)

The methyl alcohol, mixed with air, is passed over heated copper gauze, the latter serving as a catalyst.

Properties and importance. Formaldehyde is a colorless gas at room temperatures. It has a penetrating odor and is severely irritating to the eyes and the respiratory tract. It is a strong reducing agent.

Formaldehyde has a marked tendency to combine with itself, a reaction known as *polymerization*. One polymer of formaldehyde, called paraformaldehyde, is a solid that contains formaldehyde molecules in a linear chain, the length of which varies from 6 to 100 formaldehyde units.

Formaldehyde combines with proteins such as albumin or gelatin, and in high concentrations it is a protein precipitant. It is therefore highly toxic to protoplasm, which contains vitally important proteins. It also preserves and hardens tissues as a result of its action on proteins. Because of its effect on proteins it is used as an embalming agent for anatomical specimens and as a fixative for tissues in the preparation of slides of normal or pathological tissue for microscopical examination.

Formaldehyde is a powerful germicide and disinfectant. This property is due to its ability to react with and denature proteins, important constituents of bacteria. It kills both spore-forming and nonspore-forming bacteria. It is too irritating to be used on the skin; hence its principal application as a disinfectant is on inanimate objects such as surgical instruments and gloves. It is particularly effective on sputa and body excreta because it is active in the presence of considerable organic matter. Formaldehyde has received some application in the disinfecting of rooms because the gas penetrates all crevices or cracks and thus does a thorough job in destroying disease-producing organisms.

To obtain formaldehyde in a convenient form for use in the laboratory and hospital, formaldehyde gas is dissolved in water to the extent of 37%, making a solution called *formalin*.

Formaldehyde condenses with phenol to form synthetic resins. One of these resins is Bakelite, used in manufacturing articles that require high durability, such as insulating material in electrical equipment. Formaldehyde can be polymerized to form tough plastics called Delrin or Celcon.

Acetaldehyde. Acetaldehyde is prepared by the oxidation of ethyl alcohol.

$$2H-\underset{\underset{H}{|}}{\overset{\overset{H}{|}}{C}}-\underset{\underset{H}{|}}{\overset{\overset{H}{|}}{C}}-OH + O_2 \rightarrow 2H-\underset{\underset{H}{|}}{\overset{\overset{H}{|}}{C}}-\overset{\overset{O}{\|}}{C}-H + 2H_2O$$

Ethyl alcohol (ethanol)   Acetaldehyde (ethanal)

Properties and importance. Acetaldehyde has a pleasant, fruitlike odor. Like formaldehyde, it is a strong reducing agent and also has a marked tendency to polymerize. One polymer of acetaldehyde is paraldehyde, which is a compound that contains three acetaldehyde units in a ringlike structure. In medicine paraldehyde has received some use as a hypnotic or sleep-producing drug. Many people suffering from delirium tremens are quieted by the use of paraldehyde. A disadvantage of this drug is that it imparts an unpleasant odor to the breath.

KETONES, RCOR

One example of a ketone is acetone (Table 11-5).

Ketones are important in demonstrating the structure of the carbonyl group found in some sugars. The student should note that the *functional group in this series is the carbonyl group:*

$$-\overset{\overset{O}{\|}}{C}-$$

Table 11-5. An example of a ketone

Name	Formula	Structural formula
Acetone (2-propanone)	CH_3COCH_3	H—C(H)(H)—C(=O)—C(H)(H)—H

Preparation. Ketones are prepared by the *oxidation of secondary alcohols.* Thus isopropyl alcohol when oxidized yields acetone.

$$2 \begin{array}{c} CH_3 \\ \diagdown \\ C-OH \\ \diagup \\ CH_3 \end{array} \begin{array}{c} | \\ H \\ \end{array} + O_2 \rightarrow 2 \begin{array}{c} CH_3 \\ \diagdown \\ C=O \\ \diagup \\ CH_3 \end{array} + 2H_2O$$

Isopropyl alcohol (2-propanol)   **Acetone (2-propanone)**

This is a general method. The oxidation of more complex secondary alcohols yields corresponding ketones.

Properties. Ketones are of interest to us because they are an important class of substances in organic chemistry, and certain carbohydrates (discussed later) are ketones.

Acetone, the first member of the ketone series, is a colorless liquid with a pleasant odor. It is soluble in water. It is a good solvent for fats, gums, resins, paints, and varnishes. Acetone is useful in removing adhesive tape from the skin.

Acetone occurs in the blood and urine in very small amounts in healthy persons. In patients with diabetes mellitus, large amounts are often present in the blood, urine, and expired air. The increased production in diabetes mellitus is the result of incomplete metabolism of fats.

Questions for study

1. What are hydrocarbons? Name some important hydrocarbons.
2. Distinguish between saturated and unsaturated hydrocarbons.
3. What is an alkane? An alkene?
4. How is methane prepared?
5. Discuss the occurrence and importance of the hydrocarbons.
6. What is meant by the paraffin series of hydrocarbons? The alkene series? The alkyne series?
7. What is the effect of the presence of double or triple valence bonds on the properties of the compound?
8. Name three important hydrocarbon derivatives.
9. What is the formula for chloroform? Carbon tetrachloride? Iodoform?
10. What is the use in medicine of chloroform? Carbon tetrachloride? Iodoform?
11. What is an alcohol? Name three alcohols.
12. Give one general method for the preparation of an alcohol.
13. Name three complex alcohols.

Hydrocarbons, halogen compounds, alcohols, aldehydes, ketones **175**

14. What is the use in medicine of glyceryl trinitrate?
15. Distinguish between primary, secondary, and tertiary alcohols.
16. State three important chemical reactions of alcohols.
17. How is methyl alcohol prepared? Discuss its toxicity.
18. Describe the commercial method of preparation of ethyl alcohol.
19. State several physical properties of ethyl alcohol.
20. What are the uses of ethyl alcohol commercially and in medicine?
21. What is absolute alcohol? What is meant by "100 proof" alcohol?
22. What is an aldehyde? Name two aldehydes.
23. What is a general method for preparing aldehydes.
24. What is a reducing agent? Explain how aldehydes act as reducing agents. State two examples of the latter.
25. What is formaldehyde? How is it prepared?
26. What is formalin? What are its uses?
27. What is acetaldehyde? Paraldehyde? What use has been made of the latter?
28. What is a ketone?
29. State a general method for the preparation of ketones.
30. What is the importance of ketones?
31. State several physical properties of acetone.
32. What are the uses of acetone?

12 Organic chemistry—organic acids, esters, ethers, amines

ORGANIC ACIDS, RCOOH

Examples of organic acids are formic and acetic acids (Table 12-1).

The organic acids may be considered as hydrocarbons in which one or more of the hydrogens is replaced by a COOH group. The COOH group is called a carboxyl and is the functional group. The carboxyl group is made up of two components, which we have studied previously, the carbonyl group

$$-\overset{\overset{\displaystyle O}{\|}}{C}-$$

and the hydroxyl group, OH; hence the name is a contraction of the names of its two components. The carboxyl group is characteristic of all organic acids; it contains 1 hydrogen atom, which ionizes in solution and gives acid properties to the compound. An organic acid may contain either one or several COOH groups. The organic acids that contain one COOH group are called fatty acids because many of them occur in fats.

Methods of preparation. Two methods of preparing organic acids are as follows:

Table 12-1. Examples of organic acids

Name	Formula	Structural formula
Formic acid (methanoic acid)	HCOOH	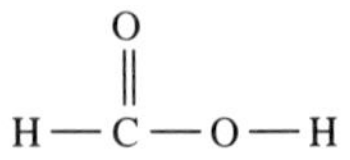 $H-\overset{\overset{\displaystyle O}{\|}}{C}-O-H$
Acetic acid (ethanoic acid)	CH_3COOH	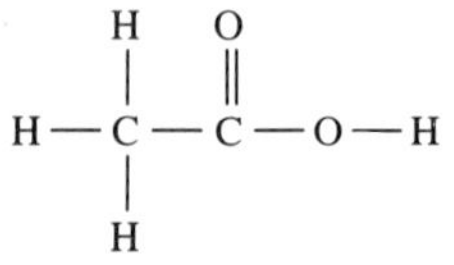 $H-\overset{\overset{\displaystyle H}{\|}}{\underset{\underset{\displaystyle H}{\|}}{C}}-\overset{\overset{\displaystyle O}{\|}}{C}-O-H$

1. When a primary alcohol is mildly oxidized, an aldehyde is formed. If the aldehyde is oxidized further, an organic acid is produced.

$$\underset{\substack{\text{Ethyl alcohol} \\ \text{(ethanol)}}}{\text{H}-\underset{\underset{\text{H}}{|}}{\overset{\overset{\text{H}}{|}}{\text{C}}}-\underset{\underset{\text{H}}{|}}{\overset{\overset{\text{H}}{|}}{\text{C}}}-\text{OH}} \xrightarrow{[O]} \underset{\substack{\text{Acetaldehyde} \\ \text{(ethanal)}}}{\text{H}-\underset{\underset{\text{H}}{|}}{\overset{\overset{\text{H}}{|}}{\text{C}}}-\overset{\overset{\text{O}}{\|}}{\text{C}}-\text{H}} \xrightarrow{[O]} \underset{\substack{\text{Acetic acid} \\ \text{(ethanoic acid)}}}{\text{H}-\underset{\underset{\text{H}}{|}}{\overset{\overset{\text{H}}{|}}{\text{C}}}-\overset{\overset{\text{O}}{\|}}{\text{C}}-\text{O}-\text{H}}$$

2. Another general method is treatment of a salt of an organic acid with a strong mineral acid.

$$\underset{\text{Sodium acetate}}{CH_3COONa} + H_2SO_4 \rightarrow \underset{\substack{\text{Acetic acid} \\ \text{(ethanoic acid)}}}{CH_3COOH} + NaHSO_4$$

General properties and reactions. The organic acids that contain less than 9 carbon atoms are liquids; those that contain more than 9 carbon atoms are solids. The lower members of the series are soluble in water; the higher members are insoluble. The organic acids are weak acids; that is, their molecules dissociate slightly, yielding a relatively small number of hydrogen ions when in solution. They exhibit the general properties of the mineral acids but not in as marked a degree, since they do not ionize to the same extent.

The reactions of organic acids with metals and bases are of interest.

Reaction with metals. The hydrogen in the carboxyl of an organic acid is ionizable. It is therefore replaceable by all metals above hydrogen in the electromotive series (p. 65). This behavior is the same as that exhibited by the hydrogen in inorganic acids except that the reaction usually takes place more slowly, since organic acids as a class are weaker than inorganic acids. Example:

$$2R-COOH + Zn \rightarrow (R-COO)_2Zn + H_2$$

Reaction with bases. Organic acids react with bases to form salts and water, a reaction similar to the action of inorganic acids with bases. Example:

$$R-COOH + NaOH \rightarrow R-COONa + H_2O$$

Some important organic acids

Formic acid (methanoic acid). Formic acid (HCOOH) is the first member of the fatty acid series. It is the strongest acid in this series. It is related to methane, and another name for this compound is methanoic acid. The structural formula is as follows:

$$\underset{\substack{\text{Formic acid} \\ \text{(methanoic acid)}}}{\text{H}-\overset{\overset{\text{O}}{\|}}{\text{C}}-\text{O}-\text{H}}$$

Formic acid is a colorless liquid at room temperatures and has an irritating odor. Its melting point is 8.4° C, and its boiling point is 100.7° C. It is highly soluble in water. It occurs in ants, bees, nettles, and pine needles. The "sting" of a bee or a nettle is due to the irritating action of formic acid liberated in the tissues.

Formic acid was prepared during the Middle Ages by the distillation of ants. One of the procedures used for obtaining formic acid was to macerate ants in water and distill the mixture. The compound received its name from the fact that it was first prepared from ants, the Latin word for ant being *formica*.

Formic acid is useful in the laboratory as a reagent and in the synthesis of organic compounds. Commercially, it is used in the tanning of hides and the manufacture of rubber.

Acetic acid (ethanoic acid). Acetic acid (CH_3COOH) is a derivative of ethane; hence another name for this compound is ethanoic acid. It may be prepared by the oxidation of ethyl alcohol to acetaldehyde, which is oxidized further to form acetic acid. This method is described on p. 177. The structural formula for acetic acid is as follows:

$$\begin{array}{c} \text{H} \quad \text{O} \\ | \quad \| \\ \text{H}-\text{C}-\text{C}-\text{O}-\text{H} \\ | \\ \text{H} \end{array}$$

Acetic acid
(ethanoic acid)

Commercially acetic acid is obtained by the oxidation of hydrocarbons. It is also prepared for commercial use by the fermentation of the sugar in fruit juices. Vinegar is made by this process. Vinegar is a dilute (3% to 6%) solution of acetic acid, which is usually made from the juice of the apple. In the formation of vinegar, sugar is first converted to ethyl alcohol by yeast; the alcohol is then oxidized to acetic acid by bacteria, called *Acetobacter,* which are in the so-called "mother" of vinegar. These bacteria occur in the air and readily get into a solution exposed to the atmosphere.

Acetic acid is a colorless liquid at room temperatures. Its melting point is 16.6° C, and its boiling point is 118° C. It is highly soluble in water. It is a weak acid; a dilute aqueous solution (0.1N) is ionized to the extent of about 1% of its molecules. A preparation containing 100% acetic acid is known as *glacial acetic acid* because of the large icelike crystals present when it is frozen.

Acetic acid is used in the manufacture of drugs, dyes, and other organic compounds. It is a valuable reagent in the chemical laboratory. As a constituent of vinegar it is an important condiment. A very dilute solution of acetic acid is a good antidote for a burn caused by an alkali.

Other organic acids. Some other organic acids of importance are *lactic acid* (2-hydroxypropanoic acid), which results from the action of bacteria on the lactose of milk and gives sour milk its characteristic acid taste; *butyric acid* (butanoic acid), the volatile substance giving rancid butter its offensive odor; *oxalic acid* (ethanedioic acid), which is valuable for cleansing purposes, such as removal of ink stains or rust; *citric acid* (2-hydroxy-1,2,3-propanetricarboxylic acid), found in fruits such as the lemon, lime, or grapefruit; *malic acid* (1-hydroxybutanedioic acid), found in apples; and *tartaric acid* (2,3-dihydroxybutanedioic acid), found in grapes, whose salt, "cream of tartar," is used in baking powders. Calcium propionate is used to retard the spoilage of bread.

ESTERS, RCOOR

An ester may be produced by the interaction of an alcohol with an acid. The formation of an ester is illustrated by the following reaction:

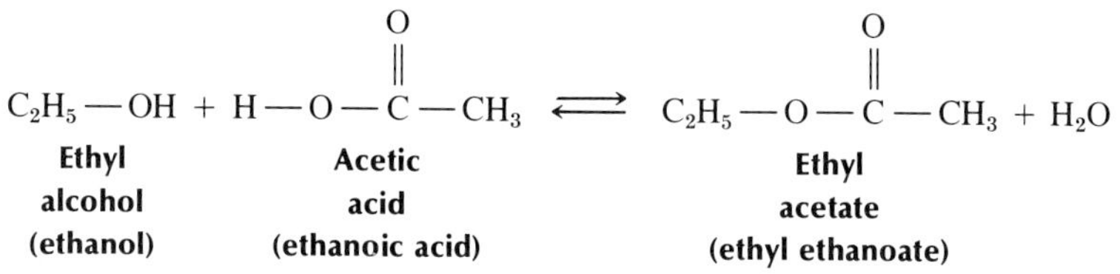

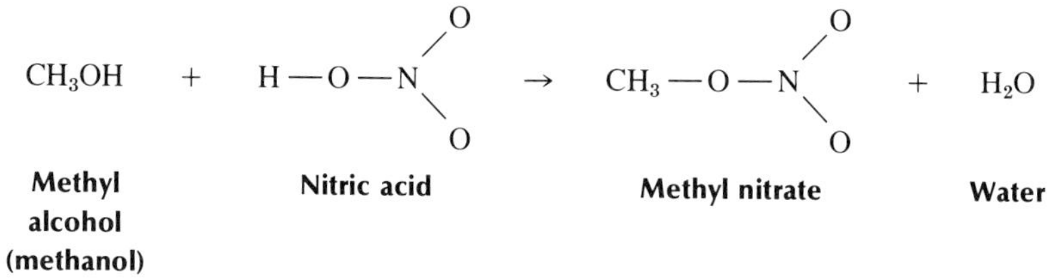

$$C_2H_5-OH + H-O-\underset{\underset{\text{Acetic acid}}{\text{(ethanoic acid)}}}{\overset{\overset{O}{\|}}{C}}-CH_3 \rightleftarrows C_2H_5-O-\underset{\underset{\text{Ethyl acetate}}{\text{(ethyl ethanoate)}}}{\overset{\overset{O}{\|}}{C}}-CH_3 + H_2O$$

Ethyl alcohol (ethanol)

$$CH_3OH + H-O-N\underset{\underset{\text{Nitric acid}}{}}{\overset{\nearrow O}{\diagdown O}} \rightarrow CH_3-O-N\underset{\underset{\text{Methyl nitrate}}{}}{\overset{\nearrow O}{\diagdown O}} + H_2O$$

Methyl alcohol (methanol)

In both of these reactions an ester and water are formed, just as a salt and water are formed in the neutralization of an acid with a base in inorganic reactions.

The esters are also called ethereal salts because they are very volatile. As a class they have a pleasant odor. The esters of organic acids are the characteristic constituents of perfumes and flavoring extracts and are, in fact, the constituents that in many cases give fruits and flowers their characteristic odors. They are widely distributed in nature, being found especially in the fruits and in the flowers of plants.

Some important esters

Isoamyl nitrite. The amyl nitrite used in medicine is isoamyl nitrite. This compound ($C_5H_{11} \cdot NO_2$) is an isomer of the normal amyl nitrite in which the carbon atoms are arranged in a straight chain. The carbon atoms of isoamyl nitrite are arranged in a branched chain.

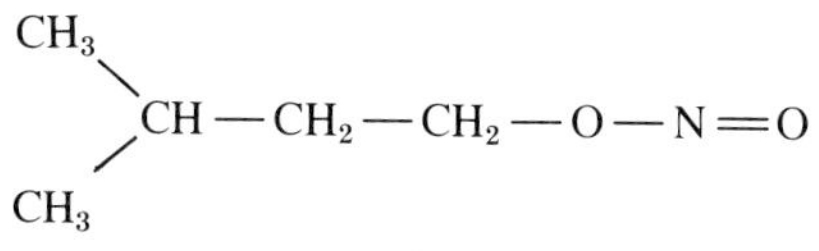

$$\underset{CH_3}{\overset{CH_3}{\diagdown}}CH-CH_2-CH_2-O-N=O$$

Isoamyl nitrite
(γ-methylbutyl nitrite)

Isoamyl nitrite is a pale yellow, volatile liquid, with the odor of banana. It has a relatively high vapor pressure; hence its vapors diffuse rapidly when the liquid is released from a container.

Isoamyl nitrite, like other nitrites, relaxes smooth muscle, especially the muscle in the walls of the smaller blood vessels. It therefore lowers blood pressure. This drug is especially effective in dilating the coronary blood vessels and has been used to relieve the paroxysms of angina pectoris.

Glyceryl trinitrate. Glyceryl trinitrate, mentioned previously, is an ester formed from the reaction of glycerol and nitric acid. Like amyl nitrite, it also is used to relax the muscles of the blood vessels.

$$\begin{array}{c} H \\ | \\ H-C-O-NO_2 \\ | \\ H-C-O-NO_2 \\ | \\ H-C-O-NO_2 \\ | \\ H \end{array}$$

**Glycerol trinitrate
(nitroglycerine)**

Esters of importance for their odors. As stated previously, certain esters of organic acids are of outstanding interest because of the odors they impart to fruits and flowers. Examples of these are

$$H-\underset{\substack{\|\\O}}{C}-O-C_2H_5$$
**Ethyl formate
(ethyl methanoate)**

$$CH_3\underset{\substack{\|\\O}}{C}-O-C_2H_5$$
**Ethyl acetate
(ethyl ethanoate)**

$$C_3H_7\underset{\substack{\|\\O}}{C}-O-C_2H_5$$
**Ethyl butyrate
(ethyl butanoate)**

$$CH_3\underset{\substack{\|\\O}}{C}-O-C_5H_{11}$$
**Amyl acetate
(1-pentanol ethanoate)**

$$C_3H_7\underset{\substack{\|\\O}}{C}-O-C_5H_{11}$$
**Amyl butyrate
(1-pentanol butanoate)**

$$CH_3\underset{\substack{\|\\O}}{C}-O-CH_2CH_2-\underset{\substack{|\\CH_3}}{CH}-CH_3$$
**Isoamyl acetate
(γ-methylbutyl ethanoate)**

$$C_4H_9\underset{\substack{\|\\O}}{C}-O-C_5H_{11}$$
**Isoamyl butyrate
(γ-methylbutyl butanoate)**

$$CH_3\underset{\substack{\|\\O}}{C}-O-C_8H_{17}$$
**Octyl acetate
(1-octanol ethanoate)**

$$C_4H_9\underset{\substack{\|\\O}}{C}-O-C_8H_{17}$$
**Octyl butyrate
(1-octanol butanoate)**

The odor of flowers and the odor and taste of fruits are largely due to certain of these naturally occurring esters or to characteristic combinations of these esters.

Fats and oils. Fats and oils are important esters. They are products of the reaction under appropriate conditions of the trihydroxy alcohol, glycerol, with fatty acids.

Other important esters. Other important esters include cellulose nitrate (nitrocellulose), used in guncotton; cellulose acetate, used to make "acetate film" for cameras, and acetate fabrics; and polyvinyl acetate polymers. Esters used in medicine include methyl salicylate (oil of wintergreen), used in liniments; and phenyl salicylate (salol), used as an intestinal antiseptic.

ETHERS, ROR

Examples of ethers are methyl and ethyl ethers (Table 12-2).

The ethers are organic oxides. They consist of 2 hydrocarbon radicals joined by an atom of oxygen. The functional group is the oxygen atom.

Methods of preparation. Ethers may be prepared as follows:

1. A general method of preparing ethers consists of treating a halogen derivative of a hydrocarbon with sodium alcoholate.

$$CH_3I + NaOCH_3 \rightarrow CH_3-O-CH_3 + NaI$$

    Methyl   Sodium   Methyl
    iodide   methylate   ether

2. The commercial method of preparing ethyl ether is by treating ethyl alcohol with sulfuric acid. In this reaction there is a removal of 1 molecule of water from 2 molecules of alcohol. The method requires specially controlled conditions with respect to amounts of sulfuric acid and ethyl alcohol used and the temperature of the reaction.

$$C_2H_5OH + HOC_2H_5 \xrightarrow{H_2SO_4} C_2H_5-O-C_2H_5 + H_2O$$

    Ethyl   Ethyl   Ethyl ether
    alcohol   alcohol

Table 12-2. Examples of ethers

Name	Formula	Structural formula
Methyl ether (methoxymethane)	CH_3OCH_3	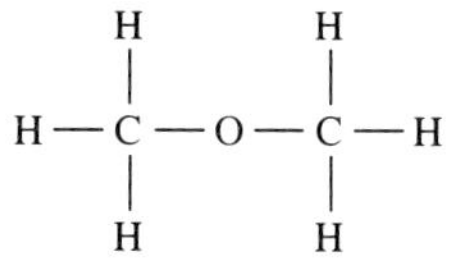 H-C(H)(H)-O-C(H)(H)-H
Ethyl ether (ethoxyethane)	$C_2H_5OC_2H_5$	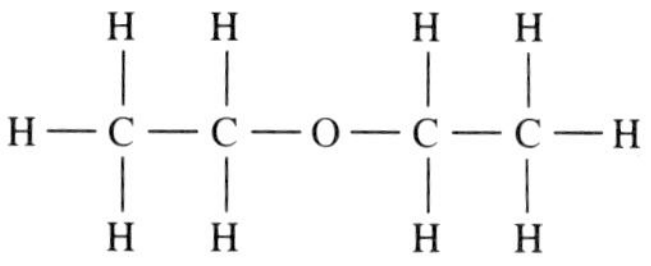 H-C(H)(H)-C(H)(H)-O-C(H)(H)-C(H)(H)-H

General properties. The ethers of lower molecular weights are colorless liquids. They are stable in the presence of strong acids and alkalies. The lower members of the series are somewhat volatile. As a class the ethers are extremely flammable.

Ethyl ether is the most important of this series of compounds because of its value as a general anesthetic. It is a colorless, volatile liquid, with a pleasant odor. Its boiling point is 34.5° C. *It is highly imflammable and is explosive when mixed with air and ignited. For this reason ether should never be used near a flame or in an atmosphere where a spark may be produced.*

Ethyl ether is an excellent solvent for fats, gums, resins, and alkaloids, as well as many inorganic substances. Aside from its combustibility, ether is generally an inactive substance chemically.

In 1842, ethyl ether was introduced into surgery as an anesthetic by Dr. Crawford W. Long at Jefferson, Georgia. Since that time it has been widely used for general anesthesia because its administration can be easily regulated, and in the absence of heart or respiratory lesions it is not toxic in the amounts used to produce the surgical state of anesthesia.

Other important ethers include divinyl ether (Vinethene), methoxyflurane (Penthrane), and methyl propyl ether (metopryl or neothyl).

$CH_2=CH-O-CH=CH_2$ $CHCl_2CF_2OCH_3$ $CH_3OC_3H_7$
Divinyl ether   **Methoxyflurane**   **Methyl propyl ether**

All three of these ethers are quite useful as general anesthetics and have some superior properties over ethyl ether.

AMINES, RNH_2

Examples of amines are methyl and ethyl amines (Table 12-3).

The amines may be regarded as organic derivatives of ammonia.

The functional group is the nitrogen atom. As shown by the formulas given in Table 12-3, in a primary amine 1 hydrogen of ammonia is replaced by an organic radical. Methyl and ethyl amines are examples of primary amines. In a secondary amine 2 hydrogens of am-

Table 12-3. Examples of amines

Name	Formula	Structural formula
Methyl amine (methylamine)	CH_3NH_2	H-C(H)(H)-N(H)(H)
Ethyl amine (ethylamine)	$C_2H_5NH_2$	H-C(H)(H)-C(H)(H)-N(H)(H)

monia are replaced by organic radicals, and in tertiary amines organic radicals replace all 3 hydrogens in ammonia.

Letting R stand for any organic radical, the formulas for primary, secondary, and tertiary amines may be written as follows:

$$R-N\begin{matrix}H\\ \\H\end{matrix} \qquad \begin{matrix}R\\ \\R\end{matrix}N-H \qquad R-N\begin{matrix}R\\ \\R\end{matrix}$$

    Primary amine   Secondary amine   Tertiary amine

A tertiary amine can react with an alkyl halide to form a compound called a *quaternary ammonium salt*.

$$R-\underset{\underset{R}{|}}{\overset{\overset{R}{|}}{N:}} \;+\; \underset{(RF,\,RBr,\,RI)}{RCl} \;\rightarrow\; \left[R-\underset{\underset{R}{|}}{\overset{\overset{R}{|}}{N{:}R}}\right]^{+}\underset{(F^-,\,Br^-,\,I^-)}{Cl^-}$$

    Tertiary   Alkyl   Quarternary
    amine   halide   ammonium salt

The ammonium ion, NH_4^+, has four replaceable hydrogens, and its organically substituted analogue is R_4N^+, the quaternary ammonium ion.

The amines are of great importance, since they are related chemically to the amino acids, which compose proteins. Only the lower members of the group of primary amines will be considered.

Preparation of methyl amine. If ammonia is mixed with hydrogen chloride, ammonium chloride is formed.

$$H-N\begin{matrix}H\\ \\H\end{matrix} \;+\; HCl \;\rightarrow\; \left[H-\underset{\underset{H}{|}}{\overset{\overset{H}{|}}{N}}-H\right]^{+}Cl^-$$

    Ammonia   Ammonium chloride

In like manner, if ammonia is treated with methyl iodide, methyl ammonium iodide is formed.

$$H-N\begin{matrix}H\\ \\H\end{matrix} \;+\; CH_3I \;\rightarrow\; \left[H-\underset{\underset{H}{|}}{\overset{\overset{CH_3}{|}}{N}}-H\right]^{+}I^-$$

    Ammonia   Methyl   Methyl ammonium
                iodide   iodide

If now the methyl ammonium iodide is treated with an alkali, methyl amine is formed.

$$\underset{\substack{\text{Methyl ammonium}\\\text{iodide}}}{\overset{\displaystyle H\quad CH_3}{\underset{\displaystyle H}{H-NI}}} + NaOH \rightarrow \underset{\text{Methyl amine}}{CH_3-N\overset{H}{\underset{H}{\diagup}}} + NaI + H_2O$$

These are typical reactions that illustrate how amines in general may be prepared.

Physical properties. At ordinary temperatures the first three members of the primary amine series are gases; those with 3 to 11 carbon atoms in their molecules are liquids; those containing more than 11 carbon atoms in their molecules are solids. The amines of low molecular weight have a fishlike odor and are highly soluble in water. The higher amines have a slight odor and are less soluble in water.

Chemical behavior. Amines react chemically with water and acids.

Reaction with water. When an amine is dissolved in water, the solution turns litmus paper blue and is therefore basic. This is due to the reaction of the amine with water to form an amine derivative of ammonium hydroxide. The complexes formed in this way are much stronger bases than ammonium hydroxide. They dissociate to a considerable extent, yielding ions of the derivative and hydroxide ions. The parallelism of the behavior of amines in water with that of ammonia in water is shown by the following equations:

$$\underset{\text{Ammonia}}{H-\underset{H}{\overset{H}{N}}-H} + H-O-H \rightleftarrows \underset{\substack{\text{Ammonium}\\\text{hydroxide}}}{H-\underset{H}{\overset{H}{N}}\diagup\overset{OH}{_{H}}} \rightleftarrows \underset{\substack{\text{Ammonium}\\\text{ion}}}{\left[H-\underset{H}{\overset{H}{N}}-H\right]^{+}} + \underset{\substack{\text{Hydroxide}\\\text{ion}}}{OH^-}$$

$$\underset{\substack{\text{Methyl}\\\text{amine}}}{CH_3-\underset{H}{\overset{H}{N}}-H} + H-O-H \rightleftarrows \underset{\substack{\text{Methyl}\\\text{ammonium}\\\text{hydroxide}}}{CH_3-\underset{H}{\overset{H}{N}}\diagup\overset{OH}{_{H}}} \rightleftarrows \underset{\substack{\text{Methyl}\\\text{ammonium}\\\text{ion}}}{\left[CH_3-\underset{H}{\overset{H}{N}}-H\right]^{+}} + \underset{\substack{\text{Hydroxide}\\\text{ion}}}{OH^-}$$

Reaction with acids. Since amines form bases in water, they will react with acids in aqueous solution to form substituted ammonium salts. This behavior is similar to that exhibited by ammonia with acids in aqueous solution. The parallelism is shown by the following equations:

$$\underset{\substack{\text{Ammonium}\\\text{hydroxide}}}{H-\underset{H}{\overset{H}{N}}\diagup\overset{OH}{_{H}}} + H-Cl \rightleftarrows H_2O + \underset{\substack{\text{Ammonium}\\\text{chloride}}}{H-\underset{H}{\overset{H}{N}}\diagup\overset{Cl}{_{H}}} \rightleftarrows \underset{\substack{\text{Ammonium}\\\text{ion}}}{\left[H-\underset{H}{\overset{H}{N}}-H\right]^{+}} + \underset{\substack{\text{Chloride}\\\text{ion}}}{Cl^-}$$

$$CH_3-\underset{\underset{H}{|}}{\overset{\overset{H}{|}}{N}}\diagdown^{OH}_H + H-Cl \rightleftarrows H_2O + CH_3-\underset{\underset{H}{|}}{\overset{\overset{H}{|}}{N}}\diagdown^{Cl}_H \rightleftarrows \left[CH_3-\underset{\underset{H}{|}}{\overset{\overset{H}{|}}{N}}\diagdown^{H}_H\right]^+ + Cl^-$$

| Methyl ammonium hydroxide | | Methyl ammonium chloride | Methyl ammonium ion | Chloride ion |

Toxicity. Most amines are toxic. When introduced into the bloodstream they have generally undesirable effects. Since foods contain proteins, which are composed of amino acids, and bacteria will degrade amino acids, forming toxic amines, poisonous substances may be formed in foods that have been kept under conditions favorable for bacterial contamination and growth. The so-called "ptomaine poisoning" was once considered to be the effect of certain toxic amines, called "ptomaines," that were formed in foods by bacterial action. However, the effect of these toxic amines is less when given by mouth than when injected into the bloodstream. This is because these substances, when absorbed from the intestinal tract, pass by way of the portal circulation to the liver. In the liver they are converted into less harmful or harmless products that are excreted in the urine.

The toxic effects of the "ptomaine poisons" now appear to be much less than was formerly believed, since it is recognized that the liver is highly efficient in detoxifying such substances. It is probable that most of the ill effects in persons with food poisoning are due to the action of pathogenic bacteria introduced into the intestinal tract in infected foods.

Questions for study

1. What is an organic acid?
2. Compare the properties of organic acids with those of inorganic acids.
3. Write the structural formula for formic acid. State several of its properties. What is the origin of its name?
4. Write the structural formula for acetic acid. State several of its properties. What are its uses?
5. Name five organic acids of importance, other than formic and acetic.
6. What is an ester?
7. How are esters prepared?
8. Write the structural formula for ethyl acetate.
9. What are the uses of isoamyl nitrite in medicine?
10. Discuss the importance of esters in relation to pleasant odors in nature.
11. What is an ether?
12. What ethers are used as general anesthetics? When was ethyl ether first used for anesthesia?
13. Mention several properties of ethyl ether.
14. What precaution is of great importance in using ether?
15. What is an amine?
16. What is the importance of the amines?
17. Distinguish between primary, secondary, and tertiary amines.
18. Mention several physical and chemical properties of amines.
19. Discuss the toxicity of amines.

13 Organic chemistry—aromatic compounds

AROMATIC HYDROCARBONS, Ar-H

Benzene. Benzene (C_6H_6) is an aromatic hydrocarbon. It is a closed-chain, six-carbon compound whose structural formula is as follows:

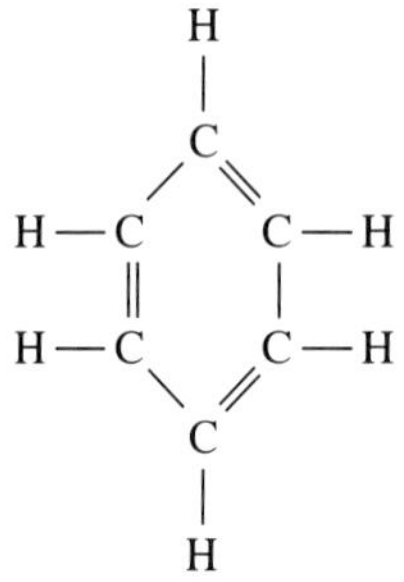

Structural formula 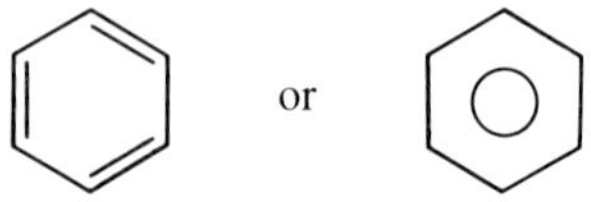   **Abbreviated structural formula**

Looking at the structural formula for benzene, the student will see that many other compounds can be made from this substance by substituting groups of atoms for one or several of the hydrogens in the ring and by joining benzene rings together. Benzene is therefore the parent substance of the aromatic compounds. Most of the aromatic compounds may be prepared with benzene as the principal starting material, just as methane serves as the starting point for the synthesis of aliphatic compounds. The same general classes found in the aliphatic division—hydrocarbons, alcohols, aldehydes, ketones, ethers, acids, esters, amines, and halides—exist in the aromatic division, the essential difference being the presence of one or more closed rings of carbon atoms in the molecule and most important, a system of alternating single and double bonds. There is also a great deal of similarity in the chemical reactions of the aliphatic and aromatic compounds.

IUPAC nomenclature. Under IUPAC rules the carbons of compounds having ring forms are also numbered. Hydrocarbon addition or substitution on a ring indicates that the lowest numbered carbon of the ring is the carbon bonded to the added or substituted group. Added or substituted groups have prefix names, and a second prefix (such as di-, tri-, or tetra-) indicates the appearance of a hydrocarbon one, two, or three times, respectively, on a ring.

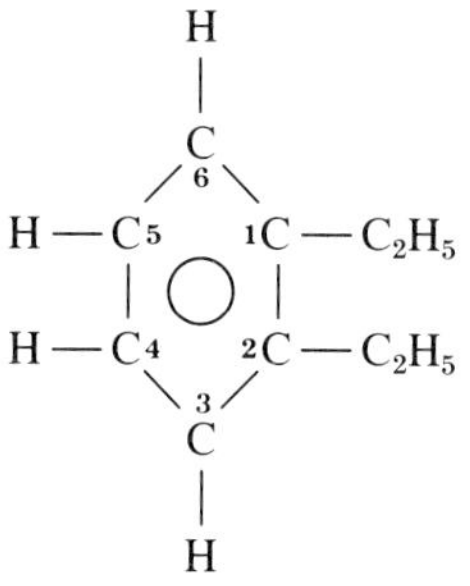

1,2-Diethylbenzene

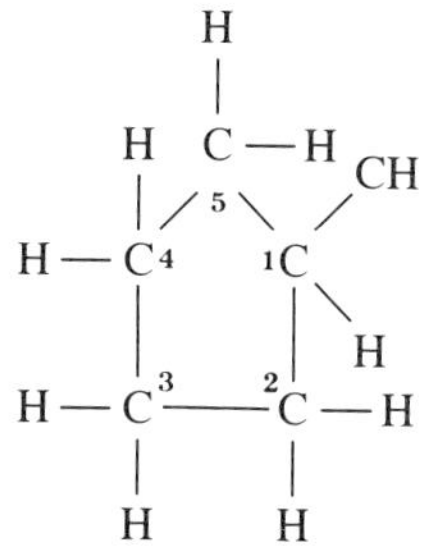

1-Methylcyclopentane

Radicals. As with the aliphatic compounds, the aromatic hydrocarbons, such as benzene, C_6H_6, form the aromatic radical, C_6H_5-, when 1 atom of hydrogen is removed. This particular radical or group is called the phenyl radical. A more general symbol would be Ar-, since this signifies any aromatic or aryl radical the same way R- signifies any alkyl or aliphatic radical. Thus we can signify benzene as one member of the aromatic hydrocarbon family of general formula Ar-H.

It is sometimes preferable to use the ring form as a hydrocarbon substitute group on a chain hydrocarbon. In this case the nomenclature is as follows:

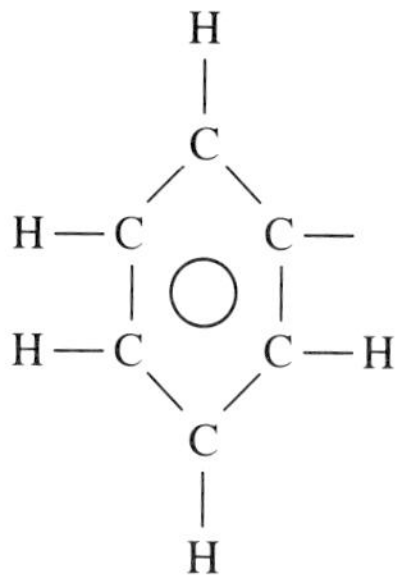

Phenyl group

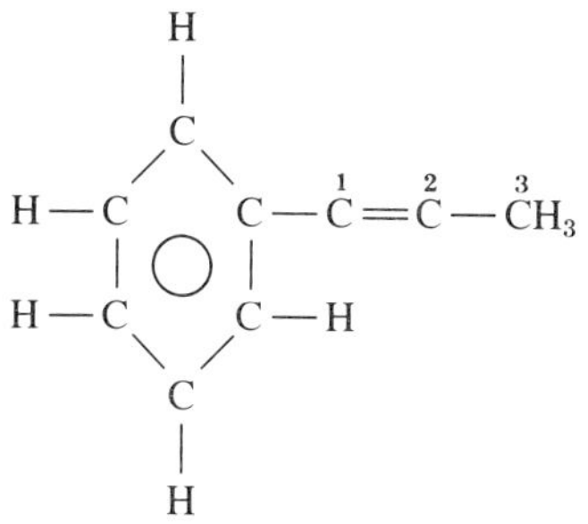

1-Phenylpropyne

Source. The great source of aromatic compounds is coal tar, which is obtained as a by-product in the manufacture of coke. In the modern gas plant, coal is heated in closed retorts, out of contact with air. By this process coke is formed and coal gas is liberated. In this procedure considerable volatile matter condenses while the gases pass through cooling pipes and form "gas liquor" or coal tar. One ton of coal yields about 100 pounds of coal tar, from which many useful compounds such as benzene, toluene, xylene, phenol, cresols, drugs, dyes, and explosives are prepared. Since World War II, petroleum has been a source of many aromatic compounds.

Preparation. Benzene can be made by heating acetylene. The preparation is accomplished practically by passing the acetylene through a red-hot tube. It may be represented as follows:

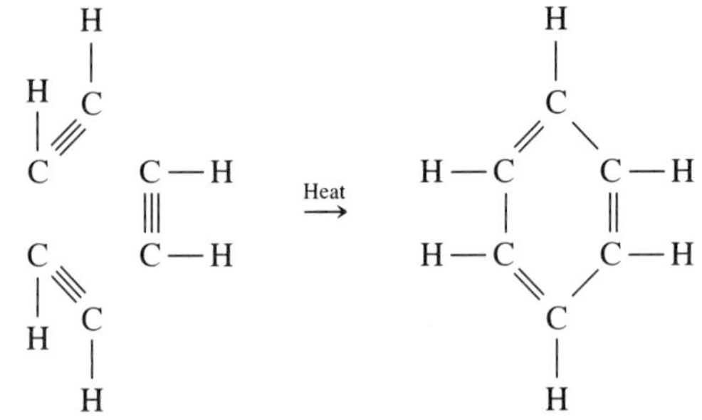

188 Roe's principles of chemistry

$$\text{Acetylene (3 molecules) (ethyne)} \xrightarrow{\text{Heat}} \text{Benzene}$$

This method is of great importance because it shows how the chemist may prepare an aromatic (ring-structure) compound from an aliphatic (straight-chain) compound.

Properties and reactions. Benzene is a colorless, volatile liquid with a characteristic aromatic odor. Its boiling point is 80° C. It is insoluble in water but highly soluble in alcohol and ether. It is a good solvent for fats.

The unbonded electrons of benzene become stabilized through resonance hybridization. The carbons and hydrogens are all on a plane. The unbonded electrons form a high electron density area that spreads evenly over the whole molecule in a "doughnut" shape. In this symmetrical hybrid molecule, all the carbon atoms are exactly alike; there are no unsaturated bond locations easily available for addition reactions as there are in the unsaturated hydrocarbons (Fig. 13-1). Benzene is fairly stable; it keeps well and is resistant to moderate oxidation. In the presence of powerful oxidizing agents, however, the ring is disintegrated, forming various products.

Under vigorous conditions, benzene will combine with hydrogen at 200° C in the presence of a nickel catalyst to form the fully saturated compound cyclohexane. Cyclohexane is an example of a carbocyclic structure.

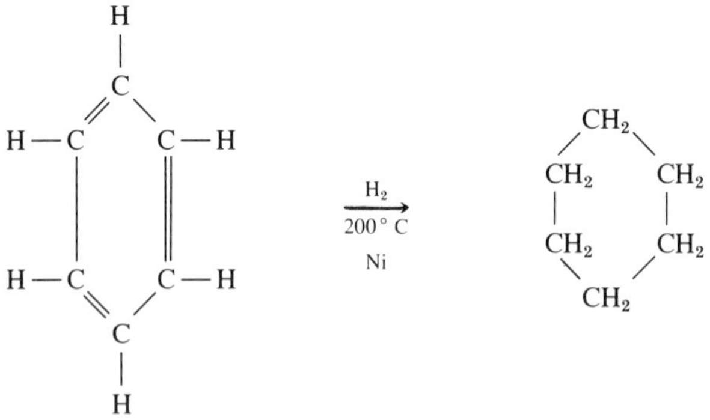

$$\text{Benzene} \xrightarrow[\text{Ni}]{\text{H}_2 \atop 200° \text{C}} \text{Cyclohexane}$$

The hydrogens attached to the benzene ring are replaceable by a number of reagents. By taking advantage of this reactivity many aromatic compounds have been prepared.

Benzene is highly flammable. Precautions should be used to avoid releasing benzene or its vapors near a flame.

Aromatic compounds 189

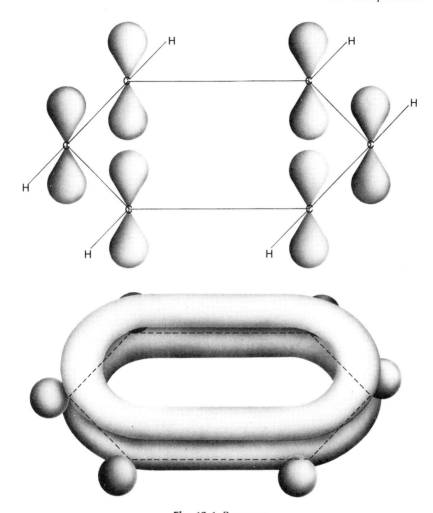

Fig. 13-1. Benzene.

It is toxic. Either inhalation of the fumes or absorption through the skin may result in serious or fatal poisoning.

Homologues of benzene. Previously it has been stated that compounds differing from each other by CH_2 or multiples of the CH_2 group belong to a homologous series and are called homologues. There are many interesting homologues of benzene. Several of them will be discussed.

By using methods that will substitute an aliphatic hydrocarbon radical for one or more of the hydrogen atoms in the benzene ring, a series of homologues may be prepared. These aliphatic groups are called *side chains*.

Methyl benzene or toluene   **Ethyl benzene**   **Propyl benzene**

Many more homologues of benzene may be prepared by lengthening the aliphatic side chain that is substituted for hydrogen or by attaching additional side chains to the ring.

Toluene. Toluene is a colorless, volatile liquid with a characteristic odor. Its boiling point is 110.8° C. It is insoluble in water and highly soluble in alcohol and ether.

Toluene occurs in coal tar and Tolú balsam, from which its name is derived. It is obtained commercially from coal tar, and it may be prepared in the laboratory by a number of methods. One laboratory method is to heat monobromobenzene with sodium and methyl bromide for several hours.

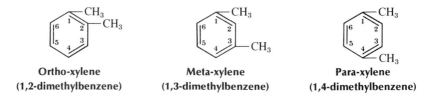

Bromobenzene   Methyl bromide   Toluene

Toluene is important as a source material for the preparation of many organic compounds in the aromatic series. Some of these are the explosive trinitrotoluene (TNT), benzaldehyde, and benzoic acid. Toluene is used as a urine preservative to prevent bacterial decomposition of sugar and other constituents of the urine when samples are collected for quantitative analysis.

The xylenes. If substitution of methyl groups for hydrogen in the benzene ring is made at two points, three homologues of benzene, which are isomers, are formed. The three products formed are ortho-, meta-, and para-xylene. Isomers are compounds that have the same number of atoms but differ in their structures.

Ortho-xylene   Meta-xylene   Para-xylene
(1,2-dimethylbenzene)   (1,3-dimethylbenzene)   (1,4-dimethylbenzene)

The xylenes are colorless liquids with a characteristic odor. They are insoluble in water but are readily soluble in alcohol and ether. They are good solvents for fatty material.

Commercial xylene is of special value in biology and pathology in the preparation of slides of tissue slices for microscopical examination. Its usefulness here is due to its ability to dissolve paraffin and also to its capacity to remove alcohol from the tissue preparation.

AROMATIC HALOGEN COMPOUNDS, Ar-X

The halogen atoms—fluorine, chlorine, bromine, and iodine—may be substituted for hydrogen atoms in the benzene ring. These halogens are the functional groups. A considerable number of halogen derivatives of benzene exist that contain one or more halogen atoms attached to the ring. Examples are chlorobenzene (C_6H_5Cl), tribromobenzene ($C_6H_3Br_3$), diiodobenzene ($C_6H_4I_2$), and fluorobenzene (C_6H_5F). Chlorobenzene will be discussed as representative of the group.

Chlorobenzene. Chlorine may replace hydrogen in the benzene molecule in the presence of a catalyst, for example, iron or aluminum chloride, forming at first monochlorobenzene.

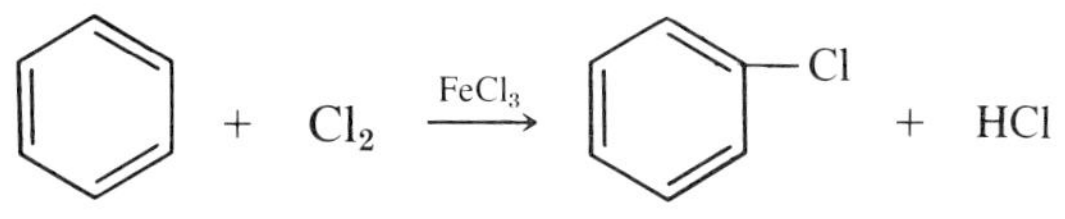

$$\text{Benzene} + Cl_2 \xrightarrow{FeCl_3} \text{Chlorobenzene} + HCl$$

By raising the temperature and continuing the chlorination, more chlorine may be substituted, forming di-, tri-, tetra-, penta-, or hexachlorobenzene. Direct halogenation of the benzene molecule may be accomplished in the same way with bromine, forming mono, di-, or polybromo- derivatives of benzene. Fluorine and iodine cannot be substituted directly onto the benzene ring; however, these elements can be substituted for the hydrogen in the ring by indirect methods.

Chlorobenzene is used in the manufacture of aromatic hydrocarbons, phenol, aniline, insecticides, and disinfectants. This compound and other halogenated derivatives are of great value as starting materials for the synthesis of many aromatic compounds.

PHENOLS AND AROMATIC ALCOHOLS, Ar-OH

Two characteristic classes of organic compounds are formed when the hydroxyl group is substituted onto the molecule of aromatic hydrocarbons. These are the phenols and the aromatic alcohols.

When an OH group is substituted onto the benzene ring, the resulting product is a phenol. There is one phenol, C_6H_5OH, that bears the name of the group. This compound is called phenol, and it is also referred to by its common name, *carbolic acid*. However, the term phenol in a general sense means any benzene derivative with one or more OH (phenolic) groups in direct combination with the benzene ring. There are many phenols.

When an OH group is substituted onto the side chain of an aromatic hydrocarbon, the resulting product is an aromatic alcohol. The simplest example of an aromatic alcohol is benzyl alcohol ($C_6H_5CH_2OH$). In this compound the OH is attached to the methyl (side chain) group. When the OH is attached to the side chain, a compound is formed that has properties resembling those of the aliphatic alcohols. The OH group thus confers quite different properties when attached to the benzene ring than when attached to the side chain of a benzene derivative. There are many aromatic alcohols.

One phenol and one aromatic alcohol, benzyl alcohol, will be discussed.

Phenol. Phenol is a derivative of benzene in which one OH group replaces a hydrogen atom of the molecule. It is prepared by boiling chlorobenzene with sodium hydroxide solution at a high temperature and elevated pressure to form sodium phenoxide and then liberating the phenol from its sodium salt by adding an acid.

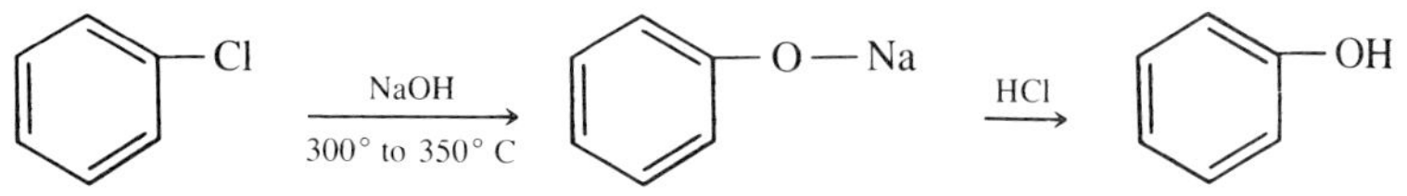

$$\text{Chlorobenzene} \xrightarrow[300°\text{ to }350°\text{ C}]{NaOH} \text{Sodium phenoxide} \xrightarrow{HCl} \text{Phenol}$$

Phenol is commonly called carbolic acid. It is a colorless, crystalline solid with a characteristic odor. Its melting point is 41° C. It dissolves in water to the extent of about 8 g per 100 ml of water at room temperature.

Phenol is appreciably acidic. When dissolved in water, the hydrogen in its OH group dissociates fairly well. It will therefore react with bases to form salts such as sodium or potassium phenoxide.

Phenol is highly poisonous. In concentrated form it will severely burn or corrode living tissues, and even in very dilute solution it is toxic to living cells. It is a powerful germicide and is used as a standard for characterizing germicidal power. The germicidal power of a given compound is often expressed as its *phenol coefficient.* This represents the germicidal capacity of the compound as compared with the potency of phenol in killing bacteria. Certain homologues of phenol (benzene derivatives having one or more OH groups) have also been used considerably as disinfectants. Some of these are the cresols, thymol, resorcinol, and hexylresorcinol; the phenol coefficients of these compounds are 2, 25, 0.3, and 58, respectively, when compared with the effect of phenol on the typhoid bacillus. Other important phenols are Lysol, hexachlorophene, picric acid, and creosote.

Benzyl alcohol. Benzyl alcohol is a colorless liquid with a boiling point of 206° C. It has a pleasant odor, which makes it desirable as a constituent of perfumes. It is slightly soluble in water. It occurs in the free state in resins and balsams, for example, balsam of Peru, and in many fragrant flowers. It also occurs as an ester of certain organic acids.

The structural formula for benzyl alcohol is as follows:

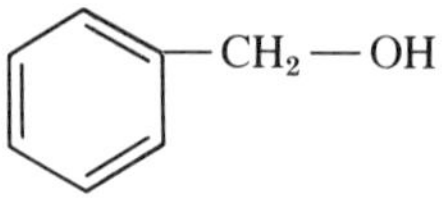

Benzyl alcohol

Benzyl alcohol is a local anesthetic. Its principal application in medicine is for easing the pain from dressing or debridement of an open wound.

AROMATIC ALDEHYDES, Ar-CHO

Aromatic aldehydes are compounds in which an aldehyde functional group (—CHO) is substituted onto an aromatic nucleus or is present in an aliphatic side chain substituted into the aromatic nucleus. Examples:

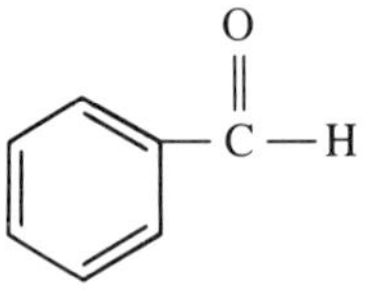

Benzaldehyde
(benzenecarbonal)

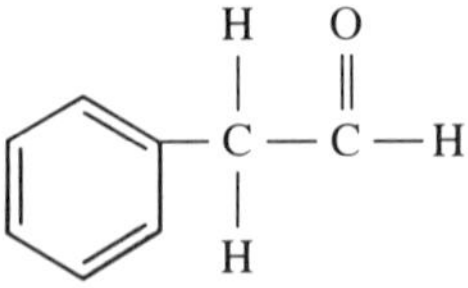

Phenylacetaldehyde

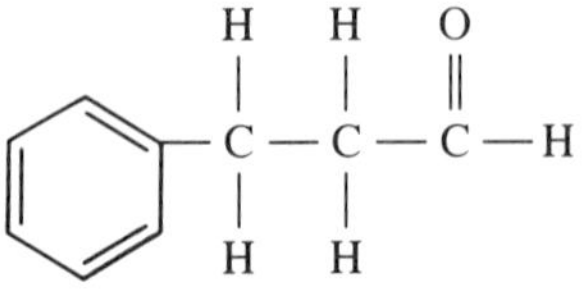

Phenyl propionaldehyde
(phenylpropynal)

Benzaldehyde (benzenecarbonal). Benzaldehyde is the simplest aromatic aldehyde. It occurs in nature in a glucoside, amygdalin, which is found in the seeds of bitter almonds and the kernels of certain fruits such as peaches and apricots. It has a characteristic odor, the odor of oil of bitter almonds. It may be obtained by the hydrolysis of amygdalin.

$$C_{20}H_{27}O_{11}N + 2H_2O \rightarrow C_6H_5-CHO + 2C_6H_{12}O_6 + HCN$$

Amygdalin Benzaldehyde Glucose Hydrocyanic acid

Benzaldehyde is a colorless, oily liquid with a boiling point of 179.5° C. It is very slightly soluble in water and readily soluble in alcohol and ether. It is a reducing agent and has other chemical properties similar to those of the aliphatic aldehydes. Benzaldehyde is used in the preparation of drugs, flavoring agents, and perfumes.

Vanillin. Vanillin occurs naturally in the vanilla bean and is produced synthetically from wood pulp waste. Vanilla extract (a solution of vanillin in ethyl alcohol) is used as a flavoring agent.

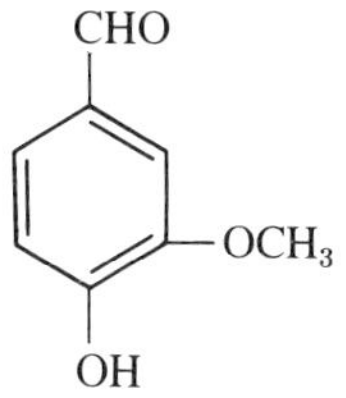

Vanillin

AROMATIC ACIDS, Ar-COOH

Aromatic acids are compounds in which a carboxyl (—COOH) group is attached to an aromatic nucleus or is present in an aliphatic side chain attached to the aromatic nucleus. Examples:

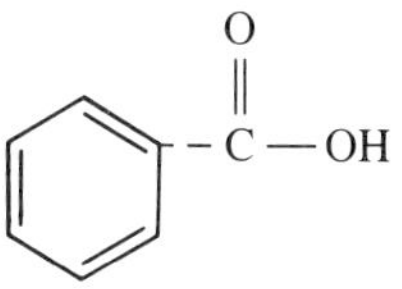

Benzoic acid Phenylacetic acid 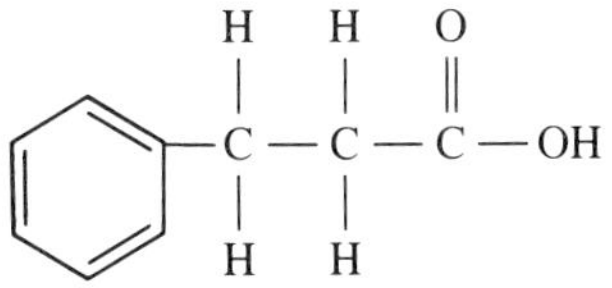 Phenylpropionic acid
(benzenecarboxylic acid)

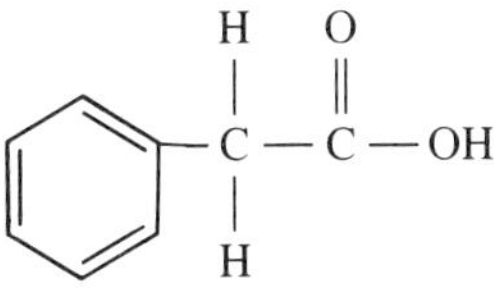

Benzoic acid. Benzoic acid occurs in nature in various balsams, for example, Tolú and Peru balsams, and as an ester in gum benzoin. It is a white, crystalline solid. It is slightly soluble in water and readily soluble in alcohol and ether.

Benzoic acid has antiseptic properties. For this reason its salt, sodium benzoate, is used as a food preservative. The standard sugar solution used in the hospital laboratory in the determination of blood or urinary sugar is glucose dissolved in a saturated water solution of benzoic acid. In such solutions, sugar keeps indefinitely because bacteria do not live in saturated benzoic acid solution.

Benzoic acid is not toxic in moderate amounts. The liver has a mechanism for conjugating it with an amino acid, glycine, to form the harmless compound, hippuric acid (C_6H_5—CO—NH—CH_2—COOH), which is excreted in the urine. The *benzoic acid liver function test* is a procedure for measuring the ability of the liver to form hippuric acid. In this test sodium benzoate is given orally or intravenously, and the urine is collected for a definite time and analyzed for hippuric acid. The amount of hippuric acid excreted in the test period is a measure of liver function.

Salicylic acid. Salicylic acid, or ortho-hydroxybenzoic acid, is of special interest because it is both an aromatic acid and a phenol, and also because two of its derivatives, methyl salicylate and aspirin, are used extensively in medicine. The structural formula for salicylic acid is as follows:

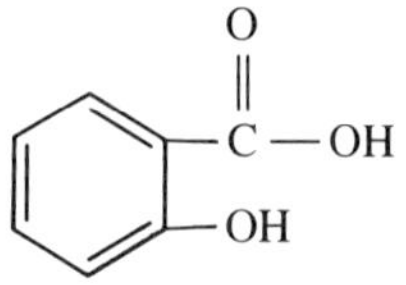

Salicylic acid

Salicylic acid is a colorless, crystalline compound that is slightly soluble in water and readily soluble in alcohol and ether. It behaves like both an aromatic acid and a phenol; it forms salts and esters through reactions of its —COOH group, and it gives color tests and other reactions typical of phenols because of its OH group.

AROMATIC ESTERS, Ar-COOR

Esters are formed by the reaction of a COOH group and an OH group (see p. 179).

Methyl salicylate. Methyl salicylate occurs in oil of wintergreen, oil of sweet birch, and other essential oils. This compound is an ester produced by the reaction of methyl alcohol with salicylic acid under appropriate conditions. It has the odor of wintergreen and is, in fact, responsible for the characteristic odor of oil of wintergreen. The structural formula for this compound is as follows:

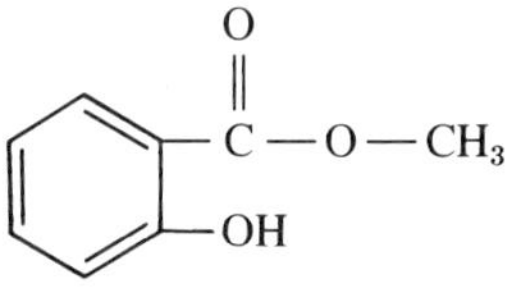

Methyl salicylate

Methyl salicylate is used as a flavoring ingredient and as a constituent of liniments. Its pharmacological effect is that of a counterirritant, a substance applied to the skin for the purpose of relieving pain in the underlying muscles or viscera.

Acetylsalicylic acid or aspirin. Acetylsalicylic acid is the chemical name of the well-known drug aspirin. This compound is an ester; that is, it is the product of the interaction of the phenol (OH) group of salicylic acid with the acid group (COOH) of acetic acid. The structural formula for aspirin is as follows:

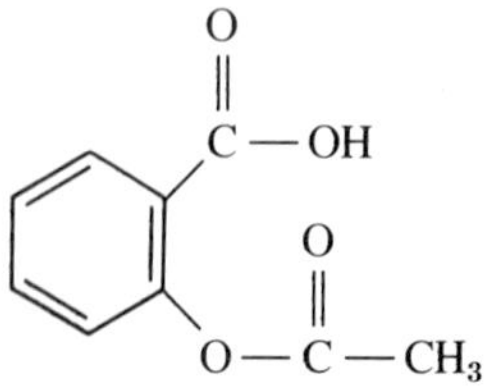

Acetylsalicylic acid or aspirin

Acetylsalicylic acid is used extensively as a remedy for the relief of pain. It is also an antipyretic, a drug that reduces fever.

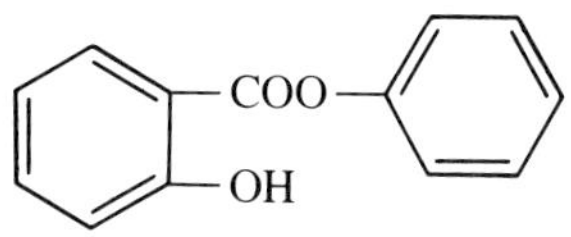

Phenyl salicylate

Phenyl salicylate. Phenyl salicylate, or salol, functions as an internal antiseptic by hydrolyzing in the small intestine to phenol and salicylic acid.

AROMATIC AMINES, Ar-NH$_2$

Aromatic amines are derivatives of ammonia in which there is a replacement of one or more of the hydrogen atoms of the ammonia by an aromatic nucleus. Representative compounds are aniline, diphenylamine, and triphenylamine.

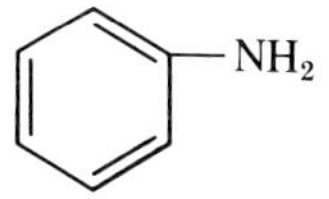

Aniline 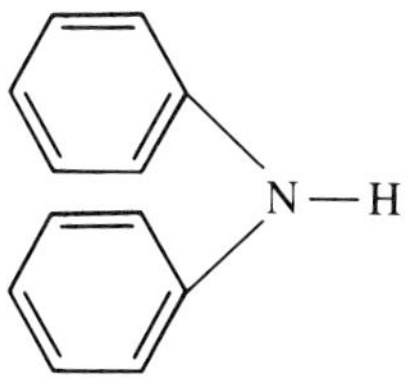   **Diphenylamine** 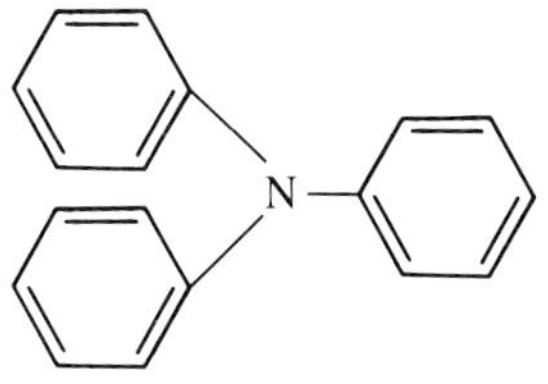   **Triphenylamine**

Aniline is a prototype of the primary aromatic amines; diphenylamine is a representative of the secondary aromatic amines; and triphenylamine is an example of tertiary aromatic amines.

Aniline. Aniline is found in small amounts in coal tar. It also occurs in indigo. Its name is derived from the Spanish word *anil,* meaning indigo, from which it was first prepared by distilling the indigo with lime. The structural formula for aniline is as follows:

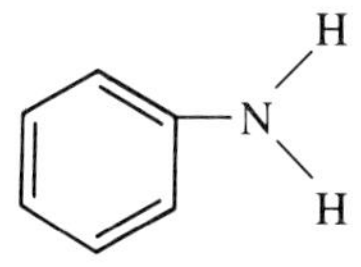

Aniline

Aniline is a colorless, oily liquid with a faint aromatic odor. Its boiling point is 184.4° C. It is slightly soluble in water. Chemically, aniline behaves like ammonia; it is basic and reacts with acids to form salts. Its behavior in general is analogous to that of the aliphatic amines. Aniline is of great importance as the starting material for the synthesis of dyes and other compounds of outstanding theoretical and practical value.

Acetanilide. One derivative of aniline of importance in medicine is acetanilide. The structural formula for this drug is as follows:

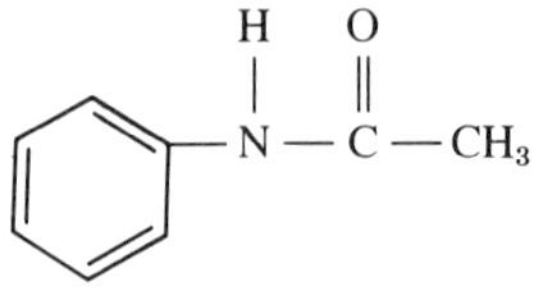

$$\underset{\text{Acetanilide}}{\text{C}_6\text{H}_5-\overset{\text{H}}{\underset{}{\text{N}}}-\overset{\text{O}}{\underset{}{\overset{\|}{\text{C}}}}-\text{CH}_3}$$

As shown, acetanilide is a compound resulting from the reaction of aniline with acetic acid; the basic amino group reacts with the acid group of acetic acid, forming water and the acetylated derivative.

Acetanilide is a white, crystalline powder, slightly soluble in water. It is used in medicine as an antipyretic and analgesic. Its effect as an analgesic is due to an elevation of the threshold to pain stimuli in the brain.

Acetophenetidin. Acetophenetidin, or phenacetin, is another derivative of aniline that is extremely useful also as an antipyretic and analgesic.

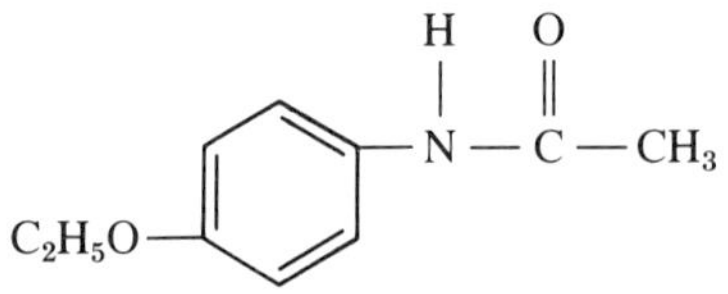

$$\underset{\text{Acetophenetidin}}{\text{C}_2\text{H}_5\text{O}-\text{C}_6\text{H}_4-\overset{\text{H}}{\underset{}{\text{N}}}-\overset{\text{O}}{\underset{}{\overset{\|}{\text{C}}}}-\text{CH}_3}$$

Saccharin. Saccharin is an imide derivative of ortho-sulfobenzoic acid. Its structure is as follows:

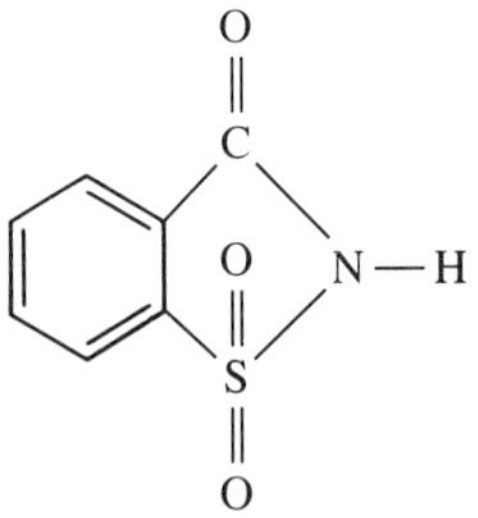

Saccharin
(ortho-sulfobenzoic acid imide)

Saccharin is a powerful sweetening agent. It has been reported to have a sweetening power from 550 to 750 times that of sucrose. This compound is used as a substitute for sugar. It does not contribute to the caloric intake or affect the blood sugar level.

The sulfonamides. The sulfonamides, or sulfa drugs, are sulfonic acid derivatives of aromatic amines. The first sulfonamide to receive major clinical use was sulfanilamide. The structural formula for this compound is as follows:

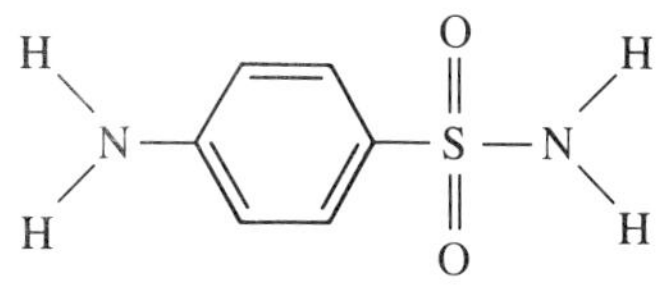

Sulfanilamide
(para-aminobenzenesulfonamide)

A drug called Prontosil, the effective component of which is sulfanilamide, was found to be effective against hemolytic streptococci in mice by Domagk in 1932. Soon thereafter a great many other sulfonamides were synthesized and applied to the treatment of bacterial infections in man. In general, the sulfonamides are effective against such organisms as *Streptococcus,* pneumococcus, gonococcus, meningococcus, *Escherichia coli, Pseudomonas,* and *Proteus vulgaris.* The formulas for some sulfonamides that are used considerably are as follows:

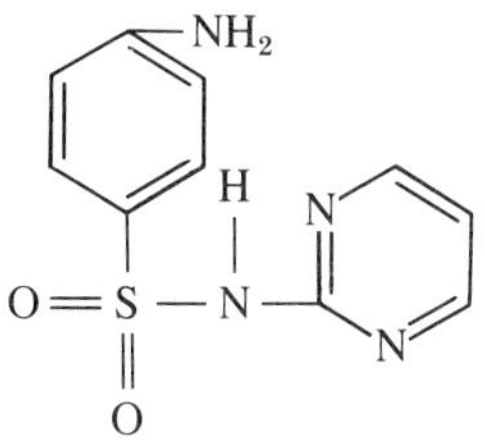

Sulfadiazine 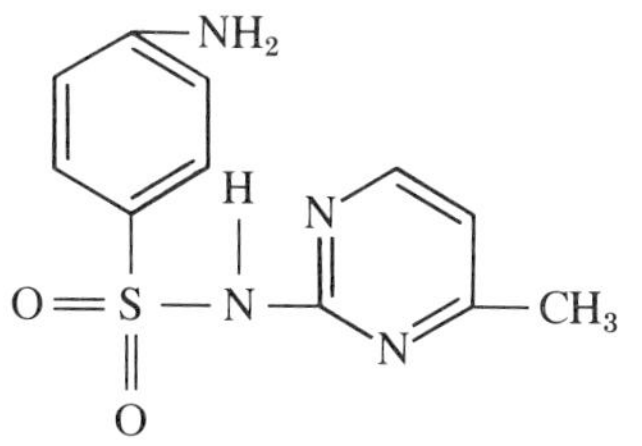   **Sulfamerazine** 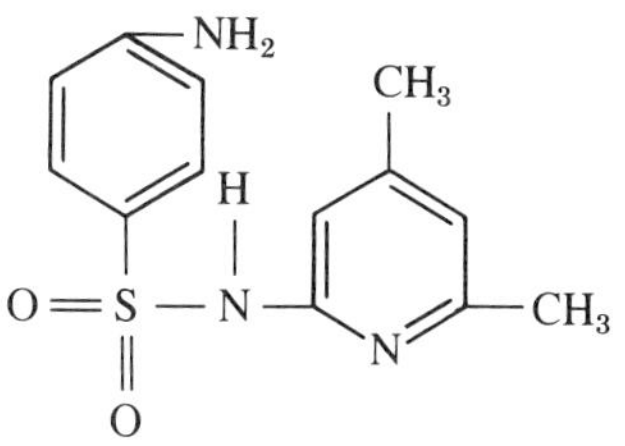   **Sulfamethazine**

One disadvantage of the sulfonamides is their relatively low solubility in water. For this reason the dosage and water intake must be adjusted to avoid depositions of crystals of the drug in the kidneys, which would result in kidney impairment. One method that has been designed to overcome this disadvantage is the use of a combination of three of the drugs (triple sulfas). This procedure lowers the amount of each drug that must be dissolved and excreted to one-third of the amount required for a therapeutic effect if any of the drugs were used alone, and since each drug has an independent solubility that is not decreased by the presence of the others in solution, the problem of using an amount of drug that will stay in solution and be excreted becomes simpler.

The production of sulfonamides reached a high level by the end of World War II (6 million pounds per year). Thereafter, the amount used decreased because of the discovery and application of the antibiotics, which have a similar antibacterial effect. The sulfonamides still receive considerable clinical use.

POLYNUCLEAR HYDROCARBONS

As stated earlier, carbon atoms have the remarkable property of being able to form a great number of stable compounds by being bonded to each other. Carbon compounds of great and widely varying chain lengths exist in the aliphatic division of organic chemistry. In the aromatic division also there is a great variety of compounds, but these compounds are found to consist of six-membered ring structures. Of great importance in the aromatic division is the existence of ring compounds in which closed rings of carbon atoms are joined at two or more places to form condensed multiple-ring structures. The multiple-ring structures are referred to as polynuclear compounds. The structural formulas of a few polynuclear hydrocarbons of special interest are as follows (double bonds have been omitted in complex structures):

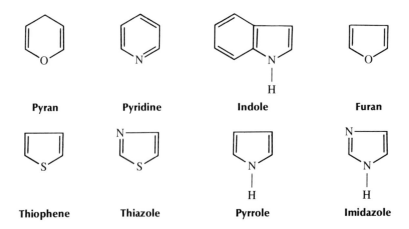

Three of these compounds (cyclopentanoperhydrophenanthrene, estrane, and androstane) are of special interest because they are related structurally to certain physiological substances (vitamins and hormones).

HETEROCYCLIC COMPOUNDS

Another division of compounds of great importance in organic chemistry consists of *closed chain structures that contain elements other than carbon in the ring*. The latter substances are called heterocyclic compounds. Structural formulas of important examples of the heterocyclic compounds follow:

These compounds are structurally related to substances of physiological importance. Reference will be made to them later in the chapters on physiological chemistry.

CARBOCYCLIC COMPOUNDS

Although carbocyclic compounds are not widely distributed, two are worth mentioning. Cyclopropane is made up of 3 carbon atoms in a closed-chain structure.

$$CH_2 \diagup\!\!\!\!\!\diagdown CH_2$$
$$CH_2$$

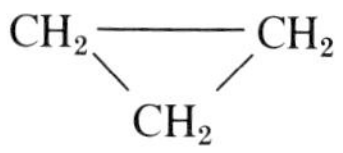

Cyclopropane

This compound is used extensively as a general anesthetic in place of ether. Care must be observed that no sparks or open flames be nearby as cyclopropane is extremely flammable.

Cyclohexane is another useful carbocyclic compound; it is composed of 6 carbon atoms in a closed-chain structure.

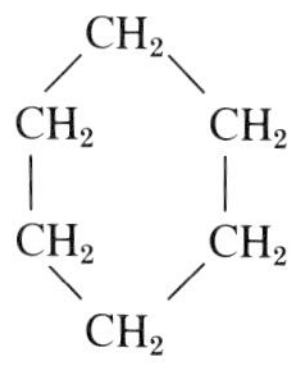

Cyclohexane

Cyclohexane's greatest use is as the starting material for making nylon.

A derivative of cyclohexane is sodium cyclamate.

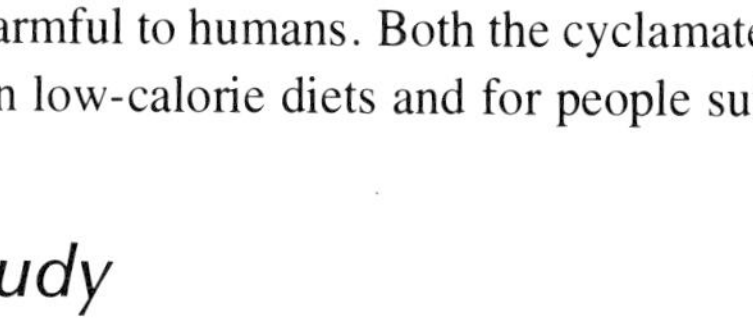

$$CH-NHSO_3Na$$

This chemical is about 30 times sweeter than sugar and is used as a noncarbohydrate sweetener. Recently the cyclamates were removed from low-calorie drinks and foods as they were believed to be harmful to humans. Both the cyclamates and saccharin have been used as sugar substitutes in low-calorie diets and for people suffering from diabetes mellitus.

Questions for study

1. What is an aromatic compound?
2. Name the subdivisions of organic compounds in the aromatic division.
3. What is an aromatic hydrocarbon? Give an example.
4. Describe the preparation of benzene. What is the importance of this preparation in organic chemistry?
5. State several properties of benzene.
6. What is a homologue in organic chemistry? Name two homologues of benzene.
7. What are the uses of toluene?
8. What is the structural difference in the three xylenes: ortho-, meta-, and para-xylene?

9. Name and give formulas for four aromatic halogen compounds. What are the uses of chlorobenzene?
10. Distinguish between the phenols and the aromatic alcohols.
11. What are the properties of phenol? What is the common name of phenol?
12. What is meant by the term "phenol coefficient"?
13. What is benzyl alcohol? What are its uses?
14. What is an aromatic aldehyde? Mention examples.
15. State several properties of benzaldehyde.
16. What are aromatic acids? Mention examples.
17. State several properties of benzoic acid.
18. What are the uses of benzoic acid in the food industry? In medicine?
19. What is salicylic acid? What is its importance?
20. What is methyl salicylate? What are its uses?
21. What is aspirin? What are its pharmacological effects?
22. What is an aromatic amine? State an example of a primary, secondary, and tertiary amine.
23. State several properties of aniline. What is its importance industrially?
24. What is acetanilide? What are its uses in medicine?
25. What are polynuclear hydrocarbons?
26. Mention several examples of polynuclear hydrocarbons.
27. What is the special significance of polynuclear hydrocarbons in physiological chemistry?
28. Mention several examples of heterocyclic compounds.
29. What is the importance of heterocyclic compounds in physiological chemistry?

14 Carbohydrates

Carbohydrates are organic compounds composed of carbon, hydrogen, and oxygen. The carbohydrate molecule contains either an aldehyde or a ketone group and also a number of alcoholic OH groups. Carbohydrates are therefore *aldehyde* or *ketone derivatives of polyhydroxy alcohols* (many OH groups). A carbohydrate containing an aldehyde group is called an aldose; one containing a ketone group is called a ketose.

Occurrence. Carbohydrates are an important class of organic compounds, of which the sugars, starch, cellulose, and glycogen are typical representatives. They are distinctly peculiar to plant life, being produced by and found chiefly in plant tissues. The woody or fibrous part of the plant is composed principally of cellulose, and the foods stored by plants in their roots, seeds, and fruits are to a large extent some form of carbohydrate. A small amount of carbohydrate is present in the tissues of animals.

Origin. Carbohydrates are produced in nature by plants. *Plants take up carbon dioxide from the air and water from the soil and, by means of the energy obtained from sunlight, combine these two compounds to form carbohydrates. This process is called photosynthesis. Chlorophyll, the green coloring matter in the leaves of plants, is necessary for photosynthesis; it makes the energy in the sun's rays available for this process.* The energy stored by plants in the form of carbohydrates thus originates from the sun.

Carbon cycle in nature. The ability of animals to form carbohydrates from amino acids or other sources is limited. Consequently animals depend to a considerable extent on plants for foods containing carbohydrates. On the other hand, plants use carbon dioxide, a waste product of animals, for the synthesis of carbohydrates and are therefore benefited by animals. Animal life is thus dependent on plant life for existence; plants in turn derive certain benefits from animals. This mutually beneficial cycle is shown by the following equations:

Process carried out by plants (photosynthesis)

$$6CO_2 + 6H_2O \xrightarrow{\text{Sunlight and chlorophyll}} C_6H_{12}O_6 + 6O_2 \tag{1}$$

From air — From soil — Simple sugar — Returned to air

Part played by animals

$$C_6H_{12}O_6 \quad + \quad 6O_2 \quad \rightarrow \quad 6CO_2 \quad + \quad 6H_2O \quad + \quad \text{Energy} \qquad (2)$$

| Sugar ingested and absorbed into tissues | Taken from air | Expired through lungs | Returned to soil and air |

Equation 1 shows that plants use carbon dioxide from the air and water from the soil to produce simple sugar, sunlight and chlorophyll being necessary to make this reaction take place. Plants also release oxygen into the air; this is the mechanism by which the oxygen content of the air is kept constant. Equation 2 indicates that the sugars (or carbohydrates) produced by plants are used by animals as food and that animals return carbon dioxide to the air and water to the air and the soil. In this connection it is to be noted that carbon dioxide and water are the same substances that were used by plants (as shown in equation 1) to form carbohydrates. These two equations therefore show that the relationship between plants and animals, with reference to carbon, hydrogen, and oxygen, is a closed cycle and that each of these two great classes of living things produces that which is necessary for the maintenance of the other. This relationship between plants and animals is known as the carbon cycle in nature.

Classification. Carbohydrates are classified on the basis of the number of saccharide (L. *saccharum,* sugar) groups in their molecules. The carbohydrate classification includes monosaccharides, disaccharides, trisaccharides, tetrasaccharides, and polysaccharides. Monosaccharides, or the simplest sugars, contain one saccharide group in their molecule (prefix *mono* means one). The monosaccharides of principal interest in this course are hexoses ($C_6H_{12}O_6$). Other carbohydrates classified under the monosaccharide group are

Table 14-1. Classification of carbohydrates

Class and examples	*Molecular formula*
I. Monosaccharides	
Pentoses	$C_5H_{10}O_5$
Ribose	
Arabinose	
Xylose	
Hexoses	$C_6H_{12}O_6$
Glucose	
Fructose	
Galactose	
II. Disaccharides	$C_{12}H_{22}O_{11}$
Sucrose	
Lactose	
Maltose	
III. Polysaccharides	$(C_6H_{10}O_5)_x \cdot (x-1)\,H_2O$
Starches	
Dextrins	
Inulin	
Cellulose	
Glycogen	
Agar	
Dextrans	
Pectins	

bioses, trioses, tetroses, pentoses, and heptoses, substances containing 2, 3, 4, 5, and 7 carbon atoms, respectively, in the molecule. The disaccharides, trisaccharides, tetrasaccharides, and polysaccharides contain two, three, four, and many saccharide units, respectively, in the molecule. The more important carbohydrates are included in Table 14-1.

Structure. The following structural formulas for glucose, fructose, and galactose will serve as examples of the structure of monosaccharides. Glucose and galatose are simple aldose sugars, and fructose is a simple ketose sugar.

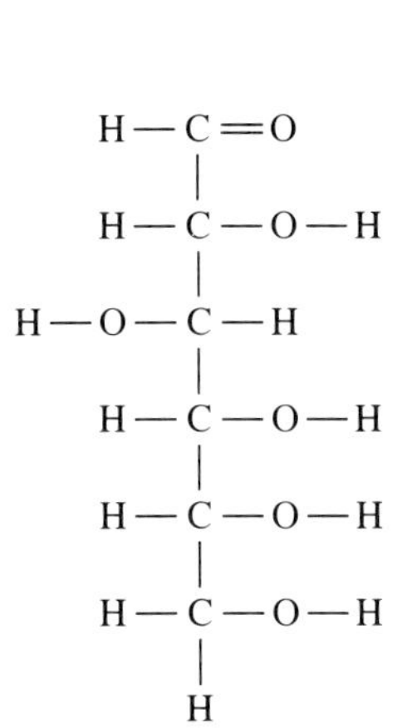

```
      H—C=O              H                     H—C=O
       |                 |                      |
      H—C—O—H          H—C—O—H                H—C—OH
       |                 |                      |
      H—O—C—H           C=O                   HO—C—H
       |                 |                      |
      H—C—O—H          H—O—C—H                HO—C—H
       |                 |                      |
      H—C—O—H          H—C—O—H                H—C—OH
       |                 |                      |
      H—C—O—H          H—C—O—H                H—C—OH
       |                 |                      |
       H                 H                      H
    Glucose           Fructose               Galactose
```

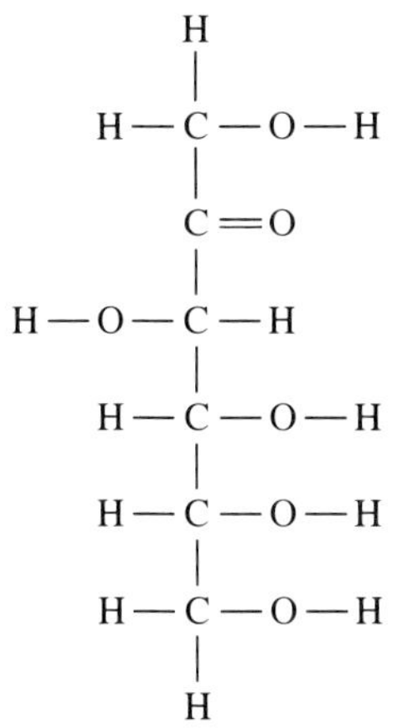

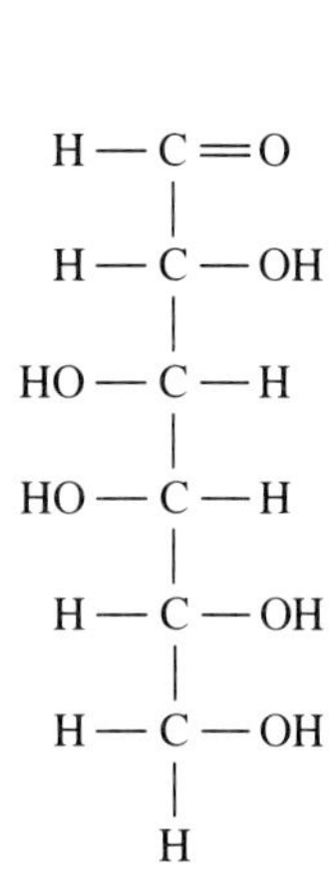

These formulas show the two representative types of structure (aldose and ketose) that are found in all carbohydrates. They illustrate the fact that carbohydrates are groups of hydrogen and oxygen atoms arranged around central carbon atoms, that there is either an aldehyde group or a ketone group in the molecule, and that hydroxyl groups are present in every case.

Actually, the preceding open-chain formulas for the three monosaccharides serve to illustrate the chemical nature rather than the true structure of these substances. The chemical nature rather than the true structure of these substances. The chemical behavior of these compounds in aqueous solution indicates that they exist principally in the form of closed-ring structures (heterocyclic) and that in solution only a very small amount of their molecules exists in the open-chain form. Taking glucose as an example, we find that the hexagonal closed-ring formulas indicate the true chemical structure of the compound.

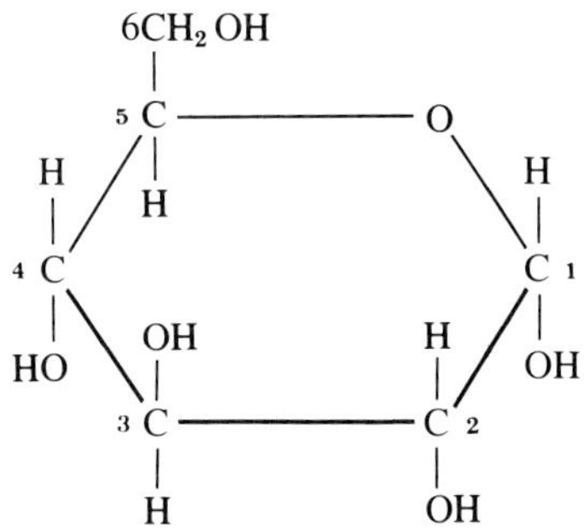

α-Glucose

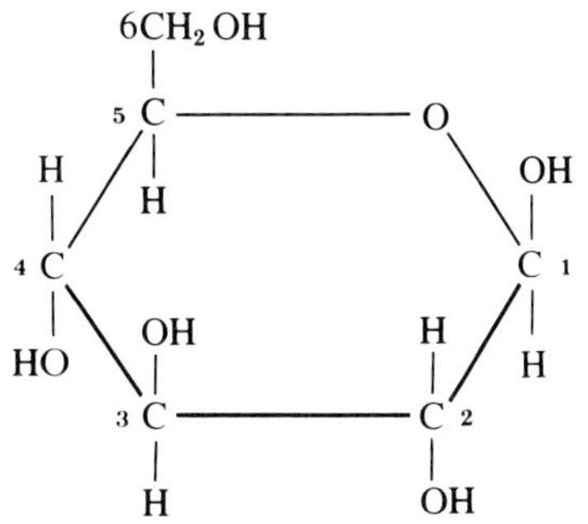

β-Glucose

In these formulas, 5 carbon atoms and 1 oxygen atom are in a closed ring. The ring must be considered as being in a plane perpendicular to the printed page, with the heavily shaded bonds closest to the reader. The H and OH atoms are on one or the other side of the plane as indicated, forming a three-dimensional structure. The arrangement of all the groups of atoms in α-glucose and β-glucose is the same except for the first carbon atoms, on which the H and OH atoms have opposite attachments.

A closed-ring structure such as that shown for glucose exists in all carbohydrates; the position of the ring in the molecule varies somewhat, but the closed ring persists.

Disaccharides are composed of 2 molecules of monosaccharides linked together. Sucrose is made up of 1 molecule of glucose and 1 molecule of fructose; maltose is composed of 2 molecules of glucose; and lactose contains 1 molecule of glucose and 1 of galactose.

The polysaccharides are composed of many molecules of monosaccharides. The number of units of monosaccharide in the molecules of various polysaccharides is not well known. The evidence indicates that this number ranges from 30 for inulin to 5000 or more for glycogen. This means that the molecular weight on inulin is approximately 5000 and that of glycogen is around 1 to 5 million.

Physical properties. The monosaccharides and disaccharides are crystalline solids. They are soluble in water. The polysaccharides are amorphous substances of varying solubilities. The dextrins and glycogen are soluble in cold water; starch, inulin, and agar are soluble in hot water; and cellulose is highly insoluble in water. The polysaccharides that can be dissolved form colloidal solutions.

Chemical properties. Chemical properties of special interest will be discussed.

Stability. The stability of carbohydrates is greatest in a neutral medium. Sugars (monosaccharides and disaccharides) decompose to some extent in acid and alkaline solutions. Polysaccharides (such as glycogen, starch, and cellulose) have a high degree of stability in alkalies.

Reducing substances (Fehling's and Benedict's tests). Carbohydrates with free aldehyde or ketone groups have the power to reduce alkaline metallic oxides (see p. 172), and this property is the basis of Fehling's, Benedict's, and other tests for sugars. Fehling's and Benedict's reagents are alkaline copper sulfate solutions. They are blue in color. When these reagents are heated with a few drops of a simple sugar solution, the blue color fades or disappears entirely, and a reddish brown precipitate of cuprous oxide is formed. This reaction is illustrated by the following equation:

$$2Cu(OH)_2 + \text{Reducing sugar} \xrightarrow{\text{Heat}} Cu_2O + 2H_2O + \text{Oxidized sugar}$$

Blue cupric hydroxide in solution + Reducing sugar → **Brown insoluble cuprous oxide** + $2H_2O$ + Oxidized sugar

This is the standard test for sugar glucose in urine. The appearance of reddish brown cuprous oxide when a few drops of urine are heated with Fehling's or Benedict's reagent is a positive indication of the presence of sugar, since sugar is the only reducing substance that occurs in appreciable amounts in the urine. All monosaccharides and all disaccharides except sucrose reduce Fehling's and Benedict's reagents.

Fermentation. Sugars undergo a chemical change known as fermentation in the presence of enzymes secreted by certain microorganisms. Fermentation is a somewhat general term and means, broadly speaking, a chemical decomposition with the evolution of gas. The products formed in the fermentation of carbohydrates vary according to the organisms that produce the reaction. The simple sugars undergo a fermentation when treated with yeast. The fermentation of sugars by yeast is known as *alcoholic fermentation,* since alcohol and carbon dioxide are the products formed. This reaction is brought about by the action of a number of specific enzymes that are present in yeast. It may be represented as follows:

$$C_6H_{12}O_6 \xrightarrow{\text{Enzymes}} 2C_2H_5OH + 2CO_2$$

Sugar   **Ethyl alcohol**   **Carbon dioxide**

Alcoholic fermentation is used commercially in making wines, beers, and other alcoholic beverages. Lactose is the only sugar that cannot be fermented, since yeast does not contain the proper enzyme.

Excessive fermentation of carbohydrates sometimes occurs in the alimentary tract through the action of enzymes secreted by bacteria, with the production of gases that are distressing. This may take place in the stomach when there is little or no free hydrochloric acid in the gastric juice. Ordinarily, the normal secretion of hydrochloric acid in the stomach acts as a disinfectant and destroys most of the bacteria that are ingested with food. In conditions in which there is little acid secreted, bacteria will multiply in the stomach and will bring about a fermentation of sugars, with gas production. Fermentation of carbohydrates also occurs in the intestines and is most marked when excess carbohydrate is present in the diet.

Hydrolysis. When boiled with dilute acids or treated with certain enzymes, the polysaccharides and the disaccharides are hydrolyzed to monosaccharides. This reaction is represented by the following equation, showing the hydrolysis of a disaccharide:

$$C_{12}H_{22}O_{11} + H_2O \rightarrow 2C_6H_{12}O_6$$

Disaccharide   **2 Monosaccharides**

Oxidation. When oxidized completely, carbohydrates yield carbon dioxide and water and liberate considerable energy. Example:

$$C_6H_{12}O_6 + 6O_2 \rightarrow 6CO_2 + 6H_2O + \text{Energy}$$

When burned either in the laboratory or in the human body, carbohydrates have an average heat value of 4 Calories per gram.

INDIVIDUAL OCCURRENCE, IMPORTANCE, AND USES OF PRINCIPAL CARBOHYDRATES
Monosaccharides

Glucose. α-Glucose is an important monosaccharide. It occurs free in the juices of fruits and in the sap of plants. Combined with itself or with other monosaccharides, it is a constituent of maltose, sucrose, lactose, starch, dextrins, glycogen, and cellulose. Glucose is the principal constituent of the commercial syrups such as corn syrup. The sugar in the blood and tissues of animals is glucose. Glucose is also called *dextrose*.

Fructose. Fructose is a monosaccharide. It is found free in the sap of many plants, in fruit juices, and in honey. It is also a constituent of sucrose and inulin. It is a very sweet sugar, being much sweeter than glucose. Fructose is also called *levulose*.

Galactose. Galactose is a monosaccharide. It is a constituent of lactose, agar, pectins, and of certain compounds in nerve tissue.

Disaccharides

Sucrose. Sucrose is a disaccharide composed of a molecule of α-glucose combined with a molecule of fructose. It occurs in the sugar cane, the sugar beet, the sugar maple tree, and in certain fruits. It is the ordinary table sugar of the daily diet and is commercially the most important disaccharide, being used in enormous quantities as a food. It is a very sweet sugar. Sucrose is also called *saccharose, cane sugar,* and *beet sugar*.

When sucrose in solution is boiled with dilute acid or treated under appropriate conditions with the enzyme invertase, it is split into its constituent monosaccharides, glucose and fructose. The product obtained in this way is called invert sugar. *Invert sugar* is an equimolar mixture of glucose and fructose prepared by the hydrolysis of sucrose. It is used extensively in candy making and in the baking industry.

Lactose. Lactose is a disaccharide consisting of a molecule of α-glucose combined with a molecule of galactose. It occurs in milk and is known as milk sugar. Lactose is synthesized in the mammary glands of mammals from the sugar of their blood, occurring in cow's milk to the extent of 4% to 5% and in human milk in amounts ranging from 6% to 8%. It sometimes occurs in the urine of women during late pregnancy or early lactation. It is not found in plants.

Maltose. Maltose is a disaccharide whose molecule consists of two α-glucose units. It is an intermediate product in the hydrolysis of starch and dextrins. It may be prepared from the dextrins by boiling them with dilute acids or by treating them with digestive enzymes such as ptyalin of the saliva. A more common name is malt sugar.

Polysaccharides

Starch. Starch is a polysaccharide consisting of α-glucose units. It is found in the seeds, tubers, fruits, and leaves of plants. It is deposited in plant tissues in granules of characteristic shape. It is a food substance of enormous importance and is the source of the dextrins, maltose, and commercial glucose preparations.

Dextrins. The dextrins are intermediate products of the hydrolysis of starch to maltose. In addition to the methods involving hydrolysis, dextrins may be prepared from starch by heating the starch at temperatures ranging from 170° to 240° C for varying lengths of time. They are colloidal substances, and their chief use is as mucilages for postage stamps, envelopes, and other purposes.

Inulin. Inulin is a polysaccharide consisting of fructose units. It occurs in the tubers of dahlias and in the Jerusalem artichoke. Inulin is not digested by the enzymes of the alimentary tract of man. It is used in clinical laboratory work as a test substance for measuring kidney function (the inulin clearance test).

Cellulose. Cellulose is a polysaccharide consisting of β-glucose units. It is the chief constituent of the walls of plant cells. It is a tough, resistant substance that gives plant tissues considerable rigidity. It thus serves an important function in the plant world as a structural or supporting substance. Wood, paper, cotton, flax, and hemp are composed principally of cellulose. Cellulose is not digested by the enzymes of the alimentary tract of man but is a desirable constituent of the diet because it gives bulk to the feces and thus tends to prevent constipation. Cellulose is a valuable material in preparing cellulose acetate, cellulose nitrate, rayon, and cellophane.

Glycogen. Glycogen is a polysaccharide consisting of glucose units. It is the form of carbohydrate that is stored in the tissues of animals. It is found in the greatest quantities in the liver and the muscles and is important as a reserve food supply in the animal body. A common name is *animal starch*.

Agar. Agar is a polysaccharide composed of galactose units. It is prepared from seaweed. It is used in medicine in the treatment of constipation since it is not digestible and will swell from the absorption of water, thus giving bulk to the feces. Agar is also used extensively in microbiology in the preparation of culture media for growing bacteria.

Dextrans. Dextrans are polysaccharides consisting of glucose units. They are synthesized from sucrose by the microorganism *Leuconostoc mesenteroides*. Dextrans differ from other polysaccharides of glucose, such as starch, dextrins, and glycogen, in the way the glucose molecules are linked to each other. They differ from each other also in the size of the molecule. Under certain conditions the bacteria synthesize dextrans with molecular weights of probably several million, and under other conditions smaller molecules are produced.

Dextrans are used in medicine as plasma volume expanders in the treatment of shock. Clinical shock is characterized by a reduced circulating blood volume. When dextran solution is injected intravenously, the dextran remains in the bloodstream for several days, being slowly excreted in the urine. While in the vascular system, dextran increases the osmotic pressure of the blood. This causes water to pass into the blood vessels from the extracellular spaces. The increase in blood volume raises the blood pressure and thus relieves the patient suffering from shock.

The dextran molecules synthesized by *Leuconostoc mesenteroides* are too large for clinical use. To make the bacterially synthesized dextrans suitable for clinical use, the product is submitted to mild, controlled hydrolysis, which shears the molecules down to a smaller size. Dextran suitable for clinical use has an average molecular weight of 70,000.

Pectin. Pectin is a term used to designate a substance or group of substances, largely carbohydrate in composition, that participate in the process involved in the preparation of fruit jellies. In the presence of fruit juices and adequate amounts of sugar, pectin serves as the agent that causes the jelly to set or assume the colloidal state of a gel.

The chemical nature of the pectins is only partly known. They are mixtures of polysaccharides and calcium-magnesium salts of a complex acid called *pectic acid*. Pectins occur in fruits, such as apples, lemons, oranges, and grapefruit, and in certain plants. Pectins are prepared for commercial use from apple and lemon residues.

• • •

In general, the carbohydrates are an important class of food substances. Of those mentioned in the foregoing discussion, all are digestible and assimilable in the human body except inulin, agar, cellulose, and dextran. Whether pectin is assimilable is not known. Cellulose can be utilized by ruminant animals (for example, the cow) by the action of bacteria that convert the cellulose to glucose. Termites also can utilize cellulose, but here again a microorganism living in the gut of the termite degrades the cellulose to glucose.

Questions for study

1. Discuss the occurrence of carbohydrates.
2. Give a definition of carbohydrates.
3. What is the origin of carbohydrates? What is photosynthesis?
4. Describe the carbon cycle in nature. What is its importance?
5. How are the amounts of oxygen and carbon dioxide in the air kept at practically constant values?
6. Discuss the true chemical structure of glucose.
7. What is a monosaccharide? Name six important monosaccharides.
8. What is a disaccharide? Name three important disaccharides.
9. What is a polysaccharide? Name eight important polysaccharides.
10. Mention several properties of carbohydrates.
11. Explain Benedict's test for sugar.
12. What is fermentation? Explain how fermentation occurs in the alimentary tract.
13. What are formed when disaccharides and polysaccharides are hydrolyzed?
14. What are the products of complete oxidation of carbohydrates? How much energy is liberated by the oxidation of carbohydrates?
15. Name the principal carbohydrates.
16. State the occurrence and importance of the following: glucose, fructose, galactose, sucrose, lactose, starch, inulin, cellulose, glycogen, agar, pectin, and dextran.
17. What carbohydrates are not digested and assimilated in the human body?

15 Proteins

The name protein comes from a Greek word meaning "to be first." Proteins are "first" in the sense that they are fundamental constituents of living matter. Every living cell contains proteins and requires a continuous supply of protein food for its existence. Proteins are therefore constituents of first importance to all living things.

Occurrence. Proteins occur in living matter or are associated with living things. They constitute a large part of the solid matter of the tissues of animals. Thus most of the solid matter of muscle, tendon, ligament, cartilage, and blood is composed of proteins, and about one-half of the solid substance of brain and nerve tissues and of bone is protein in composition. In plants, proteins are subordinate to carbohydrates as structural materials, cellulose being the chief structural substance of the plant. However, proteins are essential constituents of all plant tissues, and they occur in large quantities as storage substances in the seeds and fruits of plants. Examples of plant products having a high protein content are nuts, wheat, corn, oats, beans, and peas.

Origin. Proteins always are produced in nature by living matter and chiefly by plants. Animals may eat proteins and transform them into other types of protein, but animals have limited powers to synthesize proteins. As will be shown later, a considerable number of the building units of proteins cannot be synthesized by animals, and animals must obtain these building units from plant sources. Plants are therefore the primary protein producers of nature. In the synthesis of proteins, plants form carbohydrates first and then combine simple carbohydrate units with ammonia or other nitrogen compounds obtained from the soil to produce the various types of proteins.

Chemical nature. Proteins are organic compounds containing carbon, hydrogen, oxygen, and nitrogen. Sulfur and iodine are also elementary constituents of some proteins, whereas other proteins contain phosphorus and iron.

The protein molecule is exceedingly large. This is indicated by the formula for oxyhemoglobin $(C_{758}H_{1181}N_{207}S_2FeO_{210})_x$.

The molecular weight of oxyhemoglobin is around 68,000. The molecular weights of proteins in general range from 10,000 to 322 million. The student will get the significance of these figures by comparing them with the molecular weight of cane sugar, which is 342; of glucose, which is 180; or of sodium chloride, which is 58.5.

210 *Roe's principles of chemistry*

The "building stones" or structural units of which proteins are composed are amino acids. There are about 23 different amino acids known to exist in the various proteins. Ten of these amino acids are *essential* amino acids; that is, they are needed for proper growth, and since they cannot be synthesized by animals from chemicals in the diet, they are essential components of a proper diet. The simplest of these amino acids is glycine, or amnioacetic acid, which has the following formula:

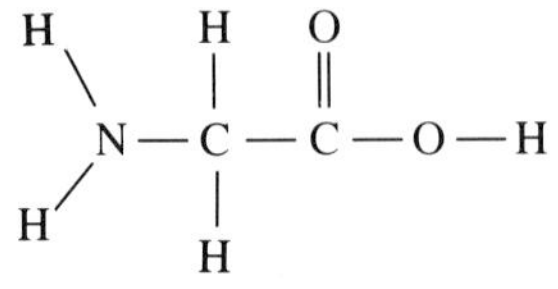

$$\underset{H}{\overset{H}{\diagdown}}N - \underset{H}{\overset{H}{\underset{|}{C}}} - \underset{}{\overset{O}{\underset{}{\overset{\|}{C}}}} - O - H$$

Amino group   Carboxyl group
Glycine
(aminoacetic acid)

Glycine is typical of all amino acids. All amino acids contain amino and carboxyl groups in their molecules. Others differ from glycine principally in that they contain more atoms in their molecules, in that they have a different arrangement of their atoms, and in that three of them contain sulfur and two contain iodine. They are soluble in water, and some of them have a sweet taste.

The formulas of ten amino acids follows:

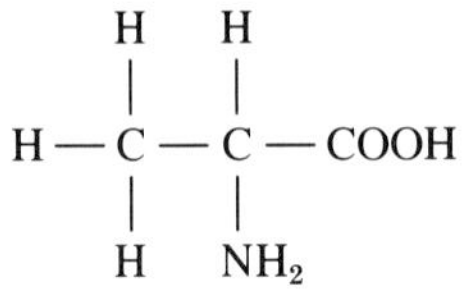

Alanine

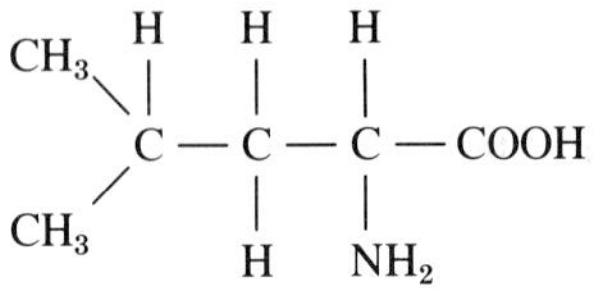

Leucine

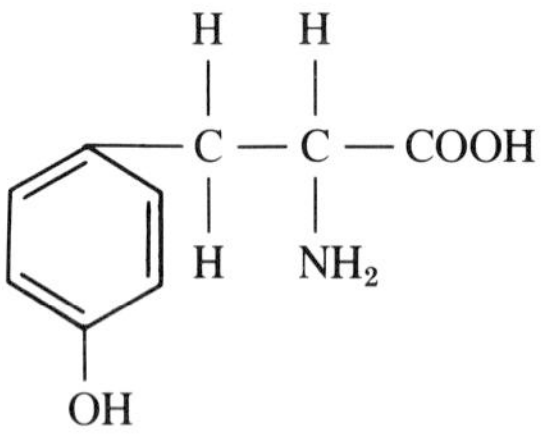

Tyrosine

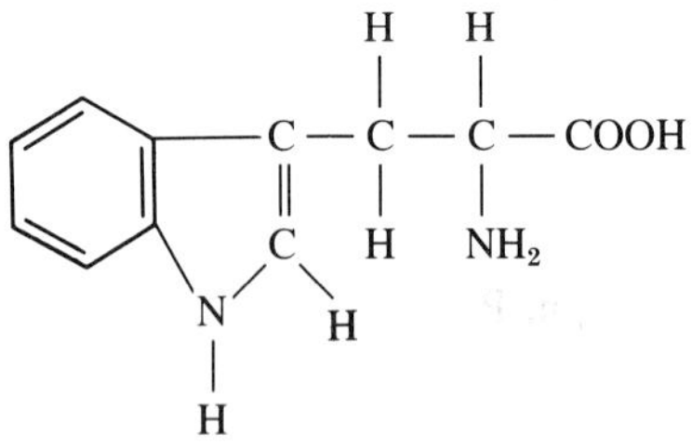

Tryptophan

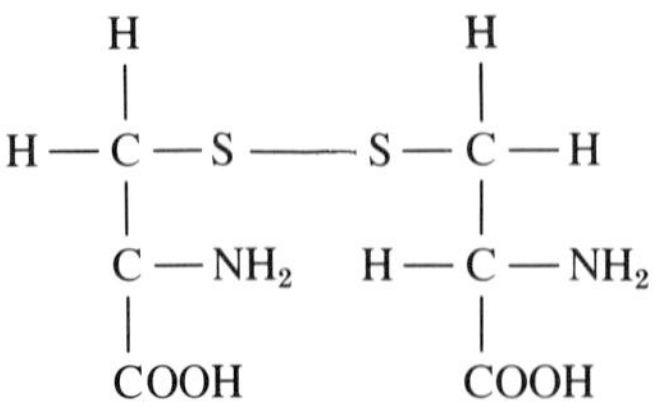

Cystine

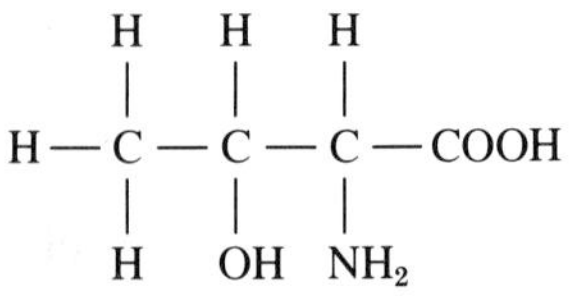

Threonine

```
    H   H  H  H  H                              H  H  H
    |   |  |  |  |                              |  |  |
H — C — C — C — C — C — COOH          HOOC — C — C — C — COOH
   /       |  |  |  |                          |  |  |
  NH₂      H  H  H  NH₂                        H  H  NH₂
         Lysine                              Glutamic
                                              acid


 H        H  H                    H  H  H  H  H
 |        |  |                    |  |  |  |  |
 C = C — C — C — COOH       H   N—C —C —C —C — COOH
 |   |    |  |              |   |  |  |  |  |
 |   |    H  NH₂            N = C  H  H  H  NH₂
 N   N—H                         |
  \ /                            NH₂
  C—H
     Histidine                       Arginine
```

With this preliminary description of amino acids, we shall proceed to define proteins. *Proteins are compounds composed of carbon, hydrogen, oxygen, nitrogen, usually sulfur, and in some instances iodine, phosphorus, and iron. Structurally their molecules consist of amino acids joined together chiefly by union of amino with carboxyl groups.* The differences in the many proteins found in nature are to be ascribed to a variation in the number, kind, and arrangement of the amino acids in their molecules.

Proteins are thus seen to be very complex substances, and we cannot hope to go very deeply into a study of their chemistry in a course of this kind. The preceding discussion, however, should impress the student with the facts that *the protein molecule is very large* and that *the fundamental units of structure of which it is composed are amino acids,* just as monosaccharides are the building stones or structural units of which starch, glycogen, and other large-molecule carbohydrates are constructed.

Structure—the peptide linkage. Proteins are large-molecule compounds composed of many amino acids. *The amino acids are linked together in long chains through a union between amino and carboxyl groups by elimination of water. This union is known as the peptide linkage.* The grouping below illustrates the peptide linkage.

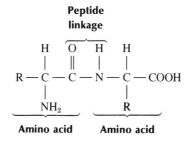

The formation of peptides is shown in Fig. 15-1. The peptide linkage is not formed as simply as indicated by direct removal of water, but indirectly by chemical reactions that are well known but need not be discussed here. As shown in Fig. 15-1, when alanine

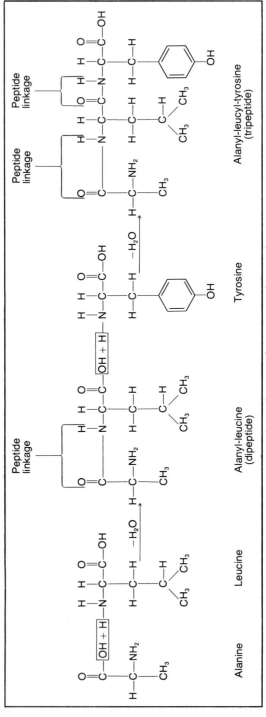

Fig. 15-1. Chart showing formation of peptides.

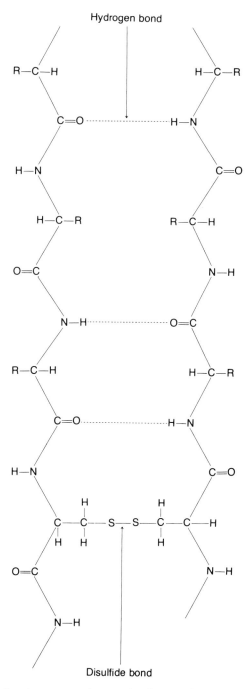

Fig. 15-2. Diagram showing two polypeptide chains of amino acids as they appear to exist in the partial protein molecule. R represents all of the amino acid except the part associated with the peptide linkage. C, N, and O atoms are in the same plane. R groups project into space to one or the other side of the plane, making a three-dimensional structure. As cross-linkages between the two chains, three hydrogen bonds and one disulfide bond are shown.

and leucine are united by a linkage between the carboxyl group of alanine and the amino group of leucine, a dipeptide, alanyl-leucine, is formed. When alanyl-leucine is united through a peptide linkage with tyrosine, the tripeptide alanyl-leucyl-tyrosine is formed. Similarly, by stepwise synthesis, amino acids may be united to form long polypeptide chains.

Polypeptides synthesized in the laboratory have been found to be similar to the peptide chains of amino acids in naturally occurring proteins. Such work, together with other evidence, has led to the conclusion that *proteins consist of long polypeptide chains of amino acids in which the chemical union between the amino acids is the peptide linkage. The polypeptide chains are held together in the molecule by cross-linkages.* Of the cross-linkages, *the disulfide (—S—S—) linkage,* shown in cystine on p. 210, is one form of bonding that holds the polypeptide chains together. Another important cross-linkage between polypeptide chains is the *hydrogen bond,* an attractive force between the hydrogen atom attached to the nitrogen of the peptide linkage (—NH) and the oxygen in the carbonyl (—CO—) group of a neighboring peptide linkage (Fig. 15-2). Other cross-linkages exist, but these will not be discussed.

An interesting example for discussion of protein structure is insulin, a hormone that has a regulatory effect on the blood sugar level. The insulin molecule consists of two polypeptide chains. In one of these chains there are 21 amino acids united by the peptide linkage, and in the other chain there are 30 amino acids combined with each other through peptide linkages. The two polypeptide chains are united with each other principally by disulfide (—S—S—) cross-linkages and are held in an arrangement in which they are parallel to each other. The parallel chains are folded in a way that produces a globular molecule. The biological effect of insulin is lost when this compound is treated by a chemical procedure that tears apart the 2 sulfur atoms in the disulfide linkage.

Shape of the protein molecule. With respect to the shape of the protein molecule, there are, in general, two types—globular and fibrous. In the globular protein molecule the polypeptide chains are folded in such a way as to form a more or less spherical structure. In the fibrous proteins long polypeptide chains are held parallel to each other, forming an elongated ellipsoidal structure. Certain treatments such as heating or placing in acid or alkali will cause the polypeptide chains of a globular protein to unfold and assume a shape more like that of a fibrous protein. Such a change is known as a *denaturation.* It has been shown by x-ray analysis that the polypeptide chains in certain fibrous proteins can be stretched to greater lengths by physical methods. The keratin in hair, for example, with suitable reagents and conditions, may be stretched, producing a "permanent wave" in the strands of hair.

Examples of globular proteins are insulin, egg albumin, lactoglobulin, poliovirus (Fig. 15-3), and hemoglobin. Representatives of the fibrous type of protein are fibrinogen of blood plasma, hair, wool, silk fibroin, tobacco maosaic virus (Fig. 15-4), and myosin of muscle fibers.

Behavior in solution—the isoelectric point. Because of their large molecular size, protein molecules are in the colloidal range. Proteins are electrolytes; their molecules contain carboxyl (—COOH) and amino (—NH_2) groups, which in solution undergo

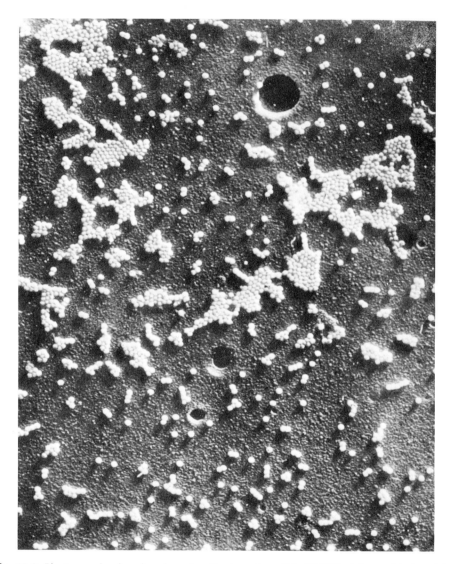

Fig. 15-3. Photograph of molecules of poliovirus, type 2 (×79,000), taken with the aid of the electron microscope. This is an example of a protein with globular molecules. (Courtesy Parke, Davis & Co. Virus Laboratory.)

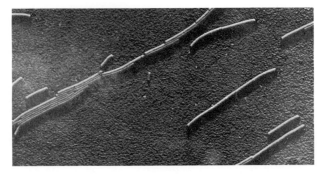

Fig. 15-4. Electron micrograph of rods of tobacco mosaic virus. The magnification (×76,000) is high enough to bring out the shapes of the individual particles. This is an example of a protein with fibrous molecules. The individual cylindrical molecules have a diameter of about 150 angstrom units and a length of about 2700 angstrom units. (Courtesy Wyckoff: Electron microscopy: technique and application, Interscience Publishers, Inc.)

ionization. Proteins therefore exist in solution as ions. These ions bear either a net positive (+) or a net negative (−) charge, depending on the pH of the solution, except at one sharply defined pH, called the isoelectric point (see p. 129). At the isoelectric point the protein ion contains an equal number of positive and negative charges. At this pH the protein ion is a dipolar ion, but it is electrically neutral because the positive and negative charges on the ion exactly balance each other. Protein ions at their isoelectric point will not move in an electric field to the positive or the negative pole because their net charge is 0, and they behave as if they had no charge at all. These ions are called *zwitterions*.

At pH values on the acid side of the isoelectric point, protein ions have a positive charge and therefore will migrate to the negative pole of an electric field. At pH values on the alkaline side of the isoelectric point, protein ions have a negative charge and will therefore pass to the positive pole.

To explain the behavior of proteins, a protein molecule will be designated as follows:

$$P\begin{matrix} \diagup COOH \\ \diagdown NH_2 \end{matrix}$$

A protein molecule thus consists of an organic skeleton designated by P and amino and carboxyl groups.

When a protein is placed in a solution with a pH that is the isoelectric point of the protein, the protein molecule is represented as follows:

$$P\begin{matrix} \diagup COO^- \\ \diagdown NH_3^+ \end{matrix}$$

At its isoelectric point the amino and carboxyl groups of the protein molecule are ionized, but the number of amino groups that are ionized is equal to the number of carboxyl groups undergoing ionization, and these groups therefore balance each other electrically. The net charge of the protein molecule at its isoelectric point is therefore 0.

If a little acid (H^+ ions) is added to a solution containing a protein at its isoelectric point, the following occurs:

$$P\begin{matrix}COO^-\\ NH_3^+\end{matrix} + H^+ \rightarrow P\begin{matrix}COOH\\ NH_3^+\end{matrix}$$

The addition of H^+ ions to an isoelectric protein molecule will suppress the ionization of the carboxyl groups and leave the resulting ion with a net positive (+) charge. This is the result at all pH values that are acid to the isoelectric point of the protein.

If, on the other hand, a little base (OH^- ions) is added to a solution containing protein at its isoelectric point, the following occurs:

$$P\begin{matrix}COO^-\\ NH_3^+\end{matrix} + OH^- \rightarrow P\begin{matrix}COO^-\\ NH_2\end{matrix} + H_2O$$

The addition of OH^- ions to isoelectric protein removes H^+ ions from the ionized amino (NH_3^+) groups and neutralizes them by forming water. The resulting protein ions are left with a net negative (−) charge. This is the condition that exists at all pH values on the alkaline side of the isoelectric point.

At the isoelectric point the protein ion shows the least activity chemically. The osmotic pressure, viscosity, and solubility of a protein also are at their minimum values at the isoelectric point.

To explain further, specific examples will be discussed. In solutions with pH values below 4.8, the protein, serum albumin, forms ions that migrate to the negative (−) pole in an electrophoresis apparatus, and therefore they bear positive charges. In solutions with a pH above 4.8, serum albumin forms ions that move to the positive (+) pole of the electrophoresis apparatus, and therefore they bear negative charges.

This knowledge has important applications in working with proteins. To bring about a chemical reaction with a protein, the protein must be placed at the right pH. Since ionic reactions occur only between oppositely charged ions, the protein must be on the side of its isoelectric point that will give it a charge opposite to the substance with which it is to react. For example, in tanning hides to make leather, tannic acid is used because it forms a tough, insoluble precipitate (leather) with the proteins of the skin. The tannate ion is negative and will react only with positive ions. In the tanning process the hides are therefore placed at a pH below the isoelectric point of the proteins of the skin to bring about a reaction with the tannic acid. This behavior may be illustrated by the following equations:

Below the isoelectric point
$$Protein^+ (ion) + Tannate^- (ion) \rightarrow Protein\ tannate$$

Above the isoelectric point
$$\text{Protein}^- \text{ (ion)} + \text{Tannate}^- \text{ (ion)} \rightarrow \text{No reaction}$$

Further examples will be helpful. Let us take, for example, the removal of proteins from blood in carrying out a chemical analysis of the blood for its sugar content. In this analysis the proteins are precipitated by tungstic acid. The compound formed by tungstate ions with protein ions, protein tungstate, is insoluble and precipitates from solution. The protein ion must bear a positive (+) charge, however, to react with the tungstate ion, which has a negative (−) charge. Hence in the blood analysis, sulfuric acid is added to place the blood proteins in a solution with a pH below their isoelectric point, where they assume a positive (+) charge and can react with tungstate (−) ions. This is illustrated by the following equations:

Below the isoelectric point
$$\text{Protein}^+ \text{ (ion)} + \text{Tungstate}^- \text{ (ion)} \rightarrow \text{Protein tungstate}$$

Above the isoelectric point
$$\text{Protein}^- \text{ (ion)} + \text{Tungstate}^- \text{ (ion)} \rightarrow \text{No reaction}$$

For another example, let us take the reaction of the ions of a metal such as silver with proteins. The ions of silver in a silver salt solution bear a positive (+) charge. Such ions can react with proteins only when the protein ions are in a solution with a pH above their isoelectric point in which they (the proteins) bear a negative (−) charge. The proteins of the body tissues are in a solution with a pH above their isoelectric point; hence their ions bear negative (−) charges. The silver ion will therefore react with the proteins of the body, forming a silver protein compound. The following equations illustrate this behavior.

Above the isoelectric point
$$\text{Ag}^+ \text{ (ion)} + \text{Protein}^- \text{ (ion)} \rightarrow \text{Ag proteinate}$$

Below the isoelectric point
$$\text{Ag}^+ \text{ (ion)} + \text{Protein}^+ \text{ (ion)} \rightarrow \text{No reaction}$$

Properties. The properties of proteins that are of special interest will be discussed.

Solubility. Proteins vary widely in their solubilities. Some are readily soluble in water as, for example, the albumins of the blood; others are soluble in dilute salt solutions, in dilute acids and alkalies, or in 70% alcohol; and some are highly insoluble in water and other solvents as, for example, the albuminoids of the skin, hair, and nails.

Large molecules. As already stated, proteins are very large molecules. They will form colloidal solutions when dissolved in water.

Reaction. Proteins are amphoteric; that is, they have both acid and basic characteristics. A protein will react with either an acid or a base, forming a protein salt. This behavior is due to the fact that the protein molecule contains free amino (NH_2) and carboxyl (COOH) groups.

Precipitation by heavy metals. Proteins are precipitated from solution by the ions of some of the heavy metals such as mercury, lead, and silver. This is the reaction that takes place when mercuric chloride is swallowed. The mercury ions of this compound combine with the proteins of the mouth, esophagus, and stomach and precipitate them from solution. In this manner a local destruction of tissue is produced. If the mercury is left in the alimentary tract, it is slowly absorbed into the blood and carried to the various internal organs and finally to the kidneys. It has a destructive action on all the tissues of the body, since all tissues contain proteins. Its most serious effect on the body is its action on the kidneys because the inflammation it produces in these organs interferes so extensively with their excretory function.

Antidote for mercury poisoning. The use of milk, eggs, and cheese as an antidote is based on the fact that these substances will react with mercury and precipitate it from solution; in this manner they substitute themselves for the proteins of the walls of the stomach and thus protect the stomach from the destructive action of this poison. A stomach tube is always introduced, and the contents of the stomach are removed as soon as possible in case of mercury poisoning even though an antidote has been used. It is important to remove the proteins that have served as antidotes; otherwise they will be digested and release the mercury, which will then be absorbed and attack the body tissues.

This reaction also explains why mercuric chloride and silver nitrate are good disinfectants. These substances combine with the proteins of bacteria, forming insoluble precipitates, and in this manner they kill bacteria.

Coagulation. Many proteins are coagulated by heat and by alcohol; that is, they are converted into an insoluble state by these agents. Coagulation is complicated chemically, and for our purposes it will be considered to mean the insoluble state of a protein that is produced by heating or by alcohol. The white of an egg is coagulated on heating.

Hydrolysis. When boiled with dilute acids or alkalies or treated with digestive enzymes, proteins are hydrolyzed to amino acids. The digestion of proteins in the human body is a series of hydrolyses in which enzymes play an important part.

Oxidation. When proteins are treated with vigorous oxidizing reagents, the carbon of their molecules is oxidized to carbon dioxide, the hydrogen is oxidized to water, and the nitrogen is oxidized to nitrates. The same end products are formed in the oxidation of proteins or amino acids in the body except those obtained from nitrogen. Oxidation in the body is too mild to oxidize nitrogen; consequently this element retains its reduced form during physiological oxidations and appears in the body excretions principally as urea or ammonium salts. The oxidation of proteins is attended by the liberation of considerable energy.

Color tests. When treated with appropriate reagents, proteins yield certain colors that are the basis of laboratory tests for these substances. The principal protein color tests used to identify the proteins are the biuret, Millon, xanthoproteic, and Hopkins-Cole tests.

BIURET TEST. When a protein solution is treated with dilute alkali and very dilute copper sulfate solution, a pink to violet solution is produced.

MILLON'S TEST. Millon's reagent is a mixture of mercuric nitrate and mercuric nitrite dissolved in nitric acid. When this reagent is mixed with a protein, a white precipitate is first produced, which turns pink on being boiled gently.

XANTHOPROTEIC TEST. When treated with nitric acid, proteins turn yellow. This color changes to a deep brown when a little alkali is added. A xanthoproteic test on the

Table 15-1. Classification of proteins

Division	Subdivision	Chief characteristics (basis of classification)	Examples
Simple proteins	Albumins	Soluble in water	Egg albumin, serum albumin
	Globulins	Insoluble in water; soluble in dilute salt solutions	Serum globulin, edestin of hemp seed
	Glutelins	Insoluble in water or salt solutions; soluble in dilute acids and alkalies	Glutenin of wheat
	Prolamins	Insoluble in water; soluble in 70% alcohol	Gliadin and zein of grains
	Albuminoids	Insoluble in water, salt solutions, dilute acids, or alkalies; hence highly insoluble	Keratin of nails and hair
	Histones	Soluble in water and dilute acids; strongly basic and insoluble in ammonia	Globin of hemoglobin
	Protamines	Strongly basic protein soluble in water; relatively small molecule	Sturin, salmin, clupein of fish's sperm
Conjugated proteins	Nucleoproteins	Compounds containing protein combined with nucleic acid	Nuclein of nucleus of cells
	Glycoproteins	Compounds containing protein combined with a carbohydrate group	Mucin of saliva
	Phosphoproteins	Compounds containing protein combined with a phosphate group	Casein of milk
	Chromoproteins	Compounds containing protein combined with a chromogenic (color-producing) nucleus	Hemoglobin of blood
	Lecithoproteins	Compounds containing protein combined with lecithin	Fibrinogen of tissues
Derived proteins	Proteans	First hydrolytic products of action of water, acids, or enzymes	Edestan
	Metaproteins	Further hydrolytic products of action of acids or alkalies	Acid metaprotein, alkali metaprotein
	Coagulated proteins	Insoluble products of action of heat or alcohol	
	Proteoses	Products of hydrolysis of proteins; soluble in water; precipitated by saturated ammonium sulfate solution; very slightly diffusible	
	Peptones	Products of further hydrolysis of proteins; soluble in water; not precipitated by saturated ammonium sulfate; readily diffusible	
	Peptides	Products of further hydrolysis of proteins; soluble in water; smaller molecule than peptones; consist of two or more amino acids linked together; more diffusible than peptones	

proteins of the skin is often performed by the student in the laboratory when nitric acid is spilled on the hands.

HOPKINS-COLE TEST. When glyoxylic acid (Hopkins-Cole reagent) and concentrated sulfuric acid are added to a solution of a protein, a pink solution is produced.

Classification. There are three major classes of proteins—simple proteins, conjugated proteins, and derived proteins. Each of these major groups contains several subdivisions. A classification of proteins is given in Table 15-1.

Functions. Proteins function as tissue builders and energy producers.

Tissue builders. Proteins are the great tissue builders of the human and animal bodies. The body selects from the proteins eaten amino acids that are appropriate for tissue construction and builds them into proteins characteristic of its own tissues. In the adult, proteins are used chiefly to renew the tissues that are constantly being broken down by the wear-and-tear processes of the body; in the young person, proteins contribute to growth as well as to the repair of the body.

Energy producers. Proteins are also an important source of energy. One gram of protein yields 4 Calories of energy when oxidized in the human body.

Questions for study

1. What is the importance of proteins?
2. Discuss fully the occurrence of proteins.
3. What is the origin of proteins in nature?
4. What elements are found in proteins? Give a definition of proteins.
5. Discuss the size of the protein molecule. How does the molecular weight of proteins compare with that of other substances?
6. What is an amino acid? Approximately how many amino acids have been found in proteins?
7. What is a polypeptide?
8. How are amino acids held together in a polypeptide chain?
9. How are the polypeptide chains held together in the protein molecule?
10. Discuss the shape of protein molecules.
11. Do proteins undergo ionization in solution?
12. At what pH values do protein ions bear a positive (+) charge? A negative (−) charge? At what pH do protein ions bear both a positive and a negative charge?
13. What is the relation of the charge on the ions of a protein to its ability to react chemically with the ions of other substances?
14. State several properties of proteins.
15. Explain why proteins are amphoteric.
16. Describe the action of mercuric chloride in a person poisoned by this substance. What are antidotes for mercuric chloride poisoning? Explain the action of antidotes for mercury poisoning.
17. How does mercuric chloride kill bacteria?
18. What are the final products of the hydrolysis of proteins?
19. How would you test for proteins in the laboratory?
20. Explain the yellow color produced on the skin when nitric acid is spilled on it.
21. Name the three principal classes of proteins.
22. What are the functions of proteins?
23. What is the caloric value of proteins?

16 Lipids

When animal or plant tissues are extracted with organic solvents such as ether, chloroform, benzene, and hot alcohol, the so-called fat solvents, and when the solvent is evaporated off, certain substances having a greasy or fatlike appearance are obtained. These substances are called lipids. The term lipid comes from a Greek word *lipos,* which means fat. Fats make up an important class in this group of compounds, but the lipids include other substances of great interest. The basis for placing these compounds in a general group is their common property of being soluble in the same solvents. They do not have much similarity in chemical nature except that nearly all lipids are esters that yield fatty acids on hydrolysis.

Properties. Lipids have the following characteristics:
1. *Lipids are soluble in ether, chloroform, benzene, and hot alcohol.*
2. *Lipids are insoluble in water.*
3. *Almost all lipids are esters that yield fatty acids on hydrolysis.*

Classification. Following is a classification of lipids:

The classes of lipids that will be discussed are fats, sterols, phopholipids, glycolipids, and waxes.

    I. *Simple lipids*—Esters of fatty acids and various alcohols.
       1. *Neutral fats:* Esters of fatty acids and the trihydroxyl alcohol, glycerol. Examples: stearin, palmitin, olein, and butyrin.
       2. *Waxes:* Esters of fatty acids and aliphatic alcohols other than glycerol. Examples: beeswax, cholesterol, esters, and esters of vitamins A and D.
   II. *Compound lipids*—Esters of fatty acids containing some characteristic group in addition to the alcohol and the fatty acids.
       1. *Phospholipids:* Esters of fatty acids containing a phosphate group and a nitrogen base. These compounds are also called phosphatides. Examples: lecithin, cephalin, and sphingomyelin.
       2. *Glycolipids or cerebrosides:* Esters of fatty acids containing a carbohydrate group and nitrogen. Examples: phrenosin and kerasin.
       3. *Sulfolipids:* Esters of fatty acids containing a sulfate group.
  III. *Derived lipids*—Derivatives obtained by hydrolysis of the substances mentioned in groups I and II.
       1. Fatty acids, saturated and unsaturated.
       2. Glycerol, monoglycerides, and diglycerides.
       3. Sterols and alcohols other than glycerol.

4. Organic bases: choline, ethanolamine, and sphingosine.
5. Miscellaneous: vitamins A, D, E, and K.

FATS

Fats are organic compounds composed of carbon, hydrogen, and oxygen. *They are esters of glycerol and fatty acids; that is, they are compounds produced by the reaction of glycerol with fatty acids.* Glycerol, also called glycerin, is an alcohol containing three OH groups. The fatty acids are organic acids containing one COOH group; they usually differ from each other by 2 or multiples of 2 carbon atoms; that is, except for rare exceptions, they all have an even number of carbon atoms. The following graphic formula for tributyrin, the glyceryl ester of butyric acid, is typical of all fats.

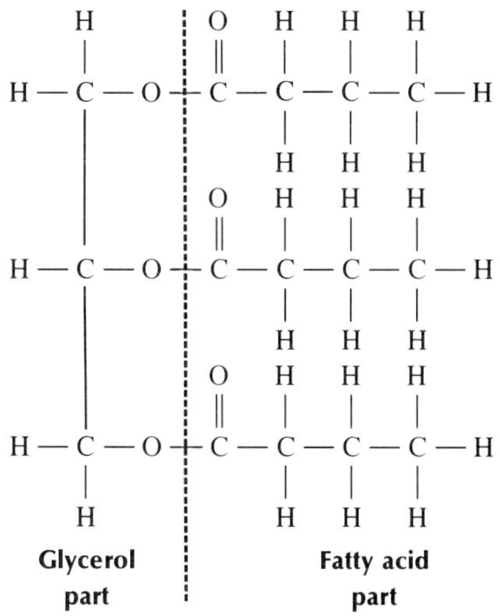

$$
\begin{array}{c}
\text{H} \\
| \\
\text{H}-\text{C}-\text{O} \\
| \\
\text{H}-\text{C}-\text{O} \\
| \\
\text{H}-\text{C}-\text{O} \\
| \\
\text{H}
\end{array}
\Bigg|
\begin{array}{c}
\text{O} \quad \text{H} \quad \text{H} \quad \text{H} \\
\| \quad | \quad | \quad | \\
\text{C}-\text{C}-\text{C}-\text{C}-\text{H} \\
| \quad | \quad | \\
\text{H} \quad \text{H} \quad \text{H} \\
\text{O} \quad \text{H} \quad \text{H} \quad \text{H} \\
\| \quad | \quad | \quad | \\
\text{C}-\text{C}-\text{C}-\text{C}-\text{H} \\
| \quad | \quad | \\
\text{H} \quad \text{H} \quad \text{H} \\
\text{O} \quad \text{H} \quad \text{H} \quad \text{H} \\
\| \quad | \quad | \quad | \\
\text{C}-\text{C}-\text{C}-\text{C}-\text{H} \\
| \quad | \quad | \\
\text{H} \quad \text{H} \quad \text{H}
\end{array}
$$

Glycerol part | **Fatty acid part**

Tributyrin

The glycerol radical is a constituent of all fats. The differences in the many fats found in nature are due to a variation in the fatty acids that enter into their composition. There are two types of fatty acids, saturated and unsaturated. Saturated fatty acids have no double bonds in their chains of carbon atoms; unsaturated fatty acids have one or more than one double bond in their carbon chains. There are about 30 fatty acids entering into the composition of the fats found in nature. Those listed in Table 16-1 are the most important.

Fats are named after the fatty acid in their molecule. For example, a fat containing butyric acid in its molecule is called tributyrin; one containing stearic acid is called tristearin; and one containing oleic acid is called triolein.

Occurrence. Fats are found in both plants and animals. In the human body all the cells contain small amounts of fat, and there are certain tissues known as "fat depots" that are specialized to store fats. The fat depots of the body are the subcutaneous tissues, the tissues around the intestines and the kidneys, and the bone marrow; these tissues may contain as much as 90% fatty substances. In plants, fats have an occurrence somewhat similar to that found in animals. All plant cells contain small amounts of fatty material,

Table 16-1. Important fatty acids that enter into the composition of fats

Fatty acid	Formula	Fat derivative
Butyric	C_3H_7COOH	Tributyrin
Caproic	$C_5H_{11}COOH$	Tricaproin
Caprylic	$C_7H_{15}COOH$	Tricaprylin
Capric	$C_9H_{19}COOH$	Tricaprin
Lauric	$C_{11}H_{23}COOH$	Trilaurin
Myristic	$C_{13}H_{27}COOH$	Trimyristin
Palmitic	$C_{15}H_{31}COOH$	Tripalmitin
Stearic	$C_{17}H_{35}COOH$	Tristearin
Oleic	$C_{17}H_{33}COOH$	Triolein

and plants also store large quantities of fats in their seeds, roots, and fruits. Cottonseeds, castor beans, peanuts, and olives are examples of storage products of plants that have a high fat content. Both plants and animals have the power to synthesize fats, and they produce them from carbohydrates and proteins.

Unsaturated fatty acids. The unsaturated fatty acids have outstanding importance in nutrition. When incorporated in the diet, they appear to have a preventive effect on the development of the atherosclerosis syndrome, which is characterized by an elevated blood cholesterol level.

In unsaturated fatty acids there are double bonds in the hydrocarbon chain. Four important unsaturated fatty acids are oleic, linoleic, linolenic, and arachidonic acids. The latter three acids are known as *essential* fatty acids. Absence of these from the diet of infants would cause a loss of weight. All three of them are found in linseed oil.

Oleic acid. Oleic acid ($C_{17}H_{33}COOH$) contains one double bond between carbon atoms 9 and 10, numbered from the carboxyl end.

$$\overset{18}{CH_3}-(CH_2)_7-\overset{10}{CH}=\overset{9}{CH}-(CH_2)_7-\overset{1}{COOH}$$

Linoleic acid. Linoleic acid ($C_{17}H_{31}COOH$) contains two double bonds, one between carbon atoms 9 and 10 and one between atoms 12 and 13.

$$\overset{18}{CH_3}-(CH_2)_4-\overset{13}{CH}=\overset{12}{CH}-CH_2-\overset{10}{CH}=\overset{9}{CH}-(CH_2)_7-\overset{1}{COOH}$$

Linolenic acid. Linolenic acid ($C_{17}H_{29}COOH$) contains three double bonds, one between carbon atoms 9 and 10, 12 and 13, and 15 and 16, respectively.

$$\overset{18}{CH_3}-CH_2-\overset{16}{CH}=\overset{15}{CH}-CH_2-\overset{13}{CH}=\overset{12}{CH}-CH_2-\overset{10}{CH}=\overset{9}{CH}-(CH_2)_7-\overset{1}{COOH}$$

Arachidonic acid. Arachidonic acid ($C_{19}H_{31}COOH$) contains four double bonds, one between carbon atoms 5 and 6, 8 and 9, 11 and 12, and 14 and 15.

$$\overset{20}{CH_3}-(CH_2)_4-\overset{15}{CH}=\overset{14}{CH}-CH_2-\overset{12}{CH}=\overset{11}{CH}-CH_2-\overset{9}{CH}=$$

$$\overset{8}{CH}-CH_2-\overset{6}{CH}=\overset{5}{CH}-(CH_2)_3-\overset{1}{COOH}$$

Physical properties. When pure, fats are colorless, odorless, and tasteless. The yellow color of many naturally occurring fats is due to the presence of pigments such as carotene or xanthophyll. In general, fats have a greasy feel and are insoluble in water but soluble in ether, chloroform, benzene, gasoline, and hot alcohol. They have a lower specific gravity than water and consequently will float on the surface of water when mixed with it.

Chemical behavior. Chemical reactions of special interest will be discussed.

Acrolein test. When a fat is heated to about 300° C, the peculiar odor of "burnt grease" is given off; this is the odor of acrolein, which is formed from the glycerol part of the fat. The formation of acrolein involves the removal of 2 molecules of water from glycerol, as indicated in the following equation.

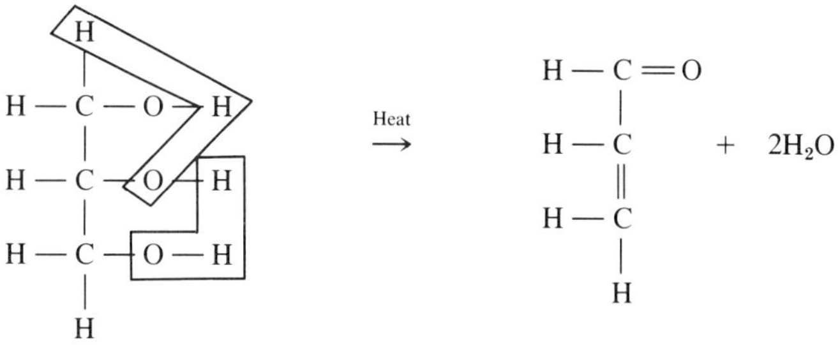

$$\text{Glycerol} \xrightarrow{\text{Heat}} \text{Acrolein} + 2H_2O$$

This reaction is favored by the presence of a drying agent such as $KHSO_4$. Hence, *a standard test for fats is to heat the fatty material with acid potassium sulfate and to note the characteristic odor of acrolein.*

Oxidation. When oxidized completely, fats yield carbon dioxide and water. This reaction takes place readily in the human body and is attended by the liberation of considerable energy. One gram of fat when burned yields approximately 9 Calories of energy, roughly 3500 Calories per pound.

Hydrolysis and saponification. When fats are boiled with acids or alkalies or treated with superheated steam, they are split into fatty acids and glycerol. If an alkali is used, it reacts with the fatty acids as soon as they are set free and forms alkali salts of the fatty acids, called soaps.

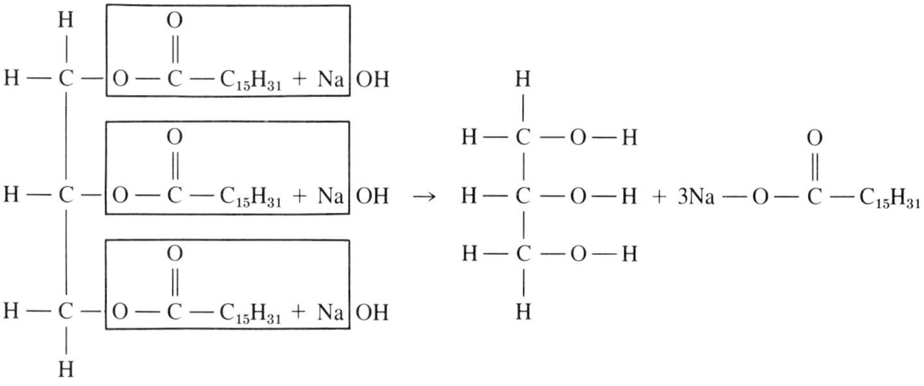

Tripalmitin + Sodium hydroxide → Glycerol + Sodium palmitate (a soap)

This reaction is called *saponification*. Fats may also be hydrolyzed by treating them with certain digestive enzymes; this reaction is the process by which fats are digested in the alimentary tract of animals.

The hydrolysis of fats by acids, steam, or enzymes is illustrated by the following equation:

$$\begin{array}{c}
\text{H} \\
| \\
\text{H}-\text{C}-\text{O}-\overset{\text{O}}{\overset{\|}{\text{C}}}-\text{R} \\
| \\
\text{H}-\text{C}-\text{O}-\overset{\text{O}}{\overset{\|}{\text{C}}}-\text{R} \;+\; 3\text{H}_2\text{O} \\
| \\
\text{H}-\text{C}-\text{O}-\overset{\text{O}}{\overset{\|}{\text{C}}}-\text{R} \\
| \\
\text{H}
\end{array}
\xrightarrow[\text{or enzyme}]{\text{Acid, steam,}}
\begin{array}{c}
\text{H} \\
| \\
\text{H}-\text{C}-\text{O}-\text{H} \\
| \\
\text{H}-\text{C}-\text{O}-\text{H} \\
| \\
\text{H}-\text{C}-\text{O}-\text{H} \\
| \\
\text{H}
\end{array}
\;+\; 3\text{R}-\overset{\text{O}}{\overset{\|}{\text{C}}}-\text{OH}$$

Neutral fat   **Glycerol**   **Fatty acid**

Emulsification. Lipids are insoluble in water and need to be emulsified to become dissolved in water solutions. Emulsification (see p. 129) is an important process. Only emulsified fats are digested to any extent in the stomach, and in the intestine the emulsification of fats by alkaline digestive juices aids greatly in their digestion.

Soaps. Soaps are sodium or potassium salts of fatty acids. They are made by boiling fats with sodium or potassium hydroxide. The reaction is as given in the discussion of hydrolysis and saponification. Most commercial soaps are sodium salts of a fatty acid. Potassium soaps are more soluble than sodium soaps but are more expensive. Hard soaps are salts of saturated fatty acids, and soft soaps are salts of highly unsaturated fatty acids. Soaps do have antiseptic properties. In addition, germicidal soaps containing hexachlorophene are marketed. Tincture of green soap is an alcohol solution of a potassium soap.

One end of the molecule of a soap, the carboxyl end, is soluble in water, and the other end, the hydrocarbon chain, is soluble in oil. Such molecules readily form oil-in-water emulsions because the hydrocarbon chain is soluble in the oil and the carboxyl end is soluble in the water. Oil dispersed in fine droplets (of colloidal size) in water containing a soap will become coated with soap films, forming an oil-in-water emulsion.

Some dirt particles that get on the skin or on fabrics are soluble in water and may be removed from the body or from clothing simply by washing with water. Most dirt particles, however, adhere to the skin or to fabrics chiefly by being dispersed in oils, which are insoluble in water. Soaps remove dirt from the skin or from fabrics by emulsifying the oils that contain dirt. After their formation the emulsified particles containing dirt are readily removed by water.

The commercial soaps in use are salts of fatty acids with hydrocarbon chains of 10 to 18 carbon atoms. If the hydrocarbon chain of the fatty acid contains less than 10 carbon atoms, the salt will not form an emulsion with oil, and if the hydrocarbon chain contains more than 18 carbon atoms, the salt is too insoluble in water to participate in the formation of an emulsion.

Detergents. The action of detergents in cleaning is similar to that described for soaps. Detergents remove dirt from surfaces by forming an oil-in-water emulsion, the dirt being present in the oil phase.

Detergents differ chemically from soaps in one respect: the water-soluble end of the molecule of a detergent is a sulfate group instead of a carboxyl group as in a soap.

Detergents are prepared by introducing a sulfate group into a hydrocarbon chain. One method consists of the addition of sulfuric acid to the double bond (unsaturated linkage) in the molecule of an unsaturated hydrocarbon. Another procedure is the treatment of a fatty alcohol with sulfuric acid. The preparation of sodium lauryl sulfate is an example of the latter method, as shown in the following equation:

$$CH_3(CH_2)_{10}CH_2OH + H_2SO_4 \rightarrow CH_3(CH_2)_{10}CH_2OSO_3H + H_2O$$
Lauryl alcohol   **Lauryl hydrogen sulfate**

$$CH_3(CH_2)_{10}CH_2OSO_3H + NaOH \rightarrow CH_3(CH_2)_{10}CH_2OSO_3Na + H_2O$$
Lauryl hydrogen sulfate   **Sodium lauryl sulfate**
                                                                    (a detergent)

A detergent, like a soap, has a dual chemical quality: one end of the molecule is soluble in oil, and the other end is soluble in water. It is this chemical quality that gives to the detergent the property of solubilizing oil particles by forming an emulsion with them that can be washed away with water.

Since detergents do not form insoluble salts with calcium or magnesium ions (as soaps do), they can be used effectively in hard water (see p. 96).

Saponification number and iodine number. All neutral fats contain the glyceryl group. They differ from each other in chain length and degree of saturation of the fatty acids present. Two chemical tests used to get information on the chemical nature of fats are the saponification number and the iodine number.

Saponification number. When fats are boiled with alkali, they are split into glycerol and salts of fatty acids. If a known amount of standardized KOH solution is used and if at the end of the boiling period the unused KOH is titrated, quantitative information on the chain length of the fatty acids present can be obtained. In carrying out such a procedure, the saponification number is derived. *The saponification number of a fat is the number of milligrams of KOH required to saponify 1 g of fat.* The saponification number of butyrin is 557 and that of stearin is 189. The fat that requires the smaller amount of KOH for saponification (stearin) has a longer chain fatty acid (stearic acid) than the fat (butyrin) that contains a shorter chain fatty acid (butyric acid) in its molecule because the long chain fatty acids weigh more, and hence there are less of them in 1 g of fat. Thus the saponification number gives information on the chain length and molecular weight of the fatty acids in a fat.

Iodine number. If a fat containing unsaturated fatty acids is placed in a solution containing iodine, the iodine will combine with the fatty acid at its points of unsaturation, that is, at its double bonds. The amount of iodine taken up is a measure of the degree of unsaturation of the fatty acids. Thus placing a weighed amount of fat in a measured amount of standardized iodine solution, letting the mixture stand for an appropriate time, and then titrating the amount of unused iodine left in the solution is a chemical procedure for determining the iodine number of an unsaturated fat. Specifically defined, *the iodine number is the number of grams of iodine absorbed by 100 g of fat*. The iodine number is a useful value for characterizing the degree of unsaturation of a fat.

Some important fats. Some important fats are butter, lard, tallow, oleomargarine, and olive oil.

Butter. Butter is a mixture of several fats. It consists of about 60% tripalmitin, 30% triolein, a little trisearin, and 7% glyceryl esters of the volatile fatty acids, which are butyric, caproic, caprylic, and capric acids.

Lard. Lard, the fat obtained from swine, is an example of the naturally occurring fats of the animal body. It consists principally of a mixture of tripalmitin, tristearin, and a small amount of triolein.

Tallow. Tallow is the fat extracted from sheep and cattle. It is a mixture chiefly of tripalmitin, tristearin and triolein.

Oleomargarine. Oleomargarine consists principally of fats of vegetable origin. When properly prepared, it is a wholesome food and is subject only to the objection that it has a lower vitamin content than the pure butter that it supplants commercially. On the recommendation of nutrition authorities the practice of fortifying oleomargarine by adding vitamin A has been widely adopted. Animal experimentation has shown that properly fortified oleomargarine has as high a nutritive value as butter. Many oleomargarines contain more polyunsaturated fatty acid than butter and in this respect can be considered superior to butter.

Olive oil. Fats that are liquids at ordinary temperatures are called oils. Olive oil is an example of this class of substances. Olive oil consists principally of triolein.

Functions of fats. Fats provide energy value and a reserve food supply. Fats also provide heat insulation and protection against mechanical injury, and they are vitamin carriers.

Energy value. Fats have a high fuel value. When oxidized in the body, they yield approximately 9 Calories of energy per gram, which is the highest energy value of all food substances.

Reserve food supply. Fats are stored in the body and are therefore an important reserve food supply.

Heat insulation. Fats are poor conductors of heat; hence the layers of fat under the skin enable the body to retain better the heat produced by the combustion of foods. This is observed in the reaction of people to extremes of temperature. Thin people suffer more from cold temperatures of winter than do fat people, and fat people complain most of the hot weather of summer.

Protection against mechanical injury. Being deposited around the internal organs and under the skin, fats fill out the body structures and protect the body against mechanical injury.

Vitamin carriers. Certain vitamins are soluble in fats. The fats of the diet therefore serve to carry into the alimentary tract of animals the "fat-soluble" vitamins and thus make these substances available to animals for their nutrition.

STEROLS

Sterol is a word derived from the Greek word *stereos,* meaning solid, and the suffix *ol,* signifying an alcohol.

The term sterol literally means solid alcohol. It is applied to a group of large-molecule alcohols that have one OH group in the molecule and that exist in the physical state of a solid. Sterols are important constituents of both plant and animal tissues. There are many sterols in nature, but we will have to confine our discussion to a consideration of one animal sterol, cholesterol.

Cholesterol. The prefix "chole" in the word cholesterol is derived from the Greek word *chole,* meaning bile, and as noted, "sterol" means solid alcohol; hence the term cholesterol literally means bile solid alcohol. Cholesterol ($C_{27}H_{45}OH$) was first prepared from gallstones. The observation that this substance is a constituent of bile led to the adoption of its name.

Cholesterol is a constituent of all animal tissues. It is present in the greatest amount in brain and nerve tissue. On a dry weight basis, 14% of the white matter of brain is cholesterol, and 10% to 15% of spinal cord is composed of this compound. The normal total cholesterol content of the blood ranges from 130 to 250 mg/100 ml of serum.

Cholesterol is abundant in egg yolk. A diet including eggs and meats therefore contains considerable cholesterol. The cholesterol that is in the food is readily absorbed from the alimentary tract. It can be synthesized by body tissues from simpler substances, and the body is therefore not dependent on food intake for its supply of cholesterol.

Cholesterol is a major constituent of bile, being present in concentrations up to 1%. In bile it is an excretory product. It is through the bile, which passes into the intestinal tract, that the body eliminates the cholesterol not needed.

Cholesterol is associated with several pathological conditions in the body. One of these is cholelithiasis, the occurrence of stones in the gallbladder. Gallstones are composed of cholesterol to the extent of 80% to 95%. Other constituents of gallstones are bile pigments, calcium carbonate, and calcium phosphate.

Another pathological condition with which cholesterol is associated is atherosclerosis. Cholesterol is deposited in the atheromatous lesions of the blood vessels that develop in advanced atherosclerosis. The relation of cholesterol to the formation of these lesions is not well known. Quite suggestive are the experiments in which atheromatous lesions are produced in rabbits by diets containing added cholesterol and in dogs by high-cholesterol diets plus a drug (thiouracil) that raises the blood cholesterol level.

The structural formula for cholesterol is as follows:

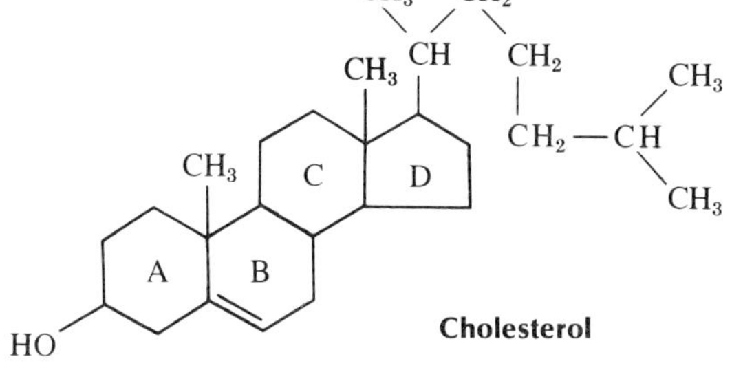

Cholesterol

230 Roe's principles of chemistry

The structure represented by rings A to D in the formula for cholesterol is known as the steroid nucleus, and compounds having this structure as the principal part of the molecule are called *steroids*. Important steroids in the animal body are the hormones of the adrenal glands, the sex hormones, the bile acids, and one of the vitamins concerned in calcification, activated 7-dehydrocholesterol (vitamin D_3). Cholesterol is the precursor or parent substance in the body of these vitally important compounds, which will be discussed later.

PHOSPHOLIPIDS

The phospholipids, also called *phosphatides,* are soluble in ether and alcohol but insoluble in acetone. They can be isolated from tissues by extracting the tissue with ether and then precipitating them from the ether solution by adding acetone. As the name suggests, these substances are lipids combined with a phosphate group. Phospholipids occur in all animal tissues, and they occur in the greatest amount in brain and nerve. Of the total lipid content of brain, which is 51% of the dry weight, phospholipids constitute one-half.

Examples of phospholipids are the lecithins found in animal tissues and in the soybean. Lecithins are degraded into their component parts by hydrolysis. When hydrolyzed by boiling with dilute acid or alkali, they yield glycerol, fatty acids, phosphoric acid, and choline. Lecithin is used in large quantities as an emulsifying agent. The formula for lecithin is as follows:

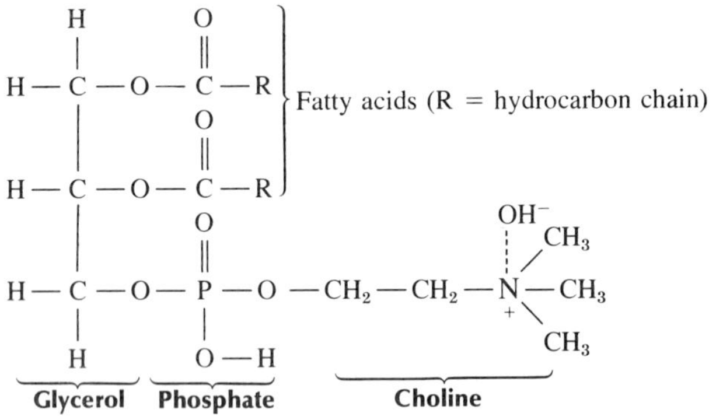

$$\underbrace{\begin{array}{c} H \\ | \\ H-C-O- \\ | \\ H-C-O- \\ | \\ H-C-O- \\ | \\ H \end{array}}_{\text{Glycerol}} \underbrace{\begin{array}{c} O \\ \| \\ C-R \\ O \\ \| \\ C-R \\ O \\ \| \\ P-O- \\ | \\ O-H \end{array}}_{\text{Phosphate}} \left. \begin{array}{c} \\ \\ \\ \\ \end{array} \right\} \text{Fatty acids (R = hydrocarbon chain)} \underbrace{CH_2-CH_2-\overset{+}{N}\overset{OH^-}{\underset{CH_3}{\overset{CH_3}{\diagup}}}}_{\text{Choline}}$$

Lecithin

Choline, one of the components of lecithin, is a nitrogenous base. It is an exceedingly important compound. Choline or its precursor must be in the diet to prevent the development of "fatty liver." This will be discussed later (p. 293). The formula for choline is as follows:

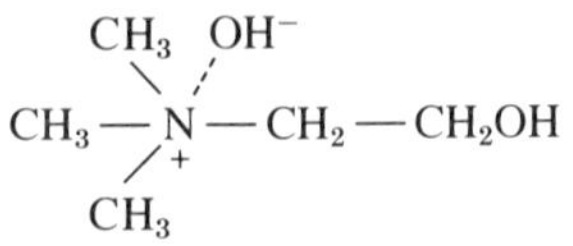

$$CH_3 - \overset{+}{N}\overset{CH_3}{\underset{CH_3}{\diagup}}\overset{OH^-}{} - CH_2 - CH_2OH$$

Choline

Choline is also important as a part of the molecule of acetylcholine, a compound resulting from the combination of choline with acetic acid. The formula for the ester, acetylcholine, is as follows:

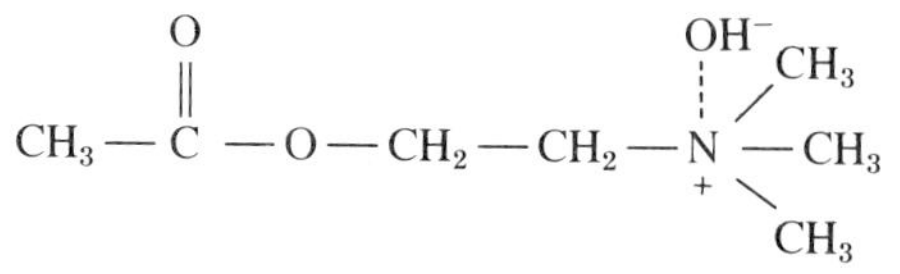

$$CH_3-\overset{\overset{O}{\|}}{C}-O-CH_2-CH_2-\overset{OH^-}{\underset{+}{N}}\diagdown_{CH_3}^{CH_3}$$

Acetylcholine

Acetylcholine is liberated at the endings of the nerves of the parasympathetic system and at the endings of nerves to muscle when these nerves are stimulated to action. Acetylcholine thus participates in the transmission of nerve impulses to the eyes, heart, blood vessels, digestive tract, bladder, and skeletal muscle.

Acetylcholine is formed by nerve tissue from choline and acetic acid through the action of an enzyme called choline acetylase. The action of acetylcholine at the nerve endings is brief because of the activity of another enzyme called cholinesterase, which splits the compound into choline and acetic acid and stops in action. Thus nature has provided one enzyme to synthesize and another to split acetylcholine, and the activity of the two enzymes in normal healthy nerve tissue is in balance.

The enzyme cholinesterase is poisoned by certain substances; for example, the drug physostigmine inactivates cholinesterase and thus permits acetylcholine to accumulate. Another compound that poisons cholinesterase is di-isopropylfluorophosphate (DFP), the so-called nerve gas, whose effects were discovered during World War II. When cholinesterase, the enzyme that exerts the normal control, is poisoned, acetylcholine accumulates and stimulates its effector organs excessively, thus producing a toxic effect.

GLYCOLIPIDS

Glycolipids are lipids that contain a sugar in place of the phosphorus in phospholipids. The sugar most often found in combination with lipid is galactose. Glycolipids occur in large amounts in brain. For this reason they have been called cerebrosides. Phrenosin, nervon, kerasin, and oxynervon are glycolipids that have been isolated from brain. Glycolipids also occur in nerve, heart, lungs, kidney, blood, muscle, and seeds. These substances are essential components of tissues, but their significance in the animal body is not known.

WAXES

Waxes are complex mixtures of esters of fatty acids and monohydroxy alcohols. Examples are lanolin, beeswax, carnauba wax, and spermaceti. Paraffin wax is not an ester but a mixture of high molecular weight hydrocarbons.

In nature, waxes serve largely as a protective coating but also in some instances as a supporting material. Lanolin, for example, is a coating over the fibers of wool. This wax is used commercially as a medium for the preparation of ointments, creams, and cosmetics. Beeswax ($C_{15}H_{31}COOC_{30}H_{61}$) is a secretion of the bee that is used for the construction of the honeycomb. The leaves and fruit of certain plants have coatings that are waxes. Carnauba wax ($C_{25}H_{51}COOC_{30}H_{61}$), for example, is a protective agent covering the leaves of the Brazilian wax palm and is widely used in floor waxes and automobile polishes. Sperm oil, taken from the head of the whale, is a mixture of oil, which is a true fat, and a wax called spermaceti. The spermaceti is used in making candles, cosmetics, and so on.

Questions for study

1. State three characteristics of lipids.
2. Discuss fully the occurrence of fats.
3. What are fats?
4. Name several fatty acids that enter into the composition of fats. Name the fats that are formed from each.
5. Mention several properties of the fats.
6. What is a chemical test for fats?
7. What products are formed when fats are oxidized completely?
8. What is the caloric value of fats?
9. What is saponification?
10. Explain how an emulsion is formed.
11. What are soaps? Detergents? Explain how soaps and detergents remove dirt.
12. What are the constituents of butter?
13. Chemically, what is lard? Tallow?
14. What is oleomargarine? Is it a satisfactory food?
15. What are the functions of fats?
16. What is a sterol?
17. What is cholesterol?
18. State the occurrence of cholesterol and its importance in the animal body.
19. With what pathological processes is cholesterol associated?
20. What is a steroid? What important biological compounds are steroids? What is the precursor in the body of these important steroids?
21. What is a phospholipid? Where do phospholipids occur?
22. What are the component parts of a lecithin?
23. What is the importance of choline in the animal body?
24. What is acetylcholine, and what function does it perform?
25. What are glycolipids? Where do they occur?
26. What are waxes?
27. State several uses of waxes.

17 Nucleoproteins, nucleic acids, and genetics

Nucleoproteins

Occurrence and structure. Nucleoproteins are constituents of all cell nuclei. They are substances of great significance in nature.

The nucleoprotein molecule consists of a protein linked to a nucleic acid. It is a very large molecule. Molecular weights of nucleoproteins range from 1.5 to 280 million.

The viruses, a group of substances that cause much disease in plants and animals, are nucleoproteins. An interesting example of a plant virus is the tobacco mosaic virus, isolated in pure crystalline form by Stanley in 1935. This virus damages the leaves of the tobacco plant. The common cold, smallpox, and poliomyelitis are examples of diseases caused by a virus. Recent work has suggested that a virus is the etiological agent in some tumors.

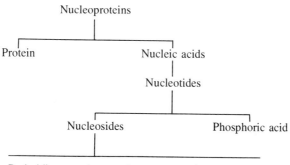

Pyrimidines, purines, and pentoses or deoxypentoses

How nucleoproteins and their components are obtained for study. Nucleoproteins are readily broken down into proteins and nonprotein nitrogenous substances called nucleic acids by treatment of nuclear material with digestive enzymes. The sources of the nuclear material used by early investigators for the preparation of nucleic acid were pus cells, the red blood cells of birds and reptiles (such cells contain nuclei), salmon sperm, and yeast. Nucleic acids are depolymerized into simpler components by treatment with digestive enzymes found in the intestinal tract. Nucleic acids are also broken down into their constituent units by boiling with dilute acid or alkali. The stepwise degradation of nucleoproteins into their basic units is illustrated by the diagram on p. 233.

Nucleic acid
COMPONENTS

The following substances are components of the nucleic acid molecule:
1. Pyrimidines, examples of which are thymine, cytosine, and uracil
2. Purines, examples of which are adenine and guanine
3. A pentose sugar, which is either ribose or deoxyribose
4. Phosphate

Pyrimidines

System of designation. The pyrimidines are heterocyclic compounds whose basic structure is a six-membered ring containing carbon and nitrogen atoms as illustrated by the parent compound, pyrimidine.

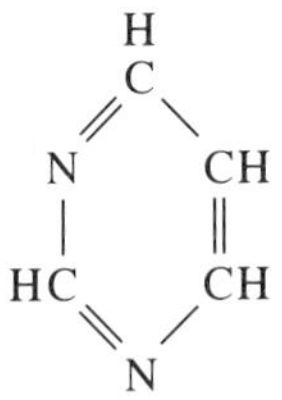

Pyrimidine

Thymine, cytosine, and uracil. Thymine, cytosine, and uracil are substituted pyrimidines found in nucleic acid. Their structural formulas are as follows:

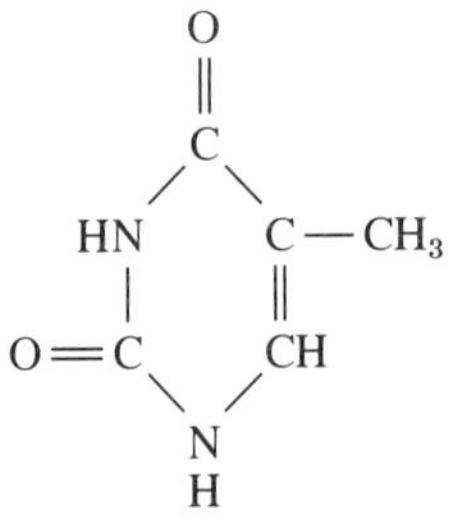

Thymine

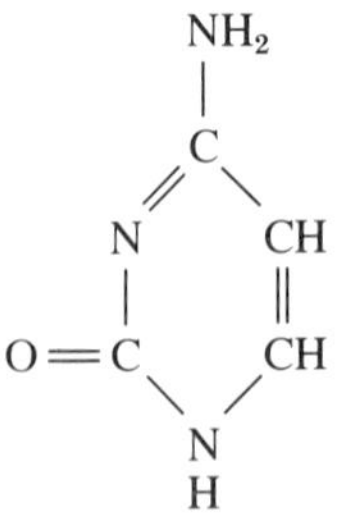

Cytosine

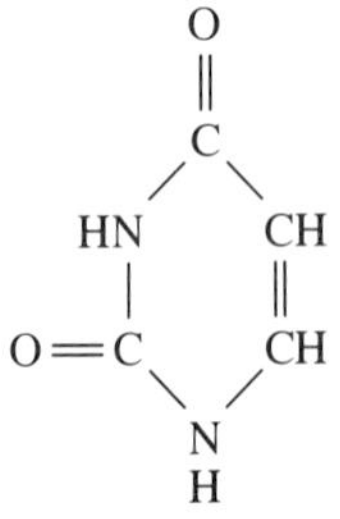

Uracil

Purines

System of designation. The parent substance, purine, consists of a pyrimidine ring attached to another heterocylic ring. The structural formula of the purine nucleus is as follows:

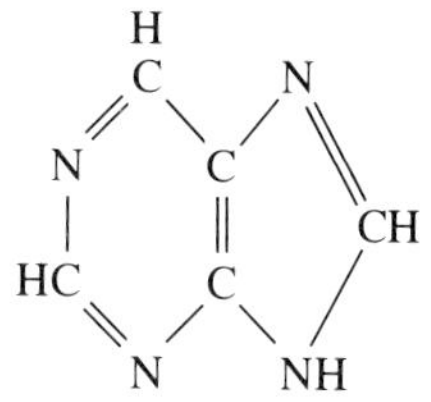

Purine

Adenine and guanine. Adenine and guanine are the principal purines found in nucleic acid. Their structural formulas are as follows:

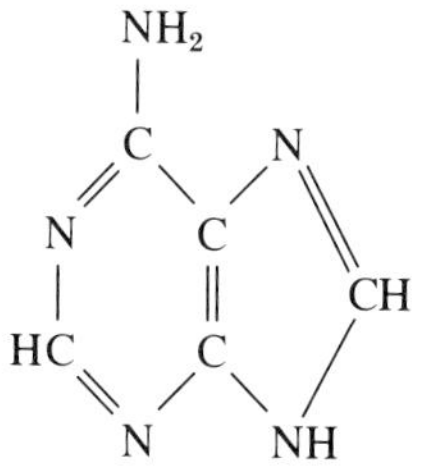

Adenine 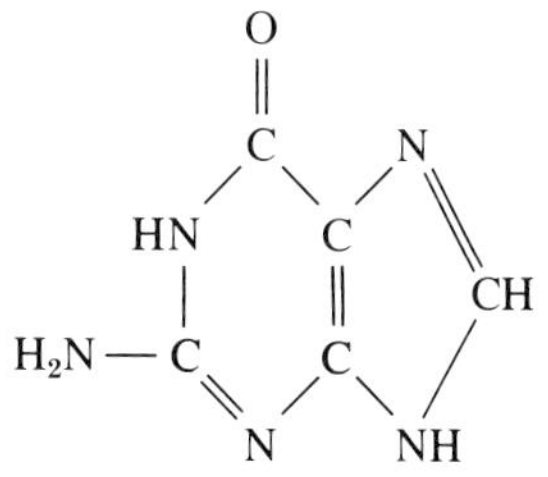   **Guanine**

Sugars in nucleic acid

Ribose and deoxyribose. One or the other of these two pentoses is a constituent of nucleic acids. Combined with a purine or pyrimidine base they enter into the composition of two exceedingly important compounds, ribonucleic acid and deoxyribonucleic acid, designated by the abbreviations RNA and DNA, respectively. The structure of these two pentoses is as follows:

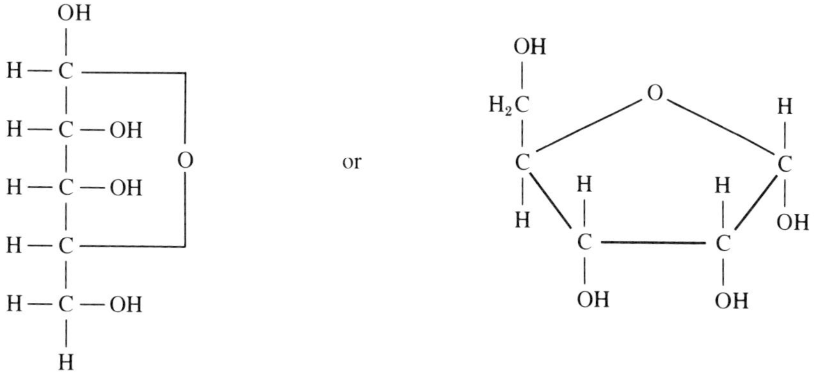

Ribose

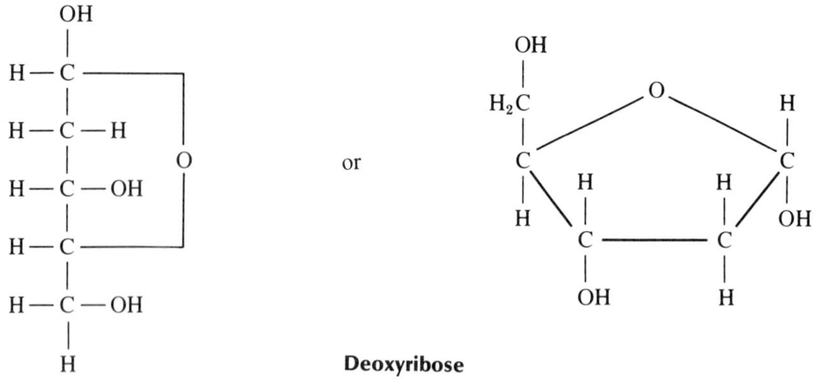

236 Roe's principles of chemistry

[Structural formulas of deoxyribose shown in open-chain and ring forms]

Deoxyribose

The only difference in the two compounds is that deoxyribose is lacking an oxygen atom.

Nucleosides

Adenosine and cytidine. Combination of a purine or pyrimidine base with ribose or deoxyribose yields a nucleoside. Examples are adenosine and cytidine. In adenosine, adenine is combined with ribose, and in cytidine, cytosine is combined with ribose. The following are the structural formulas for these two substances:

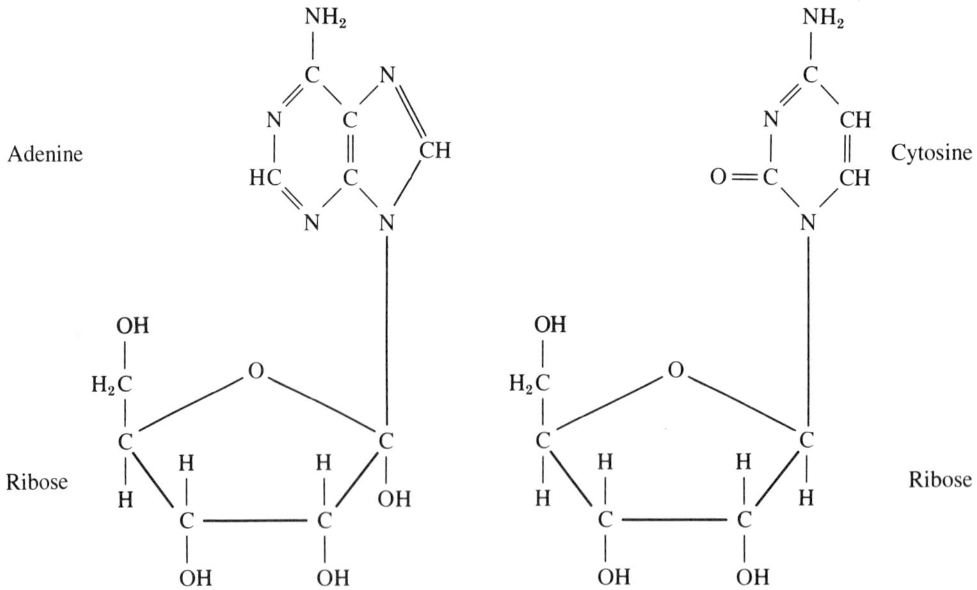

[Structural formulas of Adenosine (Adenine + Ribose) and Cytidine (Cytosine + Ribose)]

Nucleotides

Adenylic and cytidylic acids. When a nucleoside is combined with a phosphate group, the resulting compound is a nucleotide. Two important nucleotides are adenylic acid and cytidylic acid. The structure of these two substances is as follows:

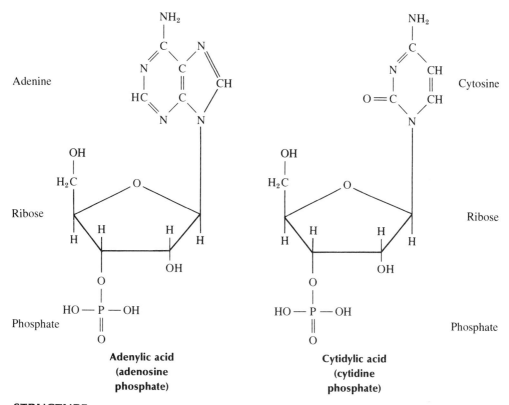

STRUCTURE

On hydrolysis, the two nucleic acids yield the following materials:

RNA	*DNA*
Phosphoric acid	Phosphoric acid
Ribose	Deoxyribose
Adenine	Adenine
Cytosine	Cytosine
Guanine	Guanine
Uracil	Thymine

Thus there are four nucleotides in RNA and four similar nucleotides in DNA. Note the only difference is in uracil and thymine. For convenience, purine bases (adenine and guanine) are symbolized by A and G, and the three pyrimidine bases (thymine, cytosine, and uracil) are symbolized by T, C, and U.

The nucleic acid molecule is a very long polymer of nucleotides linked together by phosphate groups attached to the sugar groups (Fig. 17-1). The length of the chain ranges from about 70 nucleotide units in small ribonucleic acid (RNA) molecules to an estimated 20,000 nucleotides in the deoxyribonucleic acid (DNA) molecule. Molecular weights of DNA range from about 5 to 15 million, whereas molecular weights of RNA range from 20,000 to several million.

In 1953, Dr. J. D. Watson and Dr. F. H. C. Crick proposed that the DNA molecule

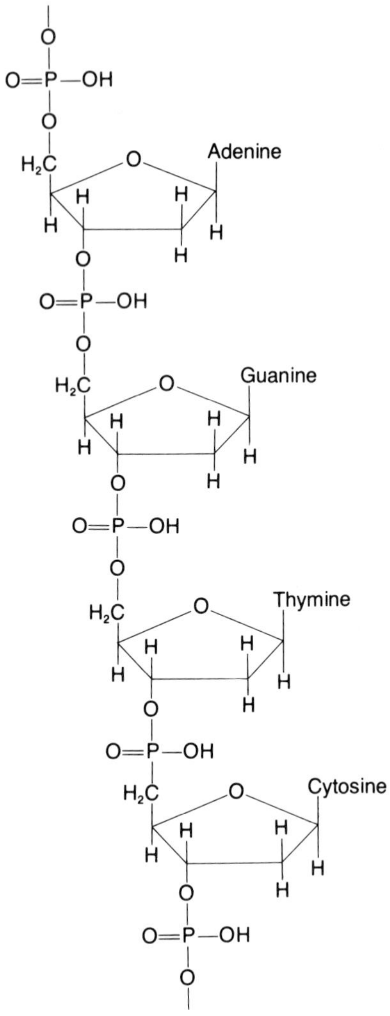

Fig. 17-1. This diagram shows four nucleotides in a deoxyribonucleic acid molecule, linked together by phosphate groups.

is composed of a double strand of nucleotides, the two strands being coiled on each other in a manner that forms a double helical structure (Fig. 17-2). In DNA the strands are held together by hydrogen bonds between their purine and pyrimidine bases, adenine being linked to thymine and guanine to cytosine. The adenine of one chain is always linked to thymine in the other chain, and guanine in one chain is always linked to cytosine in the other (Fig. 17-3). It was shown earlier (p. 214) that hydrogen bonds function in a similar way in the protein molecule, holding together peptide chains. The RNA molecule consists of a single strand of nucleotides.

Thus, if we could unwind the DNA double helix, we would find a ladderlike structure illustrated as follows:

Fig. 17-2. Helical structure of DNA. The purine and pyrimidine bases are omitted to emphasize the twisting framework. (From Trumbore, R. H.: The cell—chemistry and function, The C. V. Mosby Co.)

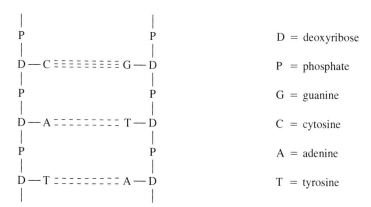

The solid lines represent chemical bonds and the dashed lines represent hydrogen bonds. Each DNA molecule differs only in the sequence of the four bases, A, C, T, and G. This sequence has become known as the genetic code.

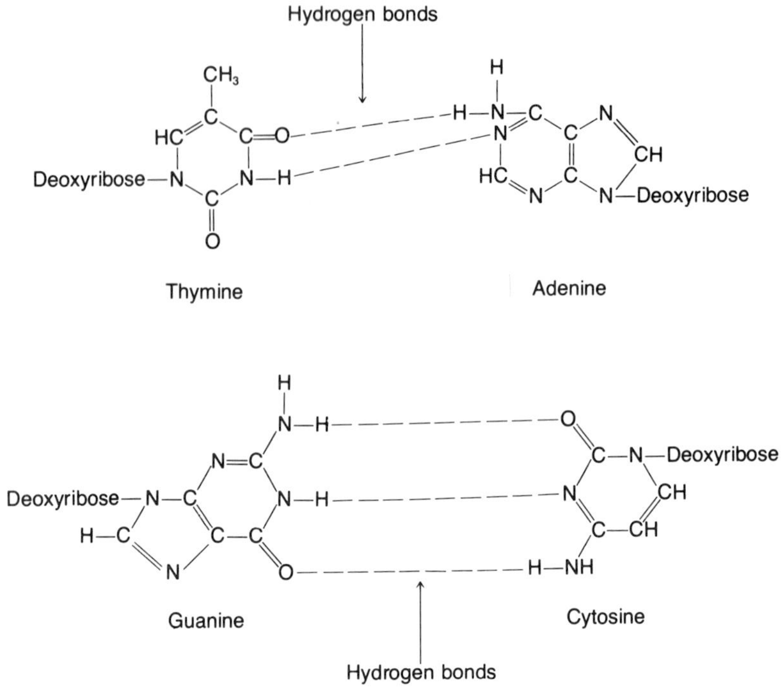

Fig. 17-3. Sections of chain of DNA showing how thymine and adenine are linked by two hydrogen bonds and guanine and cytosine are linked by three hydrogen bonds.

Genetics

All living organisms have the ability to produce a new individual that is similar to their own kind. The transmission of factors that bring about the development of similar characteristics is known as the hereditary or genetic effect. As a result of recent investigations, biological science is now in possession of remarkable information concerning the mechanisms of the transmission of hereditary characteristics.

STORAGE OF GENETIC INFORMATION

Biologists have known for a long time that heredity factors have their origin in structures called *genes,* which exist in the chromosomes of cell nuclei. Recently evidence has been obtained indicating that genes are molecules of deoxyribonucleic acid (DNA). The general concept is that one gene, or one molecule of DNA, contains the "information" or "code" that determines the synthesis of a cellular protein, or an enzyme that brings about an essential biochemical step in metabolism. This concept necessitates the existence of a great number of genes, or molecules of DNA, that differ from each other in composition. DNA is admirably adapted to such a function as it is a very long molecule and therefore capable of many variations.

In addition to its capacity to transmit genetic information, which determines biochemical behavior, the gene, or DNA molecule, is able to reproduce itself. The process is called *replication*. We will discuss replication first and then consider the *transmission of genetic information*.

REPLICATION

The quality of the DNA molecule that makes it able to reproduce itself—that makes it replicable—resides in its possession of purine and pyrimidine bases held together by hydrogen bonds. These bases are (1) adenine and thymine, which are bound to each other by two hydrogen bonds, and (2) guanine and cytosine, which are attached by three hydrogen bonds (see Fig. 17-3). The bases in each of these pairs fit each other uniquely, and they do not become attached to other bases in a similar manner.

When replication occurs, the hydrogen bonds between the two strands of the DNA molecule are ruptured and the two strands are separated. Acting as a template, or model, each separated single strand of DNA, in the presence of nucleoside triphosphates and an enzyme called DNA polymerase, brings about the synthesis of a complementary DNA strand like the one to which it was attached before the splitting occurred. Each newly synthesized strand remains united with its complementary parent strand, and the result is the formation of two molecules of DNA that are exactly like the original DNA molecule. This is called replication. The conditions that cause these changes are only slightly known; however, a separation of the DNA strands by mild heating and a partial recombination by cooling slowly (annealing) has been accomplished in the laboratory.

TRANSMISSION OF GENETIC INFORMATION

DNA primarily serves as a storehouse of genetic information in cell nuclei, but it does not function outside the cell nucleus. A mechanism is therefore necessary to transmit the genetic information stored in DNA beyond the cell nuclei. Such a mechanism is provided by the formation of messenger ribonucleic acid, known as mRNA. RNA is found in both the cytoplasm and the nucleus of the cell.

To form molecules of nucleic acid that transmit genetic information, the DNA strands split apart and serve as templates for the synthesis of copies, as in replication, except that *ribose* is substituted for the *deoxyribose* as the sugar component. The newly formed molecule is called messenger ribonucleic acid, mRNA; it has the same base sequence as existed in a parent strand of DNA and therefore has the genetic potential of the original DNA. The newly formed mRNA, because of its ability to pass through the nuclear membrane and into the cytoplasm, has the capacity to carry genetic information. In the cytoplasm the mRNA acts as a template for synthesis of protein molecules.

When the mRNA passes from the cell nucleus into the cytoplasm it becomes attached to a ribosome. The ribosome is a nucleoprotein particle consisting of protein bound to ribonucleic acid (ribosomal RNA). Ribosomes are attached to intracellular membranes. When anchored to a ribosome the mRNA promotes the synthesis of protein of a kind determined by the chemical information that came from the original DNA in the cell nucleus.

The amino acids used in the synthesis of protein are brought to the mRNA by another kind of RNA, soluble or transfer RNA, a small molecule (molecular weight of about

30,000) designated as sRNA or tRNA. sRNA combines with an amino acid and transfers it to a site of synthesis of protein on the ribosome. Presumably there is one sRNA for transferring each of the 20 amino acids to the site of protein synthesis. The synthesis of protein is accomplished by incorporation of amino acid into peptide chains under the directing influence of mRNA, in the presence of an enzyme, a synthetase. The sRNA is split off and is capable of returning with an additional amino acid. Thus the buildup of the protein chain is based on the genetic information or code obtained from the DNA within the nucleus of the cell.

• • •

In summary, it may be said that the full operation of the mechanisms involved in DNA replication and in the storage, transmission, and application of genetic information the following nucleic acids are required:
1. *Deoxyribonucleic acid, DNA:* for storage of genetic information
2. *Messenger ribonucleic acid, mRNA:* for transferring genetic information from the cell nucleus to ribosomes in the cytoplasm where it determines the kind of protein that is synthesized
3. *Transfer ribonucleic acid, sRNA or tRNA:* for combining with and carrying amino acids to the site of synthesis of proteins on the ribosomes

As noted earlier, *messenger, template,* or *informational* RNA is a single-strand nucleic acid that is synthesized in the nucleus of the cell but that diffuses into the cytoplasm and functions outside of the nucleus. It is a complementary copy of a single DNA strand and therefore contains the hereditary information that originally existed in DNA. The genetic information in mRNA is used to direct the synthesis of proteins, which are made from 20 amino acids. Another kind of RNA, transfer RNA, (sRNA or tRNA), combines with an amino acid and carries it to a site of synthesis of protein on the ribosome. There is a transfer RNA for each of the 20 amino acids.

The proteins synthesized under the directional influence of mRNA may be enzymes that catalyze vitally important metabolic processes or they may be important structural components of living tissues. Thus DNA, through its master influence on the synthesis of a genetic copy of itself, the mRNA molecule, has a determining influence on the formation and functioning of living things.

Biological science is now in possession of information that can give an insight into, and an understanding of, the way in which genetic characteristics are transmitted and made to function. This information involves the genetic code, which will now be discussed.

THE GENETIC CODE

To understand this part of our discussion, keep in mind that the functioning of DNA and mRNA in the transmission of genetic information is dependent on their possession of four bases: adenine, guanine, thymine, and cytosine. Instead of thymine, the structurally similar pyrimidine uracil in mRNA functions widely in genetic transmission. Uracil has the structure of thymine without its methyl group and therefore can function in forming hydrogen bond combinations with adenine the same as thymine does.

Nirenberg, in 1961, was the first to synthesize a compound from one of the four nucleotides. This compound, polyuridylic acid (poly-U), had the ability to function like a natural mRNA. Its synthesis was accomplished by mixing uridylic acid with a cell-free extract of colon bacilli that contained a polymerase. Polyuridylic acid, consisting of a long chain of uridylic acid molecules linked together (UUU . . .), was found to incorporate phenylalanine into a polypeptide resembling a miniature protein. Polyuridylic acid therefore has the right sequence of bases, or the right "code," for the synthesis of a proteinlike compound from phenylalanine. The only 1 of 20 amino acids that could be incorporated into protein with this compound was phenylalanine.

Since the remarkable discovery described here, there have been other polynucleotides synthesized from the four nucleotides. These also have the property of stimulating the incorporation of amino acids into protein. In fact, at least one compound has been prepared that will bring about the incorporation of each of the 20 amino acids into protein.

It is now believed that the sequence of three nucleotides or bases, called a *codon,* serves as a code for each amino acid. Each codon serves to insert a specific amino acid into the growing peptide chain to form protein. Since we have four bases, we can have 4^3, or 64, different codons. Examples of such codons are UUU, which promotes the synthesis of phenylalanine into artificial protein; UUA, which promotes the synthesis of isoleucine into protein; and AUG, which brings about incorporation of glutamic acid into protein.

Table 17-1. Genetic codons*

Amino acid incorporated into protein	U-containing codons that stimulate protein synthesis	Non-U-containing codons that stimulate protein synthesis
Alanine	CUG	CAG, CCG
Arginine	GUC	GAA, GCC
Asparagin	UAA, CUA	CAA
Aspartic acid	GUA	GCA
Cysteine	GUU	
Glutamic acid	AUG	AAG
Glutamine	UAC	AGG, AAC
Glycine	GUG	GAG, GCG
Histidine	AUC	ACC
Isoleucine	UUA, AAU, CAU	
Leucine	UAU, UUC, UGU, CCU	
Lysine	AUA	AAA
Methionine	UGA	
Phenylalanine	UUU, UCU	
Proline	CUC	CCC, CAC
Serine	CUU, UCC	ACG
Threonine	UCA	ACA, CGC, CCA
Tryptophan	UGG	
Tyrosine	AUU, ACU	
Valine	UUG	

*Groups of three purine or pyrimidine bases in a sequence in a nucleotide chain (RNA) that stimulate protein synthesis. As explained in the text, U stands for uridylic acid, A stands for adenylic acid, G stands for guanylic acid, and C stands for cytidylic acid, or their respective bases.

Some of the genetic codons, or sequences of bases, that stimulate amino acid incorporation into protein synthesis are shown in Table 17-1.

For example, the messenger RNA (mRNA) codon for methionine is UGA, which means that the corresponding DNA codon is ACT (but not TGA).

Research by Nirenberg indicated that the sequence of bases, or codons, in mRNA influences the specific amino acid entry into the protein molecule. The sRNA holds the amino acid, and apparently the sRNA is bonded to the codon of the mRNA. This allows the amino acid to be inserted in the proper sequence in the growing polypeptide chain as patterned after the original codon in the DNA.

Considering there are an estimated 20,000 nucleotides in a DNA molecule, this is enough to form about 7000 codons (three nucleotides to a codon).

If each codon is responsible for the placement of one amino acid in a growing polypeptide chain, and since there are 20 amino acids, it is easy to visualize the tremendous number of combinations possible. This is why proteins are extremely complicated in molecular structure, and yet each protein molecule is made up of the correct sequence of amino acids, all patterned after the original DNA codon.

It is possible that the codon may become scrambled in some way or be translated wrong. The result is called a mutation. Mutations can be caused by chemicals and, particularly, by excessive exposure to x-rays or nuclear radiations.

When an important enzyme (protein) is not synthesized in the body because of a defect in the DNA-RNA transfer process, genetic diseases result, hemophilia and phenylketonuria (PKU) to mention a few.

The chemistry of the genetic process is complex, obviously. It will become easier to understand as more light is shed on the mechanisms by the active research being devoted to the subject. Meanwhile the student should realize that nothing more than a general understanding of these processes and an appreciation of their function is expected from a perusal of this chapter.

Questions for study

1. Where are nucleoproteins found? What is the importance of nucleoproteins?
2. What is a pyrimidine? Name two. Where do they occur?
3. What is a purine? Name three. Where do they occur?
4. What is a nucleoside? A nucleotide?
5. What are the components of nucleic acid?
6. How are the components of nucleic acid linked together?
7. How large is the nucleic acid molecule?

Nucleoproteins, nucleic acids, and genetics

8. What is DNA? RNA?
9. What is the difference between DNA and RNA with respect to the following: (a) the pentose in each, (b) the number of strands of nucleotides in the molecule of each, and (c) the molecular weight of DNA and RNA?
10. What are genes?
11. What is the relation of DNA to the gene?
12. Describe how the DNA molecule reproduces itself. What is this process called? What is its significance?
13. Describe how genetic information is transmitted from cell nuclei to the cytoplasm.
14. Name the four nucleotides involved in the storage and transmission of genetic information.
15. What is the significance of the genetic code?

18 Enzymes

The rate of chemical reactions can often be changed by introducing a catalyst. The living cells of plants and animals contain important catalysts called enzymes. *Enzymes are proteins that function as catalysts.* They are produced by living organisms, and their activities occur chiefly in living matter, although they do not necessarily have to be in living cells to carry out their work.

Enzymes have great power to make chemical reactions take place more rapidly. For example, the enzyme rennin has been shown to have remarkable ability to promote the coagulation of the casein of milk. One part of rennin will coagulate 72 million parts of fresh raw skimmed milk in 10 minutes under favorable conditions of temperature and hydrogen ion concentration.

Other examples of enzymes are those found in yeast, which act on certain sugars, forming alcohol and carbon dioxide; ptyalin in saliva, which aids in the digestion of starch, dextrin, and glycogen; and pepsin in gastric juice, which aids in the digestion of proteins.

It is important to understand that enzymes accelerate but do not start chemical reactions. Their function is to make reactions take place faster, not to initiate them.

Properties. The most important properties of enzymes are (1) they are specific in their action, (2) they require favorable temperatures for their activity, and (3) they are active in a limited range of hydrogen ion concentration.

Specific in action. Enzymes are specific in their action; that is, an enzyme will stimulate only one chemical reaction. Pepsin digests proteins, and it will not act on carbohydrates, fats, or other substances. Rennin coagulates the casein of milk, and it will not attack any other substance. Ptyalin will act only on starch or substances that have chemical linkages like those in starch, such as dextrins or glycogen.

Dependent on favorable temperatures for activity. The most favorable temperature for the action of an enzyme is known as its optimum temperature. For the enzymes of the animal body the optimum temperature is 38° C. Above 50° C the activity of enzymes diminishes rapidly. At 70° to 80° C most enzymes are irreversibly inactivated. Only a few enzymes are not destroyed by heating above 80° C. Low temperatures retard the action of enzymes but do not destroy them; if they are placed in a favorable temperature again, they will continue their activity.

Active in a limited range of hydrogen ion concentration. Some enzymes work best in a medium that is around the neutral point; some require a dilute acid reaction; and others need a slightly alkaline medium for their activity. For example, urease is an enzyme that accelerates the conversion of urea into ammonium carbonate. Urease has varying degrees of activity within a pH range of 4 to 8.8, with optimum activity at pH 6.8 to 7.2; it is inactive below pH 4 or above pH 8.8. Pepsin is active in a pH range of approximately 1 to 5, with optimum at pH 1.5 to 2. Trypsin is effective at a pH of approximately 7 to 10, the optimum activity of this enzyme being at a pH of 8 to 9. Thus the range of pH in which an enzyme will manifest activity is limited and that in which optimum activity is exhibited is rather narrow.

Substrate. The substance on which an enzyme acts is called its substrate. For example, proteins are the substrate of pepsin, casein is the substrate of rennin, and starch is the substrate of ptyalin.

Nomenclature. The general rule in naming enzymes is to name them after their substrate, changing the suffix of the name of the substrate to *ase*. The enzyme that digests lactose is caled lact*ase,* the one that digests maltose is called malt*ase,* and the one that digests sucrose is called sucr*ase*. This rule, however, has many exceptions. Some enzymes such as pepsin, trypsin, rennin, and ptyalin have special names that were given to them before a systematic method of naming was established.

Classification. Classifying the enzymes will provide a basis for a better understanding of their action. The following classification is based on the chemical action they perform.

    I. *Hydrolytic enzymes (hydrolases)*—Enzymes that catalyze the splitting of compounds by water with the introduction of the elements of water into the products formed.
        1. *Esterases:* Enzymes that catalyze the hydrolysis of esters.
            a. *Simple esterases:* These substances act on simple esters. Example: liver esterase, which catalyzes the splitting of simple esters of short-chain fatty acids.

$$C_2H_5O-CO-C_3H_7 + H_2O \underset{}{\overset{\text{Esterase}}{\rightleftarrows}} C_2H_5OH + C_3H_7COOH$$

              Ethyl butyrate   **Ethyl alcohol**   **Butyric acid**

            b. *Lipases:* Enzymes that catalyze the hydrolysis of fats into fatty acids and glycerol. Example: the pancreatic lipase steapsin (p. 278).
            c. *Phosphatases:* Enzymes that catalyze the hydrolysis of esters of phosphoric acid. Example: glyceryl phosphatase.

$$Glyceryl-PO_3H_2 + H_2O \underset{}{\overset{\text{Phosphatase}}{\rightleftarrows}} Glycerol + H_3PO_4$$

              Glyceryl phosphate   **Phosphoric acid**

        2. *Carbohydrases:* Enzymes that catalyze the hydrolysis of complex carbohydrates into simpler ones.
            a. *Saccharidases:* Enzymes that split disaccharides and trisaccharides into mono-

saccharides. Example: sucrase, which promotes the hydrolysis of sucrose into glucose and fructose.
  b. *Polysaccharidases:* Enzymes that split polysaccharides. Example: salivary amylase, which splits starch into maltose and dextrins.
3. *Proteases:* Enzymes that hydrolyze proteins.
  a. *Proteinases:* Enzymes that split proteins into peptides and amino acids. Examples: trypsin and chymotrypsin, the action of which is shown on p. 273.
  b. *Peptidases:* Enzymes that split peptides into amino acids. Examples: carboxypeptidase, aminopeptidase, and dipeptidase, whose action is described on p. 277.
4. *Nucleases:* Enzymes that catalyze the hydrolysis of nucleic acids, liberating nucleotides. Examples: ribonuclease and deoxyribonuclease.
5. *Amidases:* Enzymes that catalyze the hydrolysis of amides. Example: urease, which catalyzes the hydrolysis of urea as follows:

$$(NH_2)_2CO + 2H_2O \xrightarrow{\text{Urease}} (NH_4)_2CO_3$$

Urea   **Ammonium carbonate**

II. *Oxidation-reduction enzymes*
  1. *Oxidases:* Enzymes that catalyze the oxidation of organic substances. Example: xanthine oxidase, found in liver and milk, which catalyzes the oxidation of hypoxanthine to xanthine and of xanthine to uric acid.
  2. *Dehydrogenases:* Enzymes that catalyze the removal of hydrogen from a substrate, for example, from fatty acids, amino acids, or intermediates in carbohydrate metabolism. A cofactor must be present to serve as an acceptor and carrier of the hydrogen. Example: lactic dehydrogenase, which in the presence of the cofactor, diphosphopyridine nucleotide, removes 2 atoms of hydrogen from lactic acid.
  3. *Catalases:* Enzymes that decompose hydrogen peroxide, liberating molecular oxygen. Example: catalase in blood.
  4. *Peroxidases:* Enzymes that decompose hydrogen peroxide or organic peroxides with a transfer of oxygen to another compound. Example: peroxidase in horseradish.
III. *Transferases*—Enzymes that catalyze the transfer of a component of the molecule of one substance to the molecule of another substance. Examples: transaminase, which catalyzes the transfer of an amino group from glutamic acid or aspartic acid to a ketogroup of another amino acid.
IV. *Lyases*—Enzymes that catalyze the breaking or the formation of a carbon chain.
  1. *Decarboxylases:* Enzymes that catalyze the liberation of CO_2 from the carboxyl of an organic acid. Example: pyruvic decarboxylase, which in the presence of cocarboxylase (thiamine pyrophosphate) removes CO_2 from pyruvic acid.
  2. *Carbonic anhydrase:* An enzyme that catalyzes the reversible combination of CO_2 and H_2O to form H_2CO_3.

$$CO_2 + H_2O \xrightleftharpoons[]{\text{Carbonic anhydrase}} H_2CO_3$$

Mechanism of action. Enzymes accomplish their catalysis by combining with the substrate or substance on which they act. The substrate, after combining with the enzyme, undergoes a chemical change and forms a new substance called the product. The complex, consisting of the product combined with the enzyme, then dissociates, yielding the free enzyme and the new product. This is illustrated by the following equations:

$$\text{E} + \text{S} \rightleftarrows \text{E}-\text{S}$$

Enzyme   Substrate   Enzyme-substrate complex

$$\text{E}-\text{S} \rightleftarrows \text{E}-\text{P} \rightleftarrows \text{E} + \text{P}$$

Enzyme-substrate complex   Enzyme product   Enzyme   Product

In more technical language, we say that enzymes accomplish catalysis by lowering the activation energy or the energy required to bring about the chemical change. The energy in this case is in the form of heat. In living cells containing enzymes, chemical reactions take place faster because the activation energy of these reactions is within the range of temperatures of these cells. The presence of enzymes in the animal body thus brings about more rapid chemical changes than would occur in the absence of these enzymes; without enzymes, chemical reactions in tissues would be too slow to maintain life.

The decomposition of hydrogen peroxide, for example, requires 18,000 calories of energy to change 1 mole of this substance into water and molecular oxygen. If colloidal platinum is added, the decomposition of the same amount of hydrogen peroxide requires only 11,700 calories. In the presence of the enzyme catalase the energy required for the decomposition of hydrogen peroxide is less than 2000 calories per mole.

When hydrogen peroxide is added to a cut wound, a rapid evolution of oxygen gas occurs because the tissues contain the enzyme catalase, which accelerates the decomposition of the hydrogen peroxide at the temperature of the wound. If hydrogen peroxide is allowed to stand in an open container at the same temperature without a catalyst, it will decompose, but at a rate that is very much slower than the rate of decomposition in the presence of catalase.

Importance. The great importance of enzymes is that they catalyze most of the chemical reactions that occur in animals and plants. Such processes as the digestion of foods, respiration, oxidations in the tissues, and metabolism in general are accomplished almost wholly through the activity of enzymes.

It is well known that applying heat will make chemical reactions take place faster. However, high temperatures cannot be used in tissues to bring about increased chemical activity because high temperatures are damaging to living cells and would be difficult to control. With their many enzyme systems, living cells can bring about increases in the rates of chemical reactions as needed—at moderate and essentially constant temperatures —and at the same time maintain control over the processes.

Coenzymes. Most enzymes require the presence of another nonprotein substance in the reaction mixture to enable them to carry out their catalytic action. When the additional factor is an organic compound, it is called a coenzyme.

In some coenzymes a vitamin is a part of the molecule. Examples are coenzyme A, which contains the vitamin pantothenic acid as a part of its molecule, and cocarboxylase, whose molecule contains the vitamin thiamine (B_1). Coenzyme A, as we shall see later, is essential for the action of certain enzymes that function in metabolism, and cocarboxylase participates with the enzyme carboxylase in the metabolic breakdown of pyruvic acid, an intermediate in the metabolism of carbohydrates.

There are other coenzymes, which structurally are not related to the vitamins. Unlike most enzymes, coenzymes are damaged very little or not at all by heating.

Activators. As previously stated, the activity of most enzymes is promoted by the presence in the reaction mixture of some other substance. When the necessary substance is an inorganic ion, the factor is called an activator. Examples of activators are the chloride ion, which promotes the action of the salivary enzyme ptyalin; magnesium ion, which accelerates the action of several enzymes active in carbohydrate metabolism; and the hydrogen ion of HCl, which converts pepsinogen of gastric juice into the active enzyme pepsin.

Inhibitors. Certain substances slow down or stop the action of enzymes. These substances are called inhibitors. Examples are heavy metals such as lead or mercury, which precipitate the enzyme (a protein) from solution; cyanide (HCN, KCN), a deadly poison in the body, which poisons some of the respiratory enzymes by combining with and thereby inactivating the vitally functioning copper and iron in the molecule; fluoride, which precipitates the magnesium ion necessary as an activator for the action of phosphorylating enzymes; and iodoacetic acid, which destroys the sulfhydryl ($-$SH) group of the enzyme that is essential for its action.

Questions for study

1. What are enzymes?
2. Give three properties of enzymes.
3. What is a substrate?
4. What is the general rule for naming enzymes?
5. Why are enzymes so important?
6. Do enzymes start chemical reactions? Explain.
7. What is a coenzyme? Name two.
8. What is an activator? Inhibitor?

19 Hormones

Hormones are internal secretions. They are produced by glands that discharge these highly important physiological substances directly into the circulating fluids of the body. In contradistinction to an external secretion, an internal secretion is not discharged through a duct into the lumen of a viscus but is secreted into the blood and conveyed by the latter and by interstitial fluids to its site of activity. The hormones have a remarkable influence on the growth, size, appearance, functioning, and general metabolism of the body. They are effective in very small quantities, and they act as regulators. These substances are called hormones, a name that comes from a Greek word meaning "to arouse to activity," because of their stimulating influence on the functioning of the body.

The glands producing internal secretions were once known as the ductless glands because of the absence from their structure of secretory ducts. They are now called the endocrine organs. The more important endocrine organs are the thyroid and the parathyroid glands, the pancreas, the adrenal glands, the pituitary gland, the ovaries, the testes, and the thymus. The biochemical significance of the secretions of each of these will be discussed (see Fig. 19-1).

THYROID GLAND

Iodothyronines. Four of the iodine-containing compounds have thyroid hormone activity. These compounds have the basic structure of thyroxin, differing only in the number of atoms of iodine in the molecule. Besides thyroxin, whose molecule contains 4 atoms of iodine, there are two triiodothyronines and one diiodothyronine. These compounds originate in the thyroid gland. A mixture of these four iodothyronines carries on thyroid function. Thyroxin is the most abundantly occurring of these compounds.

Thyroxin. Thyroxin is a secretion produced by the thyroid gland. The thyroid gland is a bilobed and highly vascular organ, situated in the neck, in front of the larynx and trachea (Figs. 19-2 and 19-2, *A*). In man the normal thyroid weighs about 20 to 25 g.

The chemistry of thyroxin is well known. It has been isolated in the pure form from the thyroid glands of animals and has also been prepared synthetically in the laboratory.

Thyroxin has a profound influence on body functions. It has a stimulating effect on growth; it is concerned in the metabolism of the skin, which becomes thick and dry when

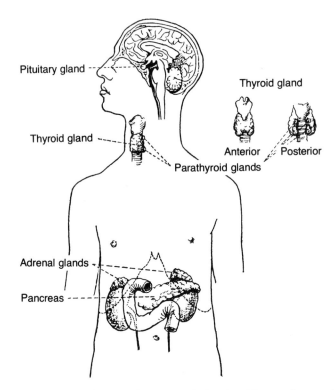

Fig. 19-1. Diagram showing location of glands of internal secretions.

the hormone is deficient, and it has a marked regulatory influence on the rate of oxidations in the body. It is extremely potent; 1 mg of thyroxin will increase the oxidations that take place in the human body 3%.

Thyroxin

Protein-bound iodine (PBI). The iodothyronines are formed and stored in the thyroid gland. These compounds are transferred from the thyroglobulin of the thyroid gland to the blood as needed. In the blood these hormones combine with proteins, forming the protein-bound iodine (PBI) of the plasma. The level of PBI in the blood is a sensitive indicator of the rate of metabolism of the body. The normal PBI level of the blood is 4 to 8 μg/100 ml of plasma. In hyperthyroidism the PBI level of the plasma is increased and in hypothyroidism the PBI level is below normal.

Thyroglobulin. Thyroglobulin is a constituent of the thyroid gland. It is a protein with an iodine content of 0.5% to 1% and a molecular weight around 700,000. When

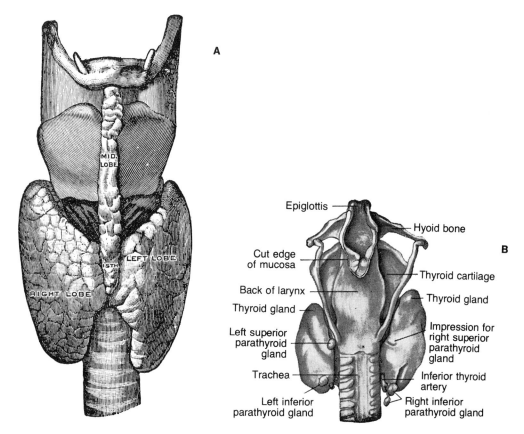

Fig. 19-2. A, Anterior view of thyroid gland. **B,** Posterior view of thyroid gland showing four parathyroid glands.

administered by mouth, it produces the characteristic effects of thyroid hormone. On hydrolysis, thyroglobulin yields the amino acids usually found in proteins, the active hormones thyroxin and triiodothyronine, and the iodine-containing amino acids monoiodotyrosine and diiodotyrosine. The monoiodotyrosine and diiodotyrosine are precursors of the thyronines. This protein apparently functions as a storage substance for the iodine-containing compounds associated with thyroid function.

Cretinism

If the thyroid gland fails to develop properly in fetal life or in childhood, an individual stunted both physically and mentally is the result. Such an individual is called a cretin, and the condition is known as *cretinism*. A cretin is a dwarfed, pot-bellied individual with coarse hair, coarse, dry skin, subnormal basal metabolism, and low mentality. Should atrophy of the thyroid gland occur during adult life, symptoms somewhat similar to those found in cretinism, with the exception of stunted growth, will result. One of these symptoms is the development of a thick, coarse, dry skin, a condition called *myxedema;* other symptoms are a reduced basal metabolism, a lowered pulse rate, a tendency to subnormal body temperature, and physical and mental sluggishness. All these abnormalities of structure and function are due to a deficient production of the iodinated hormones by the thyroid gland.

A cretin may be made to grow normally if fed preparations containing thyroid hormone before adult age is reached, and other symptoms of cretinism, such as coarse hair and skin, subnormal basal metabolism, and dwarfed intellect, may be improved or even made to disappear entirely if treatment is instituted. Likewise such symptoms of hypothyroidism in the adult as myxedema, subnormal metabolism, and physical incapacity will disappear when the subject is given a properly outlined treatment with thyroid hormone.

Simple goiter

Another form of thyroid disease is the condition known as *simple goiter*. This is a hyperplasia or increase in the cellular elements of the thyroid gland. There is also a marked increase in colloid material in the gland; hence the name *colloid goiter* is sometimes used. In this type of thyroid disease there is an increase in the size of the neck in the region of the thyroid as a result of the hyperplasia of the gland.

Simple goiter is due to a deficiency of iodine in the food and drinking water. It is endemic in regions in which there are very small amounts of iodine compounds in the soil and subsoil and consequently a low iodine content in the drinking water and in the vegetables grown in these regions. The districts of highest incidence of simple goiter are the Himalaya Mountain region of Asia, the region of the Alps in Europe, the Andes Plateau of South America, the area about the Great Lakes and St. Lawrence River, and the Cascade Mountain regions of Washington, Oregon, and British Columbia in North America. This disease is now being successfully prevented by the introduction of iodine compounds into the diet or by the ingestion of sodium iodide or other iodine salts periodically.

Hyperthyroidism

The thyroid may also manifest increased as well as subnormal activities. The condition in which there is overactivity of the thyroid gland, producing excessive amounts of thyroxin, is known as *hyperthroidism*. Its principal symptoms are an increased basal metabolic rate, a rapid and irregular heart rate, exophthalmos or bulging of the eyes, restlessness, hyperexcitability, and loss of weight. In this condition the individual is very much speeded up biologically, his vital activities taking place much too fast, in direct contrast with the hypothyroid condition in which the individual is dull and sluggish, both physically and mentally, with greatly lowered vital activities.

Effective control of the thyrotoxicosis due to hyperthyroidism is obtained by surgical removal of the thyroid gland and by oral administration of radioactive iodine. Alleviation of symptoms is also obtained by the oral administration of antithyroid drugs, such as thiourea, thiouracil, and propylthiouracil, which function by inhibiting the synthesis of thyroid hormone.

PARATHYROID GLANDS

The parathyroid glands are small structures, being about 3 to 15 mm long and 2 to 3 mm broad in man. They are situated in the neck on the posterior surface of the thyroid gland. Man usually has four of these glands, two on each lobe of the thyroid (see Fig. 19-2).

The parathyroid glands are small structures, being about 3 to 15 mm long and 2 to normal calcium content of the blood under fasting conditions is 9 to 11 mg/100 ml of serum. The particular action of the secretion of the parathyroid glands is the maintenance of this normal circulating level of calcium in the blood. If the parathyroid glands are removed from a dog, the animal's blood calcium content is lowered; it develops tetany and dies in spasms in a few days unless treatment is introduced. Tetany produced in this way is relieved by injecting an extract of the parathyroid glands, which raises the level of the blood calcium or by injecting a soluble calcium salt, which has the same effect as injecting parathyroid extract—increasing the blood calcium content. A similar influence is exerted on a normal animal by an extract of the parathyroid glands. If parathyroid extract is injected into a normal animal, the blood calcium is increased for a time to a level higher than normal. Similar experiments show that the secretion of the parathyroid glands is definitely related to the mobilization of calcium in the blood. Whether the blood calcium is low or normal, *the injection of parathyroid extract results in an elevation of the level of calcium in the blood plasma. This mobilization of calcium in the blood is brought about by a withdrawal of calcium from its storage deposits in the bones.*

Parathyroid extract has been found of value clinically in the treatment of calcium metabolism disturbances.

PANCREAS

Insulin. Insulin is a secretion of the β-islet cells of the islands of Langerhans of the pancreas (Fig. 19-3). It is a hormone that has an important part in the regulation of carbohydrate metabolism. The normal concentration of sugar in the blood under fasting conditions is from 65 to 100 mg/100 ml. The influence of insulin is to assist in maintaining this circulating level of blood sugar. Insulin accomplishes this function by aiding in the storage of glucose as glycogen in the muscles and by facilitating the oxidation of glucose. After a meal containing carbohydrates is ingested, the blood sugar is elevated to a concentration of perhaps 120 to 160 mg/100 ml of blood, which is a moderate hyperglycemia. This concentration of sugar is soon reduced to the normal level, partly by the conversion of glucose to glycogen, which is then stored, and partly by the oxidation of glucose to carbon dioxide and water. That insulin also facilitates the oxidation of glucose is shown by chemical tests (respiratory quotient studies), which reveal an increased oxidation of sugar following the injection of insulin.

Continuous *hyperglycemia* or *elevation of the blood sugar above normal* is a pathological condition. If under fasting conditions the circulating level of the blood sugar is found to be in excess of 130 mg/100 ml, there is some bodily dysfunction, which is nearly always a deficient production of insulin by the pancreas. *The disease in which*

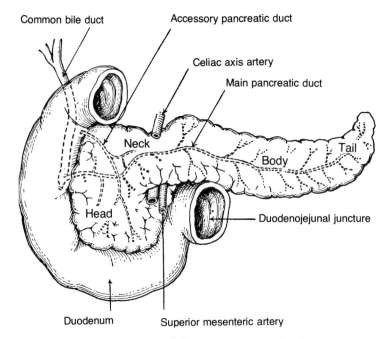

Fig. 19-3. Diagram of the pancreas and duodenum.

there is an insufficient production of insulin by the β-islet cells of the islands of Langerhans of the pancreas is known as diabetes mellitus. In this disease there is a failure to store and to oxidize glucose. It is characterized by an increased blood sugar and by glucosuria. As a result of the failure to utilize glucose, there is excess ketone body formation with marked ketonemia and ketonuria. The excretion of ketone bodies, two of which are acids, depletes the body's bases and causes an acidosis.

Insulin is prepared for clinical use by extraction of the macerated pancreases of animals (hogs, cattle, and sheep) with acid alcohol, followed by a series of chemical steps of purification. The purified product is dissolved in dilute hydrochloric acid, and its potency is determined by its action on rabbits. It is issued for use in sterile solutions of such concentrations that 1 ml contains 20, 40, 80, or 100 units, *a unit of insulin being one-third of the amount that will lower the blood sugar of a fasting 2 kg rabbit to the level at which convulsions occur within a period of 5 hours after its injection.* It is generally considered that 1 unit of insulin is necessary to bring about the metabolism of 2 g of carbohydrate in the human body.

Insulin has been obtained in a pure cystalline form. It has been found to be a protein with a molecular weight of about 6000. The chemical synthesis of insulin in the laboratory has been accomplished.

Insulin is administered in solution by subcutaneous injection and is ineffective when given by mouth because, being a protein, it is digested by proteolytic enzymes in passing through the alimentary tract. The effectiveness of insulin is reduced by the presence of infection in the body.

It is important to realize that the injection of insulin will lower the concentration of blood sugar whether the blood sugar is high or low. If insulin is given to a diabetic patient with a high blood sugar, this patient's blood sugar will be reduced, and if given to an individual with a normal blood sugar, his blood sugar will be reduced to a level below normal. *When the blood sugar becomes lower than normal, the condition is called hypoglycemia.* This sometimes occurs from an overdosage of insulin. Hypoglycemia is an undesirable and often dangerous condition. In hypoglycemia of moderate degree, such as when the blood sugar is around 50 mg/100 ml, the patient manifests such symptoms as sweating, nervousness, tremor of lips and hands, blurring of vision, and dizziness. In more advanced hypoglycemia, as when the blood sugar is 25 to 35 mg/100 ml, the patient becomes unconscious. Hypoglycemia and its attendant symptoms may be promptly dismissed by the administration of sugar either by mouth or intravenously, the intravenous method giving the more rapid response.

PITUITARY GLAND

The pituitary gland or hypophysis cerebri is a small gland weighing about 0.6 g in man. It is situated at the base of the brain in a groove in the sphenoid bone (Fig. 19-1). It consists of three parts, an anterior lobe, a posterior lobe, and a middle lobe, or pars intermedia. These three lobes are attached by a stalk, called the infundibulum, to the floor of the third ventricle of the brain, the whole structure being encased in the dura mater and pia mater, the membranes that invest the brain. The three lobes differ from each other as to embryological origin, structure, and function.

The pituitary gland may be considered the master gland of the body. Its secretions have not only important direct physiological effects but also a controlling influence on other endocrine organs.

Physiological effects of secretions of the anterior pituitary gland

Six hormones have been found in the anterior lobe of the pituitary gland. These hormones have been prepared in a highly purified form and have been found to be proteins. The effect of these secretions will be discussed.

Growth or somatotropic hormone. Of the anterior lobe secretions, a very important one is a hormone that influences growth. If a deficiency of this secretion occurs before maturity, an incomplete growth results, which in extreme cases amounts to dwarfism. An excessive secretion of this hormone produces abnormal growth. If the hypersecretion occurs before adult life, giantism results. In giantism there is unusual growth of the long bones of the body, producing a person of gigantic proportions (Fig. 19-4). If the increased secretion occurs during adult life, the principal manifestations are abnormal growth of the hands and feet and of the bones of the face, a condition known as *acromegaly.*

Growth hormone brings about one aspect of growth by increasing the rate of protein synthesis, with consequent diminution of urinary nitrogen excretion. Another phase of growth affected by the growth hormone is bone formation at the epiphyses of certain bones. The growth hormone stimulates greater phosphatase activity, bringing about the formation of longer bones, which results in giantism.

258 Roe's principles of chemistry

Fig. 19-4. Giants and dwarfs owe their abnormal stature to the overactivity or underactivity of the pituitary gland. (Courtesy Press Association, Inc., New York, N. Y.)

Gonadotropic hormones. The anterior pituitary gland produces gonadotropic hormones, substances that have a stimulating effect on the development of the genital organs. Surgical removal of the anterior pituitary gland is followed by atrophy of the testes, seminal vesicles, and prostate gland if the animal is a male and atrophy of the ovaries, fallopian tubes, and uterus if the animal is a female.

The gonadotropic secretion of the anterior pituitary gland are (1) a follicle-stimulating hormone (abbreviated designation is FSH), (2) a luteinizing hormone (LH), and (3) a luteinizing and lactogenic hormone called luteotropin (LTH). Follicle-stimulating hormone (FSH) initiates the development and stimulates the growth of the ovarian follicle in the female; in the male this hormone promotes the development and functioning of the epithelium of the seminiferous tubules of the testes, stimulating these tubules to spermatogenic activity. Luteinizing hormone (LH) in the female stimulates the ripening of the ovarian follicle, the rupture of the follicle, with release of the ovum (ovulation), and the development of the corpus luteum from the ruptured follicle. In the male, luteinizing hormone stimulates the functioning of the cells in the testes that secrete the male hormone testosterone. Luteotropin (LTH) in the female induces the corpora lutea to secrete progesterone, a female sex hormone; LTH is also necessary for the initiation of lactation in mammals at parturition. The interrelationships of these hormones in the menstrual cycle will be discussed on p. 265.

Thyrotropic hormone. Removal of the anterior pituitary gland brings about an atrophy of the thyroid gland and results in a lowered basal metabolism. The administration of proper amounts of anterior pituitary extract to an animal whose pituitary has been removed will restore the atrophied thyroid to normal size and function. Administration of excess anterior pituitary extract produces a hyperplasia or enlargement of the thyroid gland. Such experiments have shown that the anterior pituitary gland secretes a hormone that has a controlling effect on the growth and function of the thyroid gland. This substance is called the thyrotropic hormone. It appears to be a small-molecule protein.

Adrenocorticotropic hormone (ACTH). The cortex of the adrenal glands atrophies when the anterior pituitary gland is removed from an animal. Administration of anterior pituitary extract restores the adrenal glands of such an animal to normal. Thus it can be shown that the anterior pituitary gland secretes a hormone that aids in maintaining normal structure and function of the adrenal glands. This hormone is called the adrenocorticotropic hormone, ACTH, or corticotropin. ACTH aids in keeping the anatomical structure of the adrenal cortex normal and also stimulates this gland to produce the cortical hormones or corticoids—such as corticosterone and cortisone. The effects of the corticoids will be discussed later.

Relation of the anterior pituitary to carbohydrate metabolism—the diabetogenic factor. It has been found that daily injections of crude anterior pituitary extract into animals produces hyperglycemia and glucosuria. Continuous daily injection of the crude extract for several weeks produced permanent diabetes in dogs. This experimentally produced diabetes appeared to be the result of an elevation of the blood sugar to a concentration that, when maintained continuously, had a damaging effect on the β-islet cells of the pancreas (which secrete insulin) and thus resulted in a permanent diabetes.

Two of the hormones secreted by the anterior pituitary gland have a blood sugar–raising effect; these are the growth hormone and adrenocorticotropin, or ACTH.

When purified growth hormone is injected into an animal, the blood sugar level is raised. This effect is as yet unexplained, but experimental work has brought forth suggestive evidence. It has been observed that anterior pituitary extract inhibits the action of hexokinase, the enzyme that promotes phosphorylation of glucose, the first step in glucose metabolism. The presence of an excess of the growth hormone in the tissues might therefore increase the concentration of glucose in the blood and tissues by preventing the entrance of this sugar into its normal metabolic pathway.

The effect of ACTH is indirect. As mentioned previously, this hormone stimulates the adrenal coretx to produce various cortical hormones (such as corticosterone and cortisone); these, in turn, promote gluconeogenesis or the formation of glucose from noncarbohydrate precursors. Gluconeogensis takes place in the liver, and when this process is speeded up, more glucose is passed from the liver into the bloodstream, thereby increasing the level of the blood sugar. Thus ACTH, through its stimulatory effect on

Table 19-1. The amino acids of oxytocin and vasopressin

Oxytocin	Vasopressin
Cystine	Cystine
Tyrosine	Tyrosine
Isoleucine	Phenylalanine
Glutamine	Glutamine
Asparagine	Asparagine
Proline	Proline
Leucine	Arginine
Glycinamide	Glycinamide

the production of blood sugar–raising hormones by the adrenal cortex, indirectly brings about an elevation of the blood sugar level when present in the tissues in excess.

In view of the effects on the blood sugar level just described, it appears that the functioning of the anterior pituitary gland may have some relationship to diabetes mellitus. It is of further significance that some patients with acromegaly (hyperpituitarism in the adult) exhibit hyperglycemia and glucosuria.

Secretions of the posterior pituitary gland

The posterior pituitary gland secretes two hormones—*oxytocin* and *vasopressin*. Each of these hormones consists of eight amino acids held together by peptide linkages (Table 19-1).

Oxytocin has a stimulating action on smooth muscle. This action is characteristically effective on uterine muscle, causing contraction of the uterus, a physiological response that is called the oxytocic effect. This hormone also stimulates the ejection of milk from the mammary gland.

Vasopressin is an antidiuretic hormone—a substance that diminishes the excretion of water by the kidneys. Its specific action is to stimulate the kidney tubules to reabsorb water. When the water content of the body is low, the posterior pituitary gland secretes more vasopressin, which passes by way of the blood to the kidneys where it stimulates the tubules to reabsorb water more actively and thus to conserve the water supply. On the other hand, when there is an abundance of water in the body, the posterior pituitary gland secretes little or no vasopressin; consequently, less water is reabsorbed by the tubules, and a greater excretion of water by the kidneys results. *Thus the posterior pituitary gland, through its antidiuretic hormone, vasopressin, assists in the regulation of the water balance of the body.*

Vasopressin has a vasoconstrictor action on peripheral blood vessels, an effect that tends to raise blood pressure; but this pressor effect is neutralized by the action of the hormone on the heart, which decreases coronary output. Administration of this hormone does not raise the blood pressure of man under normal conditions, but under anesthesia or other experimental conditions a pressor effect may be produced by the hormone.

Posterior pituitary extract has *two clinical applications:* it is used in obstetrics and in the treatment of diabetes insipidus.

The use of posterior pituitary extract in obstetrics is based on the physiological effect of oxytocin, a powerful stimulatory action on contraction of the uterus. It is a practice among obstetricians to administer pituitary extract immiediately after the birth of the baby to control postpartum hemorrhage and to facilitate the expulsion of the placenta. Obstetricians are very conservative about the use of oxytocin to induce labor, however, because of the danger of injury to the uterus or the baby.

Disease of the posterior pituitary gland brings about severe polyuria, with marked thirst, a condition known as *diabetes insipidus*. Diabetes insipidus disease is treated by daily administrations of posterior pituitary extract. This clinical application is based on the effect of vasopressin, the antidiuretic hormone.

ADRENAL GLANDS

The adrenal glands are small, highly vascular organs, one being located just above the upper pole of each kidney (Fig. 19-1). In man the weight of one gland is about 4 g. Each adrenal body consists of two parts, the cortex and the medulla. The cortex is the outer portion of the gland, and the medulla is the central part. Each of these parts is distinct, structurally and functionally, and each produces an internal secretion of utmost importance.

Secretions of the adrenal cortex

The symptoms of deficiency and the physiological effects of the secretions of the adrenal cortex will be discussed.

Symptoms of deficiency. If the adrenal glands are removed from an animal, the animal will not live. Death is caused by the absence of hormones secreted by the cortex. Pathology of the adrenal cortex in man brings on the condition known as *Addison's disease*. Symptoms of Addison's disease are a marked bronze coloration of the skin, profound weakness, prostration, loss of appetite, nausea, and vomiting.

Chemical analyses of the blood of patients with Addison's disease have revealed a decrease in the blood sugar and the sodium, chloride and bicarbonate ions and an increase in the potassium ion and the nonprotein nitrogen of the blood. Blood analyses of animals from which the adrenal glands have been removed show similar changes in composition. After adrenalectomy, animals become dehydrated, their blood volume decreases, and their blood pressure falls. Chemical studies of the urine of adrenalectomized animals show an increased excretion of sodium chloride and water after the removal of the glands.

Physiological effects. The hormones of the adrenal cortex affect the salt balance and the carbohydrate and protein metabolism.

Effects on salt balance. One function of the hormones secreted by the adrenal cortex is regulation of the sodium ion level in the blood. These hormones stimulate the kidney tubules to reabsorb sodium. When a sufficient amount of these hormones is present, the blood sodium is maintained at a normal level; when the amount of these hormones secreted is inadequate, sodium is excreted more rapidly in the urine, and a lowered level of sodium in the blood results. Since the chloride ion is associated with the sodium ion, the urinary loss of sodium is paralleled by a corresponding loss of chloride; consequently, the adrenalectomized animal or the animal with adrenocortical disease suffers from abnormal losses of sodium chloride in the urine. Losses of sodium chloride are accompanied by losses of water from the body, changes that decrease the blood volume and thereby lower the blood pressure. An insufficiency of adrenocortical hormones in a patient thus sets up a chain of abnormal changes in the body chemistry that finally results in collapse. The adrenocortical hormones that regulate the renal excretion of sodium chloride are sometimes referred to as the mineralocorticoids.

Effects on carbohydrate and protein metabolism. Another function of the hormones secreted by the adrenal cortex is stimulation of the conversion of amino acids into glucose. This process is called *gluconeogenesis*. It takes place in the liver and is one of the mechanisms by which the liver makes available a supply of glucose to keep the blood sugar from getting low and to maintain the stores of glycogen in the muscles and liver. In adrenocortical failure the blood sugar and the glycogen in the muscles and liver are decreased to very low levels. Administration of adrenocortical hormones produces an increase in the blood sugar concentration and the glycogen content of the liver and muscles. The adrenocortical hormones that have the property of stimulating gluconeogenesis are referred to as the *glucocorticoids*.

Chemistry. A number of highly purified crystalline compounds that have biological activity have been isolated from the cortex of the adrenal glands. Some of these substances are corticosterone, deoxycorticosterone, 11-dehydrocorticosterone, 11-dehydro-17-hydroxycorticosterone (cortisone), 17-hydroxycorticosterone (hydrocortisone), and aldosterone. These compounds are all steroids, and as the names suggest, they vary only slightly from each other in structure. Functionally there is some difference in their action and potencies. Aldosterone has the greatest sodium retention and hence life-maintenance potency.

Corticosterone

Clinical applications. Outstanding clinical applications of the adrenocortical hormones are made. Cortisone, hydrocortisone, and deoxycorticosterone are prepared synthetically and are available in adequate quantities. The therapeutic uses of these drugs come under two categories: (1) replacement therapy in adrenocortical deficiency states, for example, in the treatment of Addison's disease, and (2) as palliative treatment for the relief of certain inflammatory conditions, for example, rheumatoid arthritis, rheumatic fever, and rheumatic carditis.

Secretions of the adrenal medulla

Two hormones are secreted by the adrenal medulla. These are *epinephrine* and *norepinephrine*. Other names used for epinephrine and Suprarenin and Adrenaline. Norepinephrine is also called arterenol. Epinephrine and norepinephrine belong to the class of organic compounds called the catecholamines.

The principal physiological effects of epinephrine are that it (1) *accelerates the heart rate and increases cardiac output*, (2) *causes peripheral vasoconstriction*, (3) *raises blood pressure because of its effect on the heart and its vasoconstrictor action*, (4) *relaxes the bronchial muscles*, and (5) *stimulates the conversion of glycogen to glucose in the liver (glycogenolysis), increasing the level of blood sugar.*

Norepinephrine also is a peripheral vasoconstrictor. It therefore raises blood pressure. This is its most powerful action. It has some of the other effects of epinephrine, but to a lesser degree. It does not have any action on the heart rate or cardiac output, and it does not exhibit the bronchodilatory effect of epinephrine.

Epinephrine and norepinephrine are secreted by the adrenal glands in a ratio of about 5 parts of epinephrine to 1 part of norepinephrine. The chemical structure of these two compounds is given below:

$$\text{Epinephrine} \qquad \text{Norepinephrine}$$

The principal uses of epinephrine in medicine are in (1) providing emergency aid in heart failure since it has a marked cardiac-stimulating power, (2) alleviating the spasm of bronchial asthma since it relaxes the bronchial muscles, and (3) arresting hemorrhage when it can be applied locally, especially in local operations, since it has a powerful constricting influence on the blood vessels.

THYMUS

In man the thymus is situated partly in the lower portion of the neck and partly in the upper thorax. Its size is related to the age of the individual. At birth the thymus weighs about 15 g, and at puberty it weighs around 35 g; after puberty it undergoes involutionary changes and gradually decreases in size, weighing approximately 25 g at 25 years of age and less than 15 g at the age of 60 years.

Surgical removal of the thymus gland does not result in the death of the animal, showing that the gland is not essential to maintain life.

Apparently a relationship exists between the thymus and the sex glands. Castration retards and sexual activity (mating) hastens the involution of the thymus. It has been suggested that the thymus gland influences the rate of growth and accelerates the onset of maturity. Recent research has suggested that the thymus is associated with the immune reactions of the body to foreign protein material.

FEMALE SEX HORMONES

Our consideration of the female sex hormones will include the substances estrone, estradiol, estriol, and progesterone, which are formed in the ovaries, and the gonad-stimulating substances of the anterior pituitary gland.

Estrone was the first isolated in pure form from the urine of pregnant women in 1929. In 1930, another female sex principle, estriol, was isolated from pregnancy urine. *Estriol* is similar to estrone except that it has two additional OH groups and does not have a ketone group. *Estradiol* is estrone with 2 hydrogen atoms added. It was prepared in the laboratory from estrone in 1934 and was isolated from ovarian tissue in crystalline form in 1935. *Estradiol is about ten times as potent physiologically as estrone.* Progesterone was isolated from ovarian tissue in 1934 and was also prepared by chemical methods from a plant sterol (stigmasterol) during the same year.

It seems reasonable to assume that estradiol is the parent substance that gives rise to estrone and estriol since it is more potent physiologically. Estrone, estradiol, estriol, and progesterone originate from a common precursor, cholesterol. These substances are secreted by the ovaries into the blood, and after exerting their physiological effects, they are excreted in the urine and the bile.

Estrone, estradiol, and estriol have the physiological effect of producing estrus. Estrus is the condition in lower animals that is characterized by an urge for mating. For this reason these compounds are called the "estrogenic" hormones. *In addition to estrus production these hormones have the following physiological effects on the female. They bring about a proliferation and thickening of the endometrium of the uterus and an increase in its vascularity, and they influence the gradual growth of the mammary glands during prepuberty and the accentuated development of mammary tissue during pregnancy.* In general, it may be said that the estrogenic hormones have a profound effect on the growth and functioning of the female gonads other than the ovaries. These hormones also have a determining influence on the development of the secondary sexual characteristics of the female and distribution of body fat to effect the characteristic female body structure.

The production of the ovarian hormones is influenced by the gonad-stimulating principles of the anterior pituitary gland (FSH, LH, and LTH). In the absence of these substances the ovaries atrophy and fail in their formation of hormones. The effect of the anterior pituitary on the female sex organs is thus indirect. It functions by stimulating the production of hormones by the ripening ovarian follicle and the corpus luteum in the ovary, which in turn act directly on the other female gonads. On the other hand, evidence has accumulated to show that the estrogenic hormones have a depressant action on the function of the anterior pituitary. It thus seems that there is a reciprocal relationship between the ovaries and the pituitary gland.

During pregnancy the estrogenic hormones are found in the placenta, the amniotic fluid, the umbilical cord, and in increased quantities in the blood and urine. The amount in the urine gradually increases with the progress of gestation. It rapidly falls to a normal level within four or five days after parturition.

Progesterone

Progesterone is the active constituent of the corpus luteum. This hormone induces a mucus secretory activity in the endometrium, which is essential for the implantation of the fertilized ovum in the wall of the uterus and for the development of the placenta. Gestation will not take place in the absence of this hormone. Progesterone also has a quiescent effect on the contractions of the uterus, an action opposed to that of estrone.

The relations of the female sex hormones to the cyclic events taking place in the female gonads are summarized as follows and are illustrated diagrammatically by Fig. 19-5.

Menstrual cycle

The menstrual cycle covers a period of approximately 28 days. The cycle begins with the action of the follicle-stimulating hormone (FSH) of the anterior pituitary gland. Under the influence of this hormone a primordial graafian follicle in the ovary develops into a mature follicle that expels an ovum in 11 to 16 days. During this time the follicle actively secretes estrogenic hormones. Immediately after ovulation the ruptured graafian follicle undergoes reorganization and forms the corpus luteum. The corpus luteum goes through an active functional stage lasting until about the 28 day of the cycle. This period is under the directing influence of two other gonadotropic hormones of the anterior pituitary: (1)

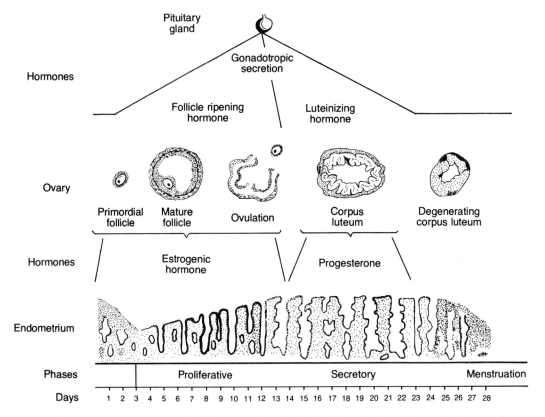

Fig. 19-5. Pituitary control of the sexual cycle in the female. See discussion in text for detailed explanation.

the luteinizing hormone (LH), which promotes the ripening of the ovarian follicle and the organization of the corpus luteum, and (2) luteotropin (LTH), which induces the secretion of progesterone by the newly formed corpus luteum. During this time progesterone is actively secreted by the corpus luteum, and there is also some secretion of estrogenic hormones, but the amount of estrogenic hormones secreted is smaller than was secreted earlier in the cycle by the developing ovarian follicle.

The first stage of the cycle is known as the follicular phase. During this time there is active production of estrogenic hormones, which brings about growth changes in the uterine endometrium. This growth consists of a proliferation of epithelial cells with a marked increase in the number of layers of epithelial tissue and rich vascular development. The estrogenic hormone also stimulates the rhythmic activity of the uterine muscle.

The second stage of the cycle is known as the luteal phase. During this period the amount of estrogenic hormone secreted is diminished and progesterone is produced in large amounts. Under the influence of progesterone, production of uterine gland secretions accelerates, and the uterine muscle relaxes, this effect being opposite to that produced by the estrogenic hormones. Progesterone thus prepares the endometrium for the implantation of a fertilized ovum.

If conception occurs, the corpus luteum persists and continues active secretion of progesterone. The placenta also secretes progesterone and thus supplements the production of this hormone, which is essential to maintain pregnancy.

If conception does not occur, the corpus luteum degenerates and the secretion of estrogenic hormones and progesterone gradually diminishes, this is followed by menstruation, the terminal stage of the female cycle, during which there is rapid destruction of endometrial tissue. After menstruation, under the influence of the follicle-stimulating hormone of the anterior pituitary gland the formation of a new graafian follicle beings, giving rise to a new cycle.

Pregnancy tests

During pregnancy there is a marked increase in the estrogenic and the gonadotropic substances of the urine. The estrogenic substances are not excreted during early pregnancy in quantities sufficient to permit a successful laboratory diagnosis of the condition by an examination of the urine for these hormones. However, within a short time after the beginning of gestation gonadotropic substances are produced in increased quantities in the organism of the pregnant woman and are excreted in amounts large enough to make possible, by an examination of the urine, an early diagnosis of pregnancy with a high degree of certainty.

Synthetic progestins—antiovulatory drugs

Progesterone, when present in quantities greater than the usual amounts secreted, has an antiovulatory effect due to inhibition of the pituitary. When given parenterally in proper dosage during days 5 to 25 of the normal menstrual cycle, progesterone inhibits ovulation. This effect is not obtained when the hormone is given by mouth unless administered in very large dosages. Certain progestins, drugs somewhat similar to progesterone in structure and having an antiovulatory effect, have been synthesized and found to be effective as contraceptives when administered orally. As examples, the formulas of two such drugs, 19-norprogesterone and a 17α-ethyl-19-nortestosterone, are given below:

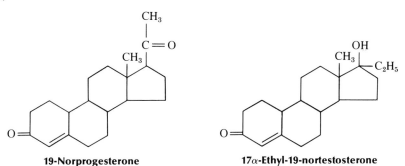

19-Norprogesterone   17α-Ethyl-19-nortestosterone

MALE SEX HORMONES

Two chemical substances, *androsterone* and *testosterone,* have a determining influence on the characteristics of the male sex. An inspection of the formulas of these two substances will show that they have considerable similarity in structure and that they do not differ much in chemical nature from the female sex hormones.

Androsterone   Testosterone

These substances are derived from cholesterol, the same precursor as that of the female hormones.

Testosterone has been isolated from the testes of animals, and androsterone has been prepared from human urine. Both hormones have been synthesized in the chemical laboratory from related chemical substances. Testosterone is about six times as potent physiologically as androsterone. Since it is more potent biologically, testosterone is probably a metabolic precursor of androsterone.

In general, *the functions of the male sex hormones are stimulation of growth of the reproductive organs of the male and maintenance of these organs in a normal condition structurally and functionally. These hormones also influence profoundly the development of the secondary sexual characteristics of the male.* In man they have masculinizing effects, such as the stimulation of growth of the beard, the production of structural changes in the larynx that bring about deepening of the voice, and the development of the masculine physique. In animals, examples of the effects of these hormones are the masculine feathering and the growth of the comb, spurs, and wattles of the cock and the growth of the antlers of the stag.

The production of male sex hormones in the testes is dependent on gonadotropic hormones of the anterior pituitary gland (FSH and LH). If the pituitary gland is removed from a male animal, the testes atrophy and production of the male hormones is diminished. The male sex characteristics are thus determined by the actions of the anterior pituitary principle and the testicular hormones. The gonadotropic hormones of the pituitary keep the testes in a normal structural and functional condition, and the testes produce secretions that maintain the normal structural and functional state of the other male sex organs and promote the development of the male sex characteristics.

Compounds having the biological effects of the male sex hormones (androgenic hormones) are synthesized by the adrenal glands of both the male and the female sexes. Male hormone therefore is excreted in the urine of the female. The amount of male hormone in the urine of the female, however, is considerably less than that in the urine of the male.

GASTROINTESTINAL HORMONES

Gastrin. There is considerable evidence indicating that the flow of gastric juice is stimulated by a hormone called gastrin. Extracts of the pyloric mucosa, when injected intravenously, were found to have a stimulating effect on gastric secretion in animals. It appears that the presence of certain food substances in the upper gastrointestinal tract leads to the formation of gastrin in the mucosa of this region. This hormone is absorbed into the bloodstream, through which it passes to the gastric glands where it stimulates the production and flow of gastric juice.

Enterogastrone. Enterogastrone inhibits the secretion of gastric juice and the motor activity of the stomach. It is formed in the mucosa of the upper intestinal tract, absorbed into the blood, and passed by way of the bloodstream to the stomach. The presence of fat and fatty acids in the upper intestinal tract is the stimulus that causes the formation of this hormone.

Secretin. Secretin is a hormone that stimulates the secretion of pancreatic juice and the flow of bile. This hormone is formed in the mucosa of the upper intestinal tract when acid chyme from the stomach enters the duodenum. It has been shown in animals that placing dilute HCl on the duodenal mucosa causes the pancreas to secrete. Secretion occurred also when HCl was placed in the duodenum of an animal in which the nerves to the pancreas had been cut. It was thus shown that nerve action is not involved in pancreatic stimulation by secretin; hence this substance is a true hormone; it must enter the bloodstream and pass by way of the blood to the pancreas to bring about secretion by that gland.

Potent secretion preparations can be prepared by extracting the duodenal mucosa with dilute HCl solution. The hormone has been obtained in crystalline form. Purified secretin is a peptide with a molecular weight around 5000. Some workers think that secretin exists in the intestinal mucosa in a precursor form, prosecretin, which becomes secretin by a mechanism requiring HCl.

Pancreozymin. Pancreozymin, another hormone that stimulates pancreatic secretion, has been prepared by acid extraction of intestinal mucosa. Its action differs from that of secretin in that it stimulates the production of a pancreatic juice richer in enzymes than that produced by secretin. Thus the flow of pancreatic juice is stimulated by two hormones, secretin and pancreozymin.

Cholecystokinin. Bile is stored in the gallbladder and is discharged intermittently through the cystic duct into the common duct. The hormone cholecystokinin has a regulatory influence on the emptying of the gallbladder. Its action brings about expulsion of bile from the gallbladder by stimulating contraction of the muscular walls of the gallbladder. This hormone is apparently formed in the mucosa of the upper intestinal tract in response to the presence of certain food substances in the intestine. Fats, fatty acids, peptones, and dilute HCl in the upper intestine give rise to the formation of cholecystokinin. Cholecystokinin is absorbed into the blood and passes through the bloodstream to the gallbladder.

Enterocrinin. Enterocrinin is the name given to a substance isolated from the intestinal mucosa. It has the property of stimulating the flow of the intestinal juices (succus entericus). This substance also increases the amount of the enzymes in the intestinal secretion. The evidence indicates that this substance is a hormone whose function is to exert a regulatory influence on intestinal secretion.

Questions for study

1. What is an internal secretion? Distinguish between internal and external secretions. Give examples of each.
2. What are the physiological effects of thyroxin? What is triiodothyronine? Thyroglobulin?
3. What is cretinism?
4. What are the symptoms of hypothyroidism in the adult?
5. What is the treatment for hypothyroidism?
6. What is the cause of simple goiter? Where is it especially prevalent? How may it be prevented?
7. What are the symptoms of hyperthyroidism? How is it treated?
8. Where are the parathyroid glands located?
9. What is the function of the parathyroid secretion?
10. What is the function of insulin?
11. What is hyperglycemia?
12. What is insulin chemically?
13. From what sources is the insulin used in medicine obtained?
14. What is a unit of insulin?
15. Why is insulin ineffective when taken by mouth?
16. What is hypoglycemia? How is it treated?
17. Where is the pituitary gland located?
18. What are the functions of the secretions of the anterior lobe of the pituitary gland?
19. What is the relation of the anterior pituitary gland to carbohydrate metabolism?
20. Discuss the chemistry of the two hormones secreted by the posterior pituitary gland.
21. What are the physiological effects of oxytocin? Vasopressin?
22. What is diabetes insipidus? What causes this disease?
23. What are the uses of posterior pituitary extract in medicine?
24. Where are the adrenal glands located?
25. What are the two parts of an adrenal gland?
26. What is Addison's disease?
27. What are the effects of the secretions of the adrenal cortex on salt balance? On carbohydrate and protein metabolism?
28. What are the clinical applications of the corticosterone group of hormones?
29. What are the principal physiological effects of epinephrine? Norepinephrine?
30. What are the uses of epinephrine in medication?
31. Where is the thymus gland located?
32. What is known regarding the functions of the thymus gland?
33. Name the female sex hormones.
34. What is estrus?
35. What are the estrogenic hormones?
36. What are the effects of the estrogenic hormones on the female?
37. What is the function of progesterone?
38. What hormone influences the production of the estrogenic hormones?
39. Describe the menstrual cycle, stating particularly the relations of the gonadotropic hormones of the anterior pituitary and the female sex hormones.
40. For what substances is the urine examined in pregnancy tests?
41. Name the male sex hormones.
42. What are the physiological effects of the male sex hormones?
43. What hormones influence the production of the male sex hormones?
44. What causes the formation of gastrin? What is its specific effect?
45. What is the physiological effect of enterogastrone?
46. What is secretin? What are its physiological effects?

47. What is the function of pancreozymin?
48. Describe the action of cholecystokinin.
49. What is the function of enterocrinin?

20 Digestion

Most foods must be digested in order that they may be absorbed from the alimentary tract and thus be made available to the tissues of the body. The food substances requiring digestion are proteins, carbohydrates (except monosaccharides), and fats. These substances exist in the form of very large molecules, and as a general rule they are not very soluble. Before they can be absorbed from the alimentary canal, many of them must be made more soluble, and in most cases their molecules must be broken down into smaller molecules. The membranes of the intestinal walls do not permit molecules as large as those of proteins, complex carbohydrates, and fats to pass through them. In the process of digestion the complex molecules of food substances are split into simpler molecules. These simpler molecules are much more soluble and will diffuse more rapidly through the walls of the intestine. Digestion may therefore be defined as *the chemical processes by which the molecules of proteins, carbohydrates, and fats are broken down into molecules that are more soluble and are small enough to pass readily through the intestinal walls into the blood and lymph.*

There are certain foods that do not require digestion. These are the inorganic salts, water, vitamins, and simple sugars or monosaccharides. The molecules of these substances are relatively small and will diffuse readily through the walls of the intestine into the bloodstream; hence digestion is unnecessary for substances of this nature.

Digestion is essentially a hydrolytic process. The agents that bring about the hydrolytic reactions are enzymes. When carbohydrates are digested, they are hydrolyzed to monosaccharides through the catalytic influence of enzymes. When fats are digested, their molecules are split by the hydrolytic action of certain enzymes, and fatty acids and glycerol are produced. The digestion of proteins is the hydrolytic degradation of protein molecules by enzymes into a series of substances with smaller molecules called proteoses, peptones, peptides, and amino acids. In each instance the end products of digestion are substances consisting of small molecules, which are quite soluble and will readily pass through the intestinal walls. This process is represented graphically in Fig. 20-1.

The digestive fluids in the human body are the *saliva,* which is secreted into the mouth by the submaxillary, sublingual, and parotid glands; the *gastric juice,* which is secreted into the stomach by glands in the walls of the stomach; the *pancreatic juice,* the external

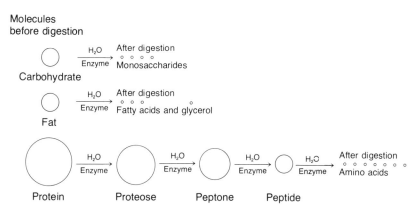

Fig. 20-1. Scheme representing change in size of molecules during digestion.

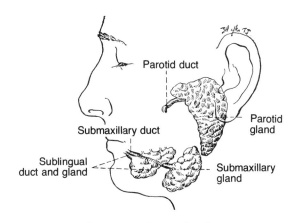

Fig. 20-2. Salivary glands.

secretion of the pancreas, which passes into the small intestine at the lower end of the duodenum; and the *intestinal juice* or *succus entericus* secreted by Brunner's and Lieberkühn's glands, which are situated in the walls of the small intestine. *Bile* is also considered a digestive fluid, since it influences the digestive process in several important respects. Bile passes by way of the common bile duct into the duodenum through an orifice common to the principal duct from the pancreas (Figs. 20-2 and 20-3).

The secretion of saliva and gastric juice is induced by a nervous reflex mechanism set in action by the sight and thought of food and by the presence of food in the mouth and stomach.

There is no hormonal stimulation of salivary secretion, but gastric secretion is promoted by a substance called *gastrin,* which appears to have hormonal action. The secretion of pancreatic juice is promoted by the hormones *secretin* and *pancreozymin.* Secretin is formed in the mucosa of the upper intestinal tract in response to the presence in the duodenum of acid chyme from the stomach. Secretin passes to the sites of its action by way of the bloodstream.

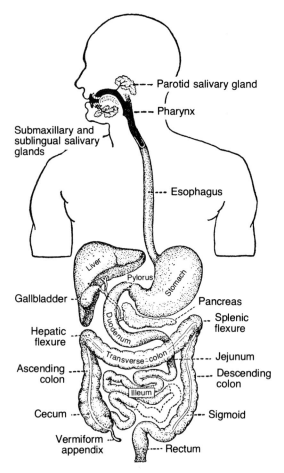

Fig. 20-3. Relationships of the different parts of the digestive system. Only a small part of the small intestine is shown. (From Turner, C. E.: Personal and community health, ed. 14, St. Louis, 1971, The C. V. Mosby Co.)

SALIVARY DIGESTION

The saliva is a fluid consisting of about 99% water, a protein called mucin, which has a lubricating function, the starch-splitting enzyme ptyalin, and small amounts of inorganic salts. Its reaction is nearly neutral, ranging from very slight acidity to very slight alkalinity. Normally about 1.5 liters of saliva are excreted in one day.

When food is taken into the mouth, it is first broken up and mixed with saliva by chewing. The saliva has the important functions of moistening the food so that it may be readily swallowed and of digesting cooked starch, dextrins, and glycogen. As soon as food is mixed with saliva, the breakdown of starch begins. The starch molecule becomes smaller by a splitting off of units of maltose; successively smaller units called dextrins are formed, which finally are split into maltose. *An enzyme in the saliva called ptyalin, or salivary* amylase, catalyzes the stepwise breakdown or digestion of starch to maltose. Starch digestion by ptyalin action will not take place unless the walls of the starch grains are ruptured by grinding or cooking. This is one of the reasons for cooking food. Ptyalin will digest starch in a medium that is very slightly acid or very slightly alkaline. The presence of the chloride ion is necessary for the action of ptyalin. The most favorable range of pH is 6.6 to 7.2. The enzyme's action is decreased by increased acidity. At a pH below 4, ptyalin action ceases. Salivary digestion therefore takes place favorably in the mouth, where the reaction is usually slightly acid or sometimes slightly alkaline. After the food has been swallowed, digestion continues in the stomach until the gastric glands secrete enough hydrochloric acid to stop the action of ptyalin.

GASTRIC DIGESTION

The gastric juice of mammals contains the enzymes pepsin, lipase, and rennin and also hydrochloric acid, mucin, and some inorganic salts. In the normal person the acidity of gastric juice increases to a peak in 30 minutes to 1 hour after ingestion of food, then diminishes slowly to a low value at the end of 2 to 3 hours. At the height of digestion the pH of normal gastric juice ranges from 1 to 2.5. About 2 to 3 liters are excreted daily.

Pepsin is a proteolytic enzyme. It acts on proteins, breaking them down to proteoses, peptones, peptides, and amino acids. The optimum pH for pepsin activity is 1.5 to 2.5.

The pepsin of gastric juice arises from an enzymic precursor or zymogen that is called pepsinogen. Pepsinogen is secreted by the chief cells, and hydrochloric acid is secreted by the parietal cells of the gastric mucosa. After pepsinogen and hydrochloric acid have passed into the lumen of the stomach and the gastric juice has become more acid than a pH of 6, the pepsinogen is converted into pepsin by the action of the hydrogen ion of the HCl. Once the process is started, it takes place more rapidly because of the presence of pepsin itself; that is, the reaction is autocatalytic, the presence of pepsin facilitating the formation of more pepsin from pepsinogen.

The lipase in gastric juice has an optimal pH around neutrality. The activity of this enzyme in the digestion of fats to fatty acids and glycerol is of minor importance in adults, since the pH of gastric juice after food ingestion by the latter rapidly shifts to a strongly acid value. Gastric lipase therefore appears to be of principal importance in digestion during infancy, since gastric acidity after food ingestion is mild in the infant.

Rennin is an enzyme that coagulates or clots milk. Rennin action is a hydrolysis or splitting of the casein of milk, forming a compound called paracasein, which in the presence of calcium ions becomes calcium paracaseinate, an insoluble curdy substance. Rennin serves a double function: it produces paracasein, whose molecule is smaller than casein, and since calcium paracaseinate is insoluble its formation slows down the passage of milk through the alimentary tract, thereby permitting a more prolonged action on this substance by other digestive enzymes. The optimum pH for rennin action is 5.4.

Rennin occurs in the gastric juice of young mammals. The evidence indicates that there is very little or perhaps not any rennin in the gastric juice of adult mammmals. Rennin action, however, does take place in the alimentary tract of the mammalian adult; but this is apparently an effect of the proteolytic enzymes, pepsin, trypsin, and chymotrypsin. These enzymes in pure crystalline form have been shown to have an action on casein similar to that of rennin.

After foods have entered the stomach, considerable digestion of starch takes place, as just noted. There is no other carbohydrate digestion in the stomach than that performed by ptyalin on cooked starch, dextrins, and glycogen, except possibly a slight hydrolysis of sucrose, maltose, or lactose by the hydrochloric acid of the gastric juice.

Gastric juice contains a variable amount of the glycoprotein mucin. This compound is secreted by the goblet cells of the mucous membranes of the alimentary tract. *The function of mucin is to serve as a lubricant to promote the passage of food along the alimentary tract, to protect the membranes against trauma by coarse food particles, and to act as a buffer against acidity in the stomach.* It is well adapted to these functions since it is a protein that will combine with acid and since it is viscid, a property that makes it cling to the walls of the stomach.

The combined action of the gastric enzymes in the presence of hydrochloric acid reduces the food to a thin liquid called chyme. As gastric digestion is completed, the acid chyme is gradually passed into the small intestine by successive, wavelike contractions of the stomach (peristalsis), which pass from the region of the fundus to that of the pylorus.

Clinical testing of gastric function. In cases of chemical imbalance in the gastric juice or failure in gastric secretion, information of value may be obtained by a gastric analysis. In such a procedure the flow of gastric juice is stimulated by giving the patient a test meal of dilute alcohol, bread, or other food or by the injection of histamine, and gastric juice is collected at intervals through a stomach tube. The material obtained is examined microscopically and chemically in the clinical laboratory.

The finding of blood in the gastric contents shows the presence of a bleeding lesion in the uppper alimentary tract. Marked excesses of bile suggest obstruction below the outlet of the common bile duct. Large numbers of yeast cells indicate that active fermentation has occurred in the stomach, a condition associated with a low secretion of hydrochloric acid. Remnants of food from a previous meal show poor emptying capacity of the stomach. In the chemical examination the factor of principal interest is the amount of hydrochloric acid secreted. Complete absence of free HCl after the administration of an effective stimulus such as histamine indicates severe pathology involving the parietal cells of the gastric mucosa. A low secretion of HCl may indicate a poor functional capacity or a gastritis. A high HCl concentration in the gastric juice is often associated with a gastric or duodenal ulcer or pyloric spasm.

An analysis for the presence or absence of free gastric acidity can be made without the inconvenience of a gastric tube. The Diagnex* test requires only the ingestion of an ion exchange resin that has been coupled with a blue dye, azure A. In the presence of gastric acidity, the blue dye is released from the resin, is absorbed into the blood, and appears in the urine within 2 hours.

---

*E. R. Squibb and Sons Division, Olin Mathieson Chemical Corp; New York, N. Y.

DIGESTION IN THE INTESTINAL TRACT

Protein digestion. In the intestine the digestion of proteins is continued through the activity of the enzymes trypsin, chymotrypsin, carboxypeptidase, aminopeptidase, and dipeptidase.

Trypsin arises from a precursor, trypsinogen, which is excreted by the pancreas. Trypsin is converted into trypsin in a medium that has a pH of 7 to 9. This reaction is autocatalytic; that is, the presence of trypsin itself accelerates the formation of more trypsin from trypsinogen. What is also very important in this connection is that the intestinal glands secrete *an enzyme called enterokinase, which very actively promotes the formation of trypsin from trypsinogen* as soon as trypsinogen appears in the intestinal tract. *Trypsin catalyzes the hydrolysis of proteins to proteoses, peptones, peptides, and amino acids. It also has a weak renninlike action on casein.*

Chymotrypsin arises as a zymogen or precursor called chymotrypsinogen, which is secreted by the pancreas. In the intestinal tract, chymotrypsinogen is converted into the active eproteolytic enzyme chymotrypsin through the action of trypsin in a medium with a pH of 6 to 10. *Chymotrypsin splits proteins into proteoses, peptones, peptides, and amino acids. It also has a renninlike action on casein.* Its rennin action is much greater than that of typsin.

Trypsin and chymotrypsin have proteolytic activity in a medium with a pH ranging from 7 to 10. Their optimum activity is at a pH of 8 to 9.

As stated previously, trypsin and chymotrypsin split proteins down to the polypeptides. It was also noted in the discussion of gastric digestion that proteins are hydrolyzed by pepsin to proteoses, peptones, and peptides. To complete the digestion of proteins, it is necessary that the peptides be split into amino acids. This is accomplished by the action of several enzymes.

One of the peptidolytic enzymes is carboxypeptidase, which is secreted by the pancreas in the inactive precursor form, procarboxypeptidase. *Carboxypeptidase attacks a peptide with an amino acid having a free carboxyl group at the end of the peptide chain. The action of this enzyme is to split off from the molecule the amino acid with the free carboxyl group and then react in a similar manner with the next amino acid in the chain* and so on until the peptide chain has been hydrolyzed into amino acids.

Another peptidolytic enzyme is aminopeptidase, which is present in the secretion of the glands of the intestinal mucosa. *Aminopeptidase acts on the end of a peptide chain that has an amino acid containing a free amino group, splitting off the latter amino acid. After having split off one amino acid with a free amino group, the action is continued in a stepwise manner until the peptide has been hydrolyzed into amino acids.* Another peptide-splitting enzyme of the intestinal juice is dipeptidase. As the name indicates, *dipeptidase splits peptides consisting of two amino acids.* This enzyme serves a useful function in splitting dipeptides that either have escaped or are resistant to the action of other peptidolytic enzymes.

Carbohydrate digestion. In the fluid mixture that is delivered into the duodenum by gastric peristalsis, there may be considerable undigested starch, also dextrin, glycogen, maltose, sucrose, and lactose. The digestion of these carbohydrates is completed in the intestinal tract through the activity of certain enzymes that will now be discussed.

The pancreas secretes a starch-splitting enzyme called amylase. The *amylase of pancreatic juice hydrolyzes starch, glycogen, and dextrins to maltose*. In the pancreatic juice and also in the intestinal juice, there is present *the enzyme maltase, which splits maltose into glucose*. In addition to maltase the succus entericus contains the enzymes sucrase and lactase. *Sucrase hydrolyzes sucrose, forming glucose and fructose. Lactase splits lactose into glucose and galactose*. Thus the undigested carbohydrates that appear in the upper intestinal tract are rapidly converted into monosaccharides by the action of certain enzymes in pancreatic and intestinal juice.

Fat digestion. In mammals most fat digestion occurs in the small intestine. The digestion of fats is brought about by the action of the enzyme steapsin (pancreatic lipase) of the pancreatic juice, on emulsified fats. *Steapsin splits fats into fatty acids and glycerol*. Since steapsin exists in aqueous solution and fats are not water-soluble, the problem of steapsin action is a matter of creating conditions that will bring about contact between the molecules of the enzyme and its substrate. This is accomplished by emulsification of the fat, bile being the emulsifying agent. Bile contains alkaline salts of the bile acids, which form thin films around minute colloidal particles of fat and thus produce an emulsion of colloidal solution of fat. These emulsified colloidal particles of fat are readily attacked by steapsin. Thus the emulsification of fats by bile serves the very important function of enabling steapsin to attack fats and split them into fatty acids and glycerol.

SUPPLEMENTARY ACTION OF ENZYMES IN THE DIGESTIVE TRACT

In the foregoing discussion it is interesting to note how the enzymes supplement the work of each other. Pepsin starts the digestion of proteins in the stomach and produces proteoses, peptones, peptides, and amino acids. In the intestine, trypsin and chymotrypsin continue the digestion of proteins and yield proteoses, peptones, peptides, and amino acids. Further digestion of the intermediate products formed from proteins by the action of pepsin, trypsin, and chymotrypsin is accomplished by carboxypeptidase, aminopeptidase, and dipeptidase, which promote the hydrolysis of peptides to amino acids. Starch is attached by ptyalin in the mouth and in the stomach for a time, but whatever starch escapes ptyalin action is digested in the intestine by amylase of the pancreatic juice. The maltose that is produced by the action of ptyalin and amylase is digested further in the intestine by maltase, which splits this disaccharide into glucose. Fat digestion is carried on effectively in the intestine by steapsin in the presence of bile, which aids in the digestive process by emulsifying the fat. The protein part of the large molecules of nucleoproteins is split off by the acidity of the gastric juice, and the nucleic acid is digested into its constituent units in the intestinal tract by muclease and phosphatase. Thus each enzyme plays an effective part in the whole process of digestion, and the final result is the production of monosaccharides, amino acids, fatty acids, glycerol, phosphates, and nucleosides. The molecules of these substances are small enough to permit absorption from the intestinal tract.

The student will be aided greatly in the study of digestion by Table 20-1, which summarizes the process.

Table 20-1. Summary of digestion

Secretion	Most favorable reaction	Enzyme	Substrate	End product
Saliva	Slightly alkaline to slightly acid, pH 6.6 to 7.2	Ptyalin (salivary amylase)	Cooked starch, dextrins, and glycogen	Maltose
Gastric juice	Acid, pH 1.5 to 2.5	Pepsin	Proteins	Proteoses, peptones, peptides, and amino acids
		Rennin	Casein	Paracasein
Pancreatic juice	Alkaline, pH 7 to 9	Trypsin and chymotrypsin	Proteins	Proteoses, peptones, peptides, and amino acids
		Amylase	Starch, dextrins, and glycogen	Maltose
		Steapsin (lipase)	Fats	Fatty acids and glycerol
		Carboxypeptidase	Peptides	Amino acids
Intestinal juice	Alkaline, pH 7 to 9	Enterokinase	Trypsinogen	Trypsin
		Aminopeptidase	Peptides	Amino acids
		Dipeptidase	Dipeptides	Amino acids
		Maltase	Maltose	Glucose
		Lactase	Lactose	Glucose and galactose
		Sucrase	Sucrose	Glucose and fructose

BILE

Bile is both a secretion and an excretion. It is a secretion in the sense that it contains substances that aid in the digestion and absorption of foods, thereby performing a physiological function. It is an excretion in that it contains a number of substances that are true waste products having no further function in the body. About 1 liter per day is emptied into the intestinal tract.

Composition. Bile consists of the following substances:

Water, 88% to 97%
Pigments: bilirubin, biliverdin
Bile salts: taurocholates, glycocholates
Cholesterol
Phospholipid
Mucin
Sex hormones
Inorganic salts
  Sodium, Potassium, Calcium, Magnesium, Ammonium
  Chlorides, Sulfates, Phosphates, Bicarbonates

Bilirubin. Bilirubin, the principal bile pigment, originates in broken-down red blood corpuscles. It has been estimated from bile excretion studies that the human body destroys about 1 trillion red corpuscles per day. The life span of human red blood cells has been found by experiment to be about 126 days. When red blood corpuscles disintegrate, their hemoglobin is broken down into globin and heme. Iron is removed from the heme, and the heme is converted into biliverdin, which is changed to bilirubin. These chemical changes take place principally in the reticuloendothelial cells of the spleen, liver, and bone marrow. The bilirubin thus formed passes from the bloodstream into the liver, where it is combined with glucuronic acid to form bilirubin diglucuronide. This substance is excreted in the bile. When obstruction in the biliary tract occurs, this pigment cannot be excreted readily, and it passes back into the bloodstream, producing obstructive jaundice, which causes the skin to become yellow.

Storage—the gallbladder. Bile may be stored for a time in the gallbladder, or it may pass directly from the liver through the bile ducts into the small intestine. It is emptied into the small intestine near the lower end of the duodenum (the ampulla of Vater), and after considerable transformation the residual products are excreted in the feces.

The gallbladder serves to store bile, which is secreted more or less continuously by the liver. In the gallbladder the bile is concentrated about five times by the loss of water. When additional bile is needed after food intake, the gallbladder is made to contract and expel bile into the intestinal tract by the hormone *cholecystokinin*. This hormone is formed in the mucosa of the upper intestinal tract in response to the presence of fatty foods. It passes to the walls of the gallbladder by way of the bloodstream.

Functions. Bile aids in the digestion and absorption of fats. It assists in fat digestion by emulsifying the fats. In this process small particles of fat become coated with a thin film of bile components, particularly the bile salts. The water-soluble, fat-splitting enzyme lipase (steapsin) can enter the film and form a lipase-fat complex. This results in a much more rapid digestion of the fat than would occur in the absence of bile.

Bile performs another critical function by aiding in the absorption of the products of fat digestion. The bile salts are the active agents in this process. They form complexes with the digestion products that are more soluble in water. The mechanism by which they aid in absorption is not understood.

If a condition arises that excludes bile salts from the intestinal tract, large amounts of lipids are excreted in the feces. Steatorrhea is the term used to designate this condition.

Questions for study

1. Why is digestion necessary?
2. What is digestion? What are the agents chiefly responsible for digestion?
3. What foods do not require digestion? Why?
4. Name the digestive fluids.
5. How is the secretion of digestive fluids brought about?
6. Describe the digestion of carbohydrates in the mouth; in the stomach; in the intestinal tract.
7. Describe the digestion of proteins in the stomach; in the intestinal tract.
8. Describe the digestion of fats in the intestinal tract.
9. What is the most favorable reaction for salivary digestion? For gastric digestion? For intestinal digestion?

10. How does bile aid in digestion?
11. Name all the enzymes that digest the following substances in the alimentary tract: carbohydrates, proteins, fats.
12. Name the end products of digestion by each of the enzymes of the alimentary tract.
13. In what respects is bile a secretion? An excretion?
14. Name the principal constituents of bile.
15. Describe the origin and production of bilirubin.
16. What is the function of the gallbladder?
17. What are the functions of bile?

21 Metabolism of carbohydrates, fats, proteins, and inorganic salts

Metabolism is a term used to indicate the chemical changes that take place in the living tissues of plants and animals. In man it includes all the *chemical changes taking place in the body after food is absorbed from the alimentary tract*. It does not include digestion. It encompasses the construction and repair of tissues, the storage of food substances in the body, the nutrition of all the cells, the conversion of the chemical energy of absorbed foodstuffs into a form that enables the body to carry on its physiological activities and to do work, and the elimination of waste products from the cells. Metabolism includes two general processes: (1) *anabolism,* the building up of body tissues and the storage of foods in the body, and (2) *catabolism,* the breaking down of body tissues and of food substances in the body.

METABOLISM OF CARBOHYDRATES

Absorption and distribution. As we have previously seen, carbohydrates are digested in the alimentary tract, forming the monosaccharides, glucose, fructose, and galactose. The monosaccharides pass through the intestinal walls into the bloodstream. *The blood carries the glucose to all parts of the body, where it either is oxidized to give heat and energy or is stored in the tissues in the form of the polysaccharide glycogen.* For a time after the ingestion of a meal, fructose and galactose may be present in the bloodstream, but these monosaccharides are soon converted into glucose or glycogen. Glucose is present in the blood and tissues at all times. Glycogen exists in the body as a storage form of carbohydrate. When the body is in a well-nourished condition, glycogen is always present in the tissues. The glycogen content of the body when the person is well fed is about 300 to 500 g or about 1 pound. The storage of glycogen is small compared with the deposition of fat in the body.

Blood sugar. Glucose, or dextrose, is the sugar of the blood. It is fermentable with yeast. The term blood sugar includes glucose and a small amount of nonfermentable substances that are not sugars. In a chemical analysis of the blood for its sugar content a small amount of nonsugar substances enters into the blood sugar value obtained, because the

Metabolism of carbohydrates, fats, proteins, and inorganic salts   283

blood sugar methods of analysis are not entirely specific for glucose. The best blood sugar methods give values that include less than 10% of nonsugar substances.

The amount of glucose in the blood of normal human subjects ranges from 65 to 100 mg/100 ml of blood. From 30 minutes to 1 hour after a carbohydrate meal has been ingested, the blood sugar rises to about 120 to 160 mg/100 ml; since glucose is rapidly converted to glycogen in the tissues the blood sugar concentration returns to the normal level of 65 to 100 mg/100 ml within 2 or 3 hours. The blood sugar thus tends to stay at a constant level in health. When an excess of glucose appears in the blood after carbohydrate ingestion, it is promptly stored in the liver and the muscles as glycogen; an elevation of the blood sugar thus leads to the formation of tissue glycogen. When the glucose of the

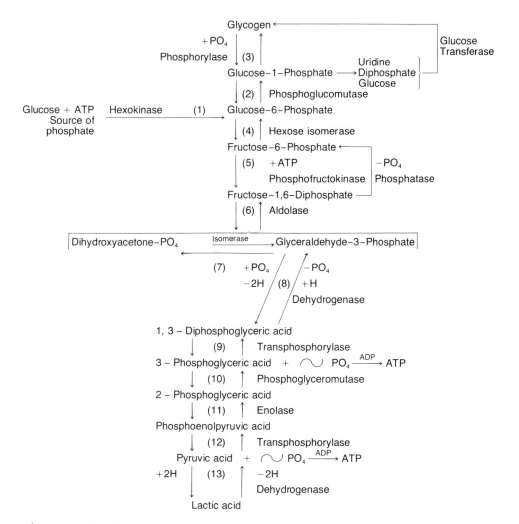

Fig. 21-1. Chart showing sequence of chemical steps in the anaerobic metabolism of carbohydrate, the process called glycolysis (Embden-Meyerhof pathway).

blood is used for metabolic processes, it is renewed from the glycogen stored in the liver. Liver glycogen in this manner serves as an immediate source of blood glucose during the fasting hours.

The hormone insulin participates in the metabolism of carbohydrates by facilitating the utilization of glucose. *The overall effect of insulin is to lower the level of the blood sugar.*

Role of the liver. The liver has the capacity to remove sugar from the blood and to store it as glycogen after a meal or to liberate glucose into the blood stream during the time that food is not being absorbed from the alimentary tract. A considerable part of the glucose delivered into the bloodstream by the liver is formed from glycogen, but the liver also has the capacity to form glucose from sugar-forming amino acids obtainable from proteins and also from fatty acids. The latter process is especially important when the glycogen content of the liver becomes low as a result of a reduced carbohydrate intake. Thus the liver possesses a mechanism by which glucose may always be supplied to the body. This is an important mechanism, since glucose is being oxidized in the body at all times to provide heat and energy for its vital processes.

The formation of glucose from glycogen in the liver is accelerated by *epinephrine* (adrenaline), a hormone secreted by the adrenal glands. This stimulation of the breakdown of glycogen by *epinephrine* is important in that it provides a mechanism by which blood sugar is quickly mobilized in times of stress when there is a special need for more blood glucose. *One effect of epinephrine in the body then is to raise the blood sugar by increasing the breakdown of glycogen to glucose in the liver.*

Intermediary metabolism. In the animal body, glucose passes through two general catabolic pathways: (1) glycolysis, which takes place without the use of oxygen and releases approximately 24,000 calories of energy per mole of glucose, and (2) the citric acid cycle, which requires oxygen and releases about 15 times as much energy as glycolysis.

Glycolysis. Glycolysis, or the Embden-Meyerhof pathway, is a series of chemical reactions by which *glycogen or glucose is transformed into lactic acid* ($CH_3 \cdot CHOH \cdot COOH$).* The sequence of chemical reactions in glycolysis is shown in Fig. 21-1. Glycolysis consists of 13 different steps, each step being catalyzed by a specific enzyme. The first step is the combination of glucose with phosphate through the action of adenosine triphosphate (ATP) and an enzyme called hexokinase. The glucose phosphate formed has the phosphate attached to the sixth carbon atom in the glucose chain and is called glucose-6-phosphate. Glucose-6-phosphate may undergo metabolic transformations leading to the formation of glycogen, or it may serve as the starting material for a series of degradative steps that result in the formation of pyruvic or lactic acids.*

Glycolysis is an anaerobic process; each step takes place without oxygen. This is one of its great advantages. The body can obtain energy from glycolysis quickly, without waiting for a supply of oxygen to be carried to the cells. Glycolysis also performs the essential function of transforming glucose into pyruvic acid, the compound that enters the citric

---

*If the pH of the body is considered, the lactic acid would be present as the lactate ion; acetic acid would be present as acetyl coenzyme A, and pyruvic acid would be present as the pyruvate ion.

acid cycle, through which the metabolism of carbohydrates is continued with the use of oxygen, an aerobic process.

The other monosaccharides, fructose and galactose, are phosphorylated in the liver by ATP and a specific enzyme. After phosphorylation they are transformed into glucose-6-phosphate through the action of other enzymes. These sugars thus enter the glycolytic cycle and follow the same metabolic pathway as glucose.

The 13 steps in glycolysis may be summarized by the following overall reaction:

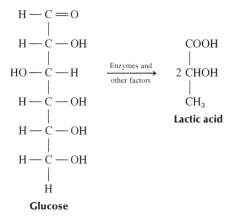

The details of the reactions in glycolysis shown in Fig. 21-1 are not to be learned. This chart is intended as a reference to give the student an overall concept. It is desirable to note that glycolysis starts with glucose or glycogen and ends with lactic acid, that a specific enzyme catalyzes each reaction, that phosphorylation and dephosphorylation play important parts, and that where the arrows point both ways the reactions are reversible. The latter statement needs clarification. If a large dose of lactic acid is given to an animal, the lactic acid will undergo the metabolic transformations shown by the chart, but in the reverse direction (from the bottom to the top of the page), and will be stored as glycogen. Thus, if more lactic acid or pyruvic acid (or other utilizable carbohydratelike material) is introduced into the animal's body than is needed by the tissues for their immediate activities, these substances are conserved by being stored as glycogen.

Citric acid cycle. The citric acid, or Krebs, cycle is an oxidative pathway of metabolism. It makes use of the atmospheric oxygen transported to the cells by the blood. It is a remarkable series of chemical reactions by which maximal amounts of energy are obtained from the foodstuffs that pass through the cycle. A specific enzyme catalyzes each reaction in the series.

The citric acid cycle is shown in Fig. 21-2. It begins with the condensation of oxalacetic acid, with acetic acid derived from pyruvic acid. The pyruvic acid comes from the glycolysis of carbohydrate and also from certain amino acids. The product of this condensation is citric acid. Citric acid is transformed successively into the following metabolites: cis-aconitic acid, isocitric acid, oxalosuccinic acid, α-ketoglutaric acid, succinic acid, fumaric acid, malic acid, and, finally, oxalacetic acid, which was one of the starting materials. The student need not be disturbed by this array of organic compounds; in

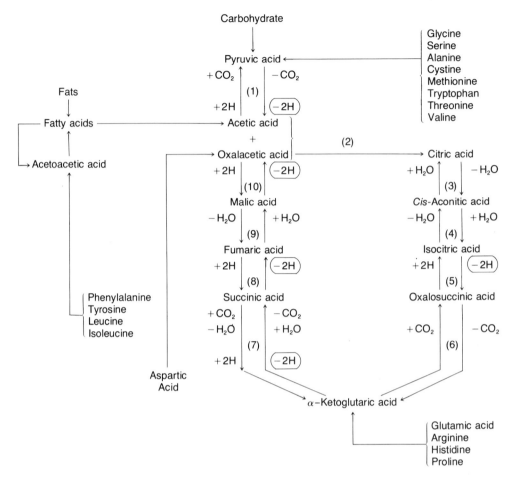

Fig. 21-2. Citric acid, or Krebs, cycle includes the ten numbered steps. The chart shows how this cycle functions as a common pathway for the metabolism of carbohydrates, fats, and the amino acids of proteins.

fact, for this course it is not desirable to learn them. It can be readily seen, however, that this series of reactions is a closed cycle; it begins with oxalacetic acid as one of the starting materials and it ends with oxalacetic acid. At five places in the citric acid cycle (steps 1, 5, 7, 8, and 10) hydrogen (encircled) is removed from the compound being metabolized, and this hydrogen is made to combine with oxygen to form water through the activity of a number of enzymes. This union of hydrogen with oxygen in the living cells is an energy-producing reaction. The student will recall from the study of inorganic chemistry that hydrogen burns with a very hot flame. *The amount of energy obtained from oxidizing the hydrogen from carbohydrates is the same as that produced by burning hydrogen in the laboratory; the living cells, however, keep these reactions so well controlled that the body temperature is maintained at around 38° C.*

In three of the steps in the citric acid cycle (1, 6, and 7) CO_2 is given off. These steps are a further degradation of the compound being metabolized, and they account for the

CO_2 exhaled in the breath. *The total amount of energy obtained from the oxidation of carbohydrates is approximately 4 Calories per gram.*

In each passage of metabolites through the citric acid cycle an amount of H_2 and CO_2 is released that will account for 1 molecule of pyruvic acid. *The citric acid cycle is thus a metabolic pathway that brings about the oxidation of 1 molecule of pyruvic acid with each turn of the cycle.* It is therefore a complete and continuous cycle of chemical events that serves to accomplish the oxidation of carbohydrates.

Occurrence of glycolytic sequence and citric acid cycle. The reactions in the two charts (Fig. 21-1 and 21-2) showing metabolic pathways of carbohydrates apply most completely to muscle and liver; however, these reactions take place in heart, brain, kidney, and other tissues perhaps to almost as great an extent. What is also of great interest is that the reactions of glycolysis and the citric acid cycle apply rather widely to plant life. Furthermore, the reactions indicated have been observed to take place to a large extent in bacteria, protozoa, and the tissues of lower forms of animal life.

Undoubtedly, marked variations from the outlined pathways of metabolism will be observed as further investigations are made, and of course it is not desirable to try to make one generalized scheme apply too widely to such complex systems as are found in living tissues. However, the outlined charts enable the student to get an overall concept of the chemical reactions by which living matter utilizes carbohydrates; when rightly understood, these charts are useful educational tools.

Common metabolic pathway. The citric acid cycle occupies a central position in the metabolic processes of the animal body. *As previously noted, the citric acid cycle is the pathway of oxidation of carbohydrates. It is also a common metabolic pathway through which pass products formed in the intermediary metabolism of fats and proteins.* As indicated in Fig. 21-4, fats undergoing metabolism yield fatty acids that form acetic acid, the same compound that arises from pyruvic acid of carbohydrate origin. Also, as shown in the diagram in Fig. 21-2, eight amino acids, while undergoing metabolism, give rise to pyruvic acid and enter the cycle at step 1. Four amino acids undergo metabolic transformations that yield α-ketoglutaric acid and gain admission to the cycle at step 7; four other amino acids give rise to acetoacetic acid and enter the cycle as fatty acids; and the amino acid, aspartic acid, loses its amino group and enters the cycle as oxalacetic acid. Thus the citric acid cycle is an exceedingly important metabolic pathway because it is the central passageway of oxidation for the three foodstuffs—carbohydrates, fats, and proteins.

Other pathways of carbohydrate metabolism. In addition to the glycolytic sequence just described there are other pathways of carbohydrate breakdown that begin with the initial steps in glycolysis but follow a different course.

However, the preceding discussion of glycolysis and the citric acid cycle is considered an adequate presentation of carbohydrate metabolism for this course.

ATP—storage and release of energy. A part of the energy released by the oxidation of carbohydrates, fats, and proteins appears in the form of heat and serves to maintain the temperature of the body. However, in the mammalian body about two-thirds of the energy resulting from the oxidation of foodstuffs is trapped and stored in certain organic

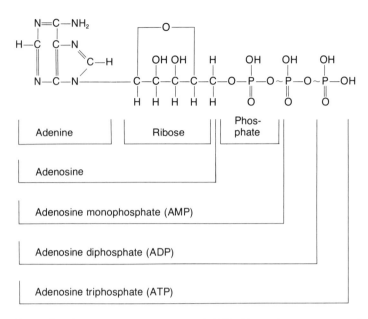

Fig. 21-3. Chart showing component parts of ATP. Note the presence of two high-energy phosphate bonds designated by a curling bond (∼).

phosphate compounds. These compounds are known as *high-energy phosphate compounds*. The storage of energy is effected by the phosphate group, and the valence bond connecting this phosphate group with the other part of the molecule is known as a *high-energy phosphate bond*.

The principal high-energy phosphate compound in living tissues is adenosine triphosphate or ATP. The structure of ATP is shown in Fig. 21-3. This compound contains three phosphate groups connected to each other, two of which are high-energy groups. The high-energy phosphate bond is designated by a curled hyphen (∼).

When, for example, energy is needed in living tissues, to enable muscles to contract, to transmit a nerve impulse, or to promote a metabolic process, an enzyme splits off one of the high-energy phosphate groups from adenosine triphosphate (ATP) and releases the stored energy for the use of the living process that needs it. When one high-energy bond of ATP is split, ADP (adenosine diphosphate) is formed, inorganic phosphate is set free, and about 12,000 calories (12 kilocalories or large Calories) of energy per mole of ATP are made available for living processes. However, if this system is to keep working, the ATP must be re-formed. This is accomplished by the energy released from the oxidation of foods. The ADP formed when energy is released for body processes is readily converted back to ATP in the presence of inorganic phosphate by the energy provided from oxidation of foodstuffs. The latter process is called *oxidative phosphorylation*. Thus ATP is formed from ADP and inorganic PO_4 by the energy derived from the oxidation of foods and is split into ADP and inorganic PO_4 when living cells use energy to perform their functions. These relationships are summarized by the following equations:

Metabolism of carbohydrates, fats, proteins, and inorganic salts

$$ATP \rightarrow ADP + PO_4 + \text{Energy for work (8 to 12 kilocalories)}$$
$$ADP + PO_4 + \text{Energy from oxidation} \rightarrow ATP$$

As shown in the preceding material, *ATP is a remarkable substance. It serves first in the trapping and storage of energy made available in the living cells by oxidation of foodstuffs, and later it serves in the release of this energy when needed by the living tissues.* ATP molecules are present in great numbers in all living cells. They are charged with energy and are ready to release their energy for living processes as needed. The process of breaking down and building up ATP, involving exchanges of energy, is continuous in living matter, and it is the central mechanism by which energy is obtained and utilized for living processes.

Abnormal metabolism. The mechanism of the production of glucosuria will be discussed.

Diabetes mellitus. As has already been stated, insulin is necessary for the utilization of glucose. When insulin is not secreted in sufficient quantities, the condition known as *diabetes mellitus* exists. This disease is characterized by an inability to utilize glucose, giving a high blood sugar (150 to 1500 mg/100 ml of blood under fasting conditions); as a consequence of the failure to oxidize sugar, there is an increased production of "ketone bodies": acetoacetic acid, β-hydroxybutyric acid, and acetone. As a result of increased sugar and ketone bodies in the blood, sugar and ketone bodies will pass through the kidneys and appear in the urine in greatly increased quantities.

The patient with diabetes mellitus has polyuria or an increased output of urine because a greater volume of water is needed to bring about the excretion of the increased load of sugar that is presented to the kidneys for excretion. The patient with diabetes has polydipsia, or increased thirst, because of the increased water requirement of the tissues due to the polyuria. The patient has an increased appetite and more or less weakness because of the failure to obtain energy from the oxidation of sugar.

Causes of glucosuria. Glucosuria occurs when the concentration of sugar in the blood is increased above the normal level. *The sugar level of the blood at which sugar is excreted in the urine is called the sugar threshold.* The sugar threshold in most subjects ranges from 140 to 190 mg of glucose per 100 ml of blood.

Some subjects do not show glucosuria until the blood sugar level is 200 to 300 mg/100 ml; these individuals are said to have a high sugar threshold. Somewhat rarely, subjects may be observed who exhibit glucosuria when the blood sugar is in normal limits. The excretion of sugar in the urine when the blood sugar is normal is a condition known as *renal diabetes.*

The most frequent cause of glucosuria is diabetes mellitus. The increased blood sugar level in this condition causes the excretion of sugar in the urine. *Other pathological conditions in which glucosuria may occur are hyperthyroidism, hyperactivity of the anterior pituitary gland, and hyperactivity of the adrenal cortex.* In these conditions the blood sugar does not increase to the high concentrations present in diabetes mellitus, but a level exists that is above the sugar threshold; consequently glucosuria occurs.

Glucosuria may occur in normal subjects. The ingestion of a large amount of sugar may raise the blood sugar level above the sugar threshold and cause glucosuria. The occurrence of glucosuria as a result of high ingestion of sugar or rapidly digestible carbohydrate is called *alimentary glucosuria*.

Emotional disturbances such as fright may also cause hyperglycemia and consequent glucosuria. The mechanism here is the liberation of epinephrine from the adrenal medulla, which raises the blood sugar by stimulating glycogenolysis or the breakdown of glycogen into glucose in the liver.

METABOLISM OF FATS

Absorption and distribution. The products of the digestion of fats in the intestinal tract are glycerol, fatty acids, and partially digested glycerides containing one or two fatty acids in the molecule. To what extent the partially digested glycerides are absorbed has not been determined. The glycerol is water soluble and is readily absorbed into the portal blood. Most of the fatty acids released in the intestinal tract by the action of digestive enzymes on fats are absorbed by the lymphatic system. These products enter the central lacteals of the intestinal villi and pass by way of anastomosing lymph vessels into the thoracic duct. The latter vessel drains into the bloodstream by an opening into the venous system at the junction of the left subclavian vein with the left internal jugular vein.

The fluid in the lymphatic vessels during the digestion and absorption of a fatty meal is milky white in appearance because of the presence of minute droplets of fat called *chylomicrons*. This fluid is called *chyle*.

There is evidence that fatty acids of chain length less than 10 carbon atoms (butyric, caproic, and caprylic acids, which occur in the fats of butter) are absorbed predominantly into the portal blood rather than into the lymphatic system. However, most of the fatty acids arising from fat digestion pass into the lymphatic system before entering the blood.

The blood finally carries the absorbed fats to all parts of the body, where they either are stored as fatty tissue or are oxidized to give heat and energy.

Glycerol metabolism. When fed to an animal, glycerol gives rise to the formation of liver glycogen. This indicates that the glycerol part of the fat molecule is metabolized by the same pathway as carbohydrates. Glycerol is phosphorylated and enters the glycolytic pathway at step 8 of Fig. 21-1, after which it could either form glycogen or undergo degradation to carbon dioxide and water. In undergoing catabolism it would pass through the succession of steps in the glycolytic sequence leading to the formation of pyruvic acid and then as pyruvic acid would enter the citric acid cycle where oxidation to carbon dioxide and water occurs.

β-Oxidation of fatty acids. The oxidation of the fatty acid part of the fat molecule is a stepwise procedure that begins at the carboxyl end of the fatty acid. *A series of reactions occurs, which includes the removal of hydrogen and a splitting off from the fatty acid chain of the two-carbon fragment, the acetyl group. These metabolic steps are called β-oxidation because the oxidation process centers around the β or second carbon atom from the carboxyl carbon.*

The β-oxidation is shown diagrammatically in Fig. 21-4. As indicated there, fatty acid combines with coenzyme A. This combination requires ATP to furnish the energy for

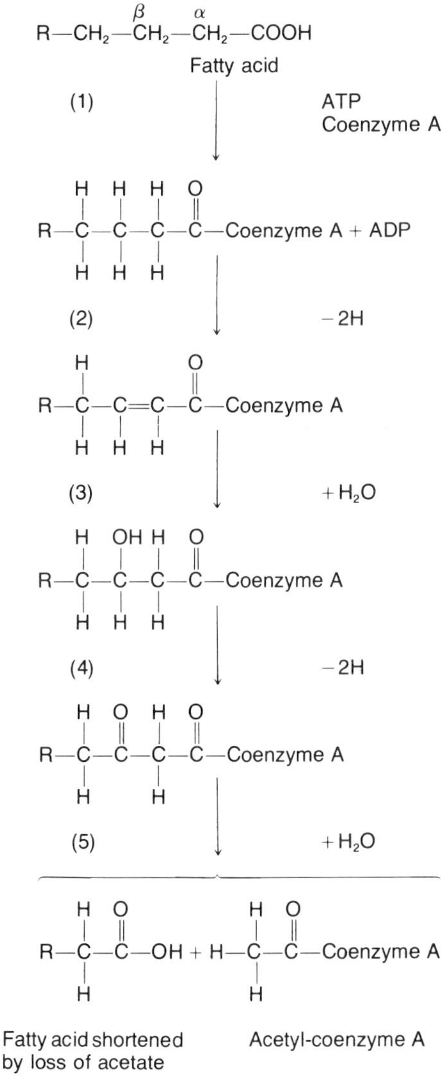

$$R-CH_2\overset{\beta}{-}CH_2\overset{\alpha}{-}CH_2-COOH$$
Fatty acid

(1) | ATP
Coenzyme A

$$R-\underset{H}{\overset{H}{C}}-\underset{H}{\overset{H}{C}}-\underset{H}{\overset{H}{C}}-\overset{O}{\underset{\|}{C}}-\text{Coenzyme A} + \text{ADP}$$

(2) | −2H

$$R-\underset{H}{\overset{H}{C}}-\underset{H}{\overset{}{C}}=\underset{H}{\overset{}{C}}-\overset{O}{\underset{\|}{C}}-\text{Coenzyme A}$$

Fig. 21-4. Chart showing β-oxidation of fatty acid.

(3) | +H$_2$O

$$R-\underset{H}{\overset{H}{C}}-\underset{H}{\overset{OH}{C}}-\underset{H}{\overset{H}{C}}-\overset{O}{\underset{\|}{C}}-\text{Coenzyme A}$$

(4) | −2H

$$R-\underset{H}{\overset{H}{C}}-\overset{O}{\underset{\|}{C}}-\underset{H}{\overset{H}{C}}-\overset{O}{\underset{\|}{C}}-\text{Coenzyme A}$$

(5) | +H$_2$O

$$R-\underset{H}{\overset{H}{C}}-\overset{O}{\underset{\|}{C}}-OH + H-\underset{H}{\overset{H}{C}}-\overset{O}{\underset{\|}{C}}-\text{Coenzyme A}$$

Fatty acid shortened by loss of acetate | Acetyl-coenzyme A

the reaction. Coenzyme A remains attached to the fatty acid throughout the series of degradative steps. Each step in the sequence is catalyzed by a specific enzyme. At two places in the sequence, steps 2 and 4, 2 hydrogen atoms are removed from the fatty acid molecule. These hydrogen atoms are transferred to the respiratory enzyme system, which ultimately brings about their union with oxygen to form water and liberate energy. In the fifth step of the sequence the complex is split hydrolytically between the α- and β-carbon atoms. This shortens the chain by 2 carbon atoms and releases a molecule of acetyl coenzyme A. The process is repeated on the shortened fatty acid chain, a two-carbon fragment (acetyl group) being split off from the carboxyl end each time the sequence of reactions occurs until the fatty acid is used up.

The β-oxidation of fatty acids is the most active, quantitatively, in the liver. It takes place to a smaller extent in kidney, smooth muscle, adipose tissue, and probably in other tissues.

Further oxidation of fatty acids. The acetyl coenzyme A complex formed by β-oxidation of fatty acids condenses with oxalacetic acid (from carbohydrate metabolism) to form citric acid, as shown in the citric acid cycle (Fig. 21-2). In the citric acid cycle the acetyl group (from fatty acid breakdown) is oxidized to carbon dioxide and water. The final stages of fatty acid metabolism are therefore the same as the final steps in carbohydrate metabolism.

The oxidation of fats in the human body liberates 9 kilocalories (9 large Calories or 9000 small calories) of energy per gram of fat. The energy produced by the oxidation of fats is utilized to maintain body temperature and for storage as ATP, which is used to supply energy for mechanical work and physiological processes.

Metabolism of ketone bodies. The ketone bodies are acetone, acetoacetic acid, and β-hydroxybutyric acid. Their formulas are as follows:

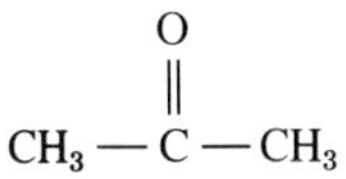

$$CH_3-\overset{\overset{O}{\|}}{C}-CH_3 \qquad CH_3-\overset{\overset{O}{\|}}{C}-CH_2-COOH \qquad CH_3-\overset{\overset{H}{|}}{\underset{|}{C}}\overset{}{H}-CH_2-COOH$$
$$\text{Acetone} \qquad\qquad \text{Acetoacetic acid} \qquad\qquad \beta\text{-Hydroxybutyric acid}$$

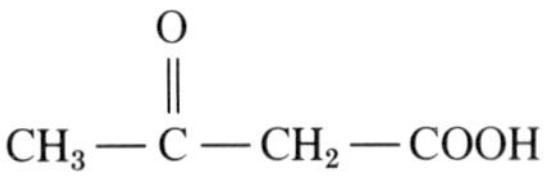

These compounds are formed in the oxidative breakdown of fatty acids in the liver. The acetyl coenzyme A complex produced by β-oxidation of fatty acids condenses with itself to form acetoacetic acid. The process is represented diagrammatically as follows:

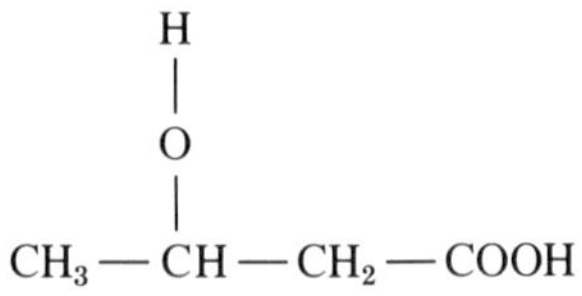

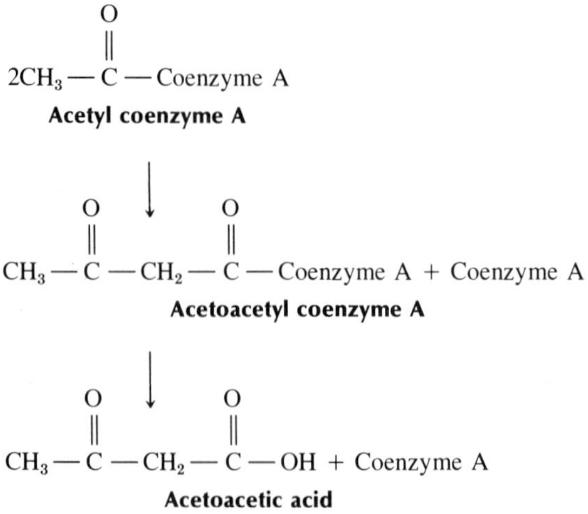

$$2CH_3-\overset{\overset{O}{\|}}{C}-\text{Coenzyme A}$$
$$\textbf{Acetyl coenzyme A}$$

$$\downarrow$$

$$CH_3-\overset{\overset{O}{\|}}{C}-CH_2-\overset{\overset{O}{\|}}{C}-\text{Coenzyme A} + \text{Coenzyme A}$$
$$\textbf{Acetoacetyl coenzyme A}$$

$$\downarrow$$

$$CH_3-\overset{\overset{O}{\|}}{C}-CH_2-\overset{\overset{O}{\|}}{C}-OH + \text{Coenzyme A}$$
$$\textbf{Acetoacetic acid}$$

The other ketone bodies, β-hydroxybutyric acid and acetone, are formed from acetoacetic acid. The ketone bodies are normal products of metabolism. They are formed by the liver and are utilized principally by the muscles. When the ketone bodies are delivered to

Metabolism of carbohydrates, fats, proteins, and inorganic salts

the muscles by the blood, they undergo further metabolic breakdown and are finally oxidized to carbon dioxide and water by way of the citric acid cycle, thereby yielding considerable energy. In a normal animal receiving a diet properly balanced with respect to carbohydrate and fat, the amount of ketone bodies formed by the liver does not exceed the quantity the muscles can oxidize. If, however, the diet contains too little carbohydrate or too much fat, more ketone bodies will be formed than the muscles (and other tissues) can utilize, and the excess of ketone bodies produced is excreted in the urine; also, if the animal has a deficient insulin secretion (diabetes mellitus), the carbohydrates are not metabolized correctly, and this will give rise to excess ketone body formation. The factor that prevents the formation of excess ketone bodies by the liver is the amount of carbohydrate utilized. Anything that interferes with the utilization of adequate amounts of carbohydrate, such as a low-carbohydrate diet, starvation, or insulin deficiency, will cause a shift of metabolic processes in the liver to a greater ketone body production. A high-fat diet limits the amount of carbohydrate available to the animal and also gives rise to increased production of ketone bodies. Some ketone bodies are formed from the amino acids leucine, isoleucine, tyrosine, and phenylalanine, but this is a small source compared with the amount obtainable from fatty acid catabolism.

From this discussion it is clear that the fatty acids are ketone-forming or ketogenic substances and that carbohydrates are antiketogenic substances. The balance between the amounts of carbohydrate and fat in the diet is important. If the ratio of fat to carbohydrate becomes too high, excess ketone bodies are formed. The highest ratio of fat to carbohydrate in the diet that will not result in excess ketone body formation is 1.5 g of fat to 1 g of carbohydrate. Thus, if 1 g of carbohydrate is utilized while not more than 1.5 g of fat are being metabolized, excess acetone body formation does not occur.

As noted previously, the ketone bodies are normal intermediary products in the catabolism of fatty acids. They become harmful or pathological when produced in greater quantities than the muscles can oxidize, and the excesses are excreted in the urine. The pathological effect is produced by two of the ketone bodies that are acids and require neutralization by bases. If the ketone bodies are oxidized in the muscle or other tissues, the basic ions remain in the body, but if these substances are excreted in the urine, they carry the bases along with them, and these basic constituents are lost from the body. Excess ketone body formation thus brings about an exhaustion of the bases of the animal's body, a process that produces an acidosis. This is a serious condition. Previous to the discovery of insulin, acidosis was the cause of death in about two-thirds of all persons with diabetes mellitus. Now diabetic acidosis can be relieved or prevented by the subcutaneous administration of insulin, which brings about correct metabolism of carbohydrates and thus prevents excess ketone body formation from fat. From this discussion it is clear that a diet properly balanced with respect to carbohydrate and fat is important in preventing a diabetic acidosis, and the same dietary balance is also essential to prevent acidosis in a normal subject.

Fatty liver. The normal fat content of the liver ranges from 4% to 8%. The fat content of the liver may be increased by improper diet, by disease, and by poisons. When rats or mice are placed on a high-fat diet or a low-protein diet, the animals develop the lesion known as "fatty liver." In this condition the liver contains from 15% to 25% fat. A somewhat similar condition results from starvation, from carbon tetrachloride or chloroform poisoning, and from eating foods with an excess of cholesterol. Depancreatized dogs, in which diabetes is controlled by the administration of insulin, will develop fatty livers unless raw pancreas, lecithin, or choline is added to the diet. There is also a tendency toward the development of fatty liver in uncontrolled diabetes mellitus.

Fatty liver of dietary origin is due to the absence from the diet of certain substances called *lipotropic agents* that enter into the formation of phospholipid. Choline and the amino acid methionine are such substances. Methionine is a donor of methyl groups that enter into the formation of choline, and choline is a part of the molecule of certain phospholipids. An example of a phospholipid is lecithin, the formula for which is given below. Phospholipids are important in fat metabolism because they are water-soluble compounds that serve as carriers of fatty acids. When there is a low supply of phospholipids in the body, fats are not transported from the liver with facility. A failure in the formation of phospholipid because of a dietary deficiency of choline or methionine thus results in an accumulation of fat in the liver.

Since fatty liver is an abnormal condition in itself and since this condition is a forerunner of cirrhosis of the liver, it is important to prevent its development. The administration of the lipotropic agents, choline and methionine, is advantageous in preventing or treating fatty liver, for reasons already given. Regardless of its cause, fatty liver is prevented or alleviated by a diet high in protein and carbohydrate and low in fat. A high-protein diet is effective in preventing fatty liver because proteins contain the amino acid methionine, which serves as a donor of methyl groups to form choline. A diet containing liberal amounts of carbohydrate is important because it serves to build up the glycogen stores of the liver and to supply the energy needs of the body from carbohydrate; when this is done, the liver is spared the necessity of utilizing greater amounts of fat.

$$\begin{array}{c} H \\ | \\ H-C-O-\text{Fatty acid} \\ | \\ H-C-O-\text{Fatty acid} \\ | \\ O \quad\quad H \quad H \quad OH^- \\ \| \quad\quad\quad | \quad | \quad\quad\!\diagup CH_3 \\ H-C-O-P-O-C-C-N-CH_3 \\ | \quad\quad\quad | \quad\quad | \quad | \quad\!\!+\!\diagdown \\ H \quad\quad\quad O \quad\quad H \quad H \quad\quad CH_3 \\ | \\ H \quad\quad\quad\quad \text{Choline} \end{array}$$

Lecithin

Obesity. Obesity is a bodily condition characterized by overweight. Obesity is present when the body weight exceeds to some extent that of the normally accepted standard weight for the individual's height and skeletal structure. This condition is considered unhealthy when the individual is 20% overweight.

Obesity results from the ingestion of more food than is needed to meet the requirements of the body for heat, energy, and maintenance of the structure of the tissues. In other words, obesity occurs when the caloric intake is greater then the caloric requirement. Under these conditions the excess calories bring about the deposition of fat in the elastic compartments of the body, and obesity is the result. One gram of fat is deposited for each 9 Calories of food intake in excess of the body's need for maintenance and energy.

It has been estimated that in the United States obesity exists in 15% of the population and that 5 million people are pathologically overweight.

Obesity is always the result of a caloric intake in excess of the body's needs. This takes place somewhat subtly during aging as the need for calories becomes less with decreasing activity. Certain pathological conditions such as hormonal imbalances may cause an increased appetite, which leads to an intake of food greater than needed, with consequent obesity.

The appetite for food is under the control of a center in the hypothalamus, a region in the third ventricle of the brain. This center has been damaged experimentally in animals by the administration of gold salts, with the resultant production of an enormous appetite (polyphagia) and marked obesity. It has also been observed that damage to other areas in the hypothalamus produces a diminished appetite (aphagia), which causes cachexia, the condition opposite to obesity.

METABOLISM OF PROTEINS

Anabolism. Proteins are digested to amino acids by the action of the proteolytic enzymes of the alimentary tract. The amino acids pass readily through the walls of the intestine into the bloodstream. During the digestion and absorption of a protein meal the amino acid content of the blood increases, but the blood rapidly distributes these absorbed protein products to all the tissues of the body, and the amino acid concentration of the blood soon returns to the normal fasting level.

Since the body tissues are composed largely of proteins and since there is a constant wear and tear on these tissues, it is important for the body to be supplied continuously with amino acids from which proteins may be constructed. *The amino acids obtained from the proteins of the diet are therefore utilized for the important function of tissue building. These substances may be used for the construction of new body tissues as in the growth of young persons, or they may be used for the repair of tissues in both adult and young persons. Amino acids are also used in the formation of enzymes and hormones. Proteins are therefore absolutely essential for the maintenance of life.* If proteins are not present in the diet, amino acids are withdrawn from the less essential parts of the body and are used to nourish the more vital parts such as the heart and the brain. In starvation the muscles and peripheral portions of the body lose protein most rapidly, and the heart and brain are the last to show a decrease in their protein content. A protective or "sparing" metabolism thus exists in the body, but an individual cannot live long on a protein-free diet, even though an abundance of carbohydrates and fats is eaten.

Unlike carbohydrates and fats, proteins are not normally stored in the bodies of adults. Practically the only instances of protein storage are in the growing young person, in underweight convalescents from wasting diseases, in individuals recovering from starvation, and in individuals undergoing vigorous physical exercise who may to a limited extent store proteins for a short period of time. With these exceptions, the anabolic function of proteins is confined to the renewal of body tissues, in which case storage does not occur.

Proteins are converted to carbohydrates in the human body, especially when there is a deficiency of carbohydrates in the diet or when the body is unable to burn carbohydrates properly, as in persons with diabetes mellitus. The amount of conversion varies according to the diet up to a maximum of 58%. Hence in calculating the amount of carbohydrates in a diet for a patient with diabetes, 58% of the protein allowed is considered as available carbohydrate.

Catabolism. The primary function of amino acids in the animal body is the construction of tissue proteins, but these substances also serve an additional useful purpose by being oxidized to yield energy. The body tissues formed from amino acids are not static in composition. Actually there is a rapid metabolic turnover of amino acids and of the proteins formed from them in the animal body. The rapidity of this chemical turnover has been demonstrated through experiments in which amino acids labeled with isotopes were fed to animals; by analyzing tissues for the labeled atoms, investigators have been able to show that there is in the animal's body a rapid turnover of amino acids and proteins.

When amino acids absorbed into the body (1) are in excess of the needs for tissue construction, (2) are not of the correct type for tissue building, or (3) are discarded in the metabolic turnover processes, they are degraded by a series of chemical reactions into carbon dioxide, water, urea, and ammonium salts. The chemical degradation of proteins in the animal body yields 4 kilocalories (4 large Calories) of heat per gram of protein.

Two catabolic processes, deamination and transamination, that apply to the metabolism of practically all amino acids will be discussed.

Deamination. Deamination is a reaction in which the α-amino group (the one next to the carboxyl) is removed from an amino acid. The reaction takes place in probably all animal tissues but to the greatest extent in the liver. It requires oxygen and is catalyzed by an enzyme, an amino acid oxidase. There are many amino acid oxidases in animal tissues. Deamination occurs as one step in the catabolism of nearly all amino acids. The overall reaction, using a general formula for an amino acid, is as follows:

$$\underset{\text{Amino acid}}{R-\underset{\underset{NH_2}{|}}{\overset{\overset{H}{|}}{C}}-COOH} + \tfrac{1}{2}O_2 \xrightarrow{\text{Amino acid oxidase}} \underset{\text{Keto acid}}{R-\overset{\overset{O}{\|}}{C}-COOH} + NH_3$$

In this reaction a keto acid derivative of the amino acid is formed and ammonia is released. Since ammonia is a toxic by-product, it must be converted to nontoxic products and eliminated.

A small amount of the ammonia liberated by oxidative deamination combines with acid radicals (such as sulfate and phosphate) to form ammonium salts. Most of the ammonia released by this reaction in the presence of CO_2, certain amino acids, and enzymes form urea, $NH_2 \cdot CO \cdot NH_2$. Urea formation takes place in the liver only. In the formation of urea and ammonium salts, the nitrogen of amino acids is converted into a form that is highly soluble in water and therefore readily excreted in the urine.

FATE OF KETO ACID. *In part, or as the total fragment, the keto acid formed by oxidative deamination of amino acid may enter the citric acid cycle and be oxidized to carbon dioxide and water; it may enter the glycolytic pathway and form glucose and glycogen; or it may enter into the biosynthesis of fatty acids.*

Transamination. Transamination is a metabolic process in which there is a transfer of an amino group from an amino acid to a keto acid. The process is shown using the amino acid, which donates an amino group to the keto acid, pyruvic acid, which serves as a recipient of an amino group.

$$HOOC-CH_2-CH_2-\underset{\underset{NH_2}{|}}{CH}-COOH + CH_3-\underset{\underset{O}{\|}}{C}-COOH \rightleftarrows$$

Glutamic acid   **Pyruvic acid**

$$HOOC-CH_2-CH_2-\underset{\underset{O}{\|}}{C}-COOH + CH_3-\underset{\underset{NH_2}{|}}{CH}-COOH$$

α-Ketoglutaric acid   **Alanine**

In this reaction there is a transfer of the amino group from glutamic acid to pyruvic acid with the resulting formation of α-ketoglutaric acid and another amino acid, alanine. The reaction is reversible and is catalyzed by an enzyme called glutamic pyruvic transaminase. The vitamin pyridoxine is a coenzyme in this reaction. Transaminases are found in practically all animal tissues and also in bacteria and plants.

Glutamic acid can donate an amino group to keto acids other than pyruvic acid. In fact, glutamic acid has been found to serve as a donor of an amino group to the corresponding keto acid of nearly all amino acids.

The role of α-ketoglutaric acid, serving as a recipient of an amino group, is illustrated by the same equation, read in the reverse direction, from right to left. It has been found that α-ketoglutaric acid in the presence of an extract of liver, kidney, heart muscle, skeletal muscle, or certain plants will, under appropriate conditions, accept an amino group from all amino acids except lysine and threonine, which do not participate in transamination. When aminated, α-ketoglutaric acid becomes glutamic acid, which, as shown, may continue the transamination process by aminating other keto acids.

Transamination occurs widely in plant and animal life. The purpose of transamination is to provide the variety of amino acids needed by the tissues by utilizing the amino groups of amino acids present in greater quantities than needed, to synthesize amino acids that may not be present in adequate amounts. Transamination serves to bring about a rapid interchange of amino groups between amino acids and keto acids and thus to make available the kind of amino acids needed by the tissues.

There are two transaminases of clinical importance—serum glutamic oxalacetic transaminase (SGOT) and serum glutamic pyruvic transaminase (SGPT). Transaminases are found in all tissues of the body but in much greater quantities in some organs than in others.

Heart muscle is especially rich in glutamic oxalacetic transaminase. When acute injury to the heart occurs, as in myocardial infarction, glutamic oxalacetic transaminase is released from the damaged cells into the bloodstream, and the blood level of this enzyme is markedly elevated for several days after the injury has occurred. The determination of the SGOT level is therefore of diagnostic value in acute heart disease.

Serum glutamic pyruvic transaminase is found in all body tissues, and this enzyme occurs in the greatest amount in liver. Both SGPT and SGOT are elevated in the blood as a result of acute liver damage, but SGPT is elevated to a greater extent. Determinations of SGPT and SGOT are therefore of diagnostic value in certain diseases of the liver. (See Table 24-2.)

METABOLISM OF INORGANIC SALTS

The inorganic salts do not contribute energy to the body, but they are necessary for correct metabolism. About 25 g of inorganic salts are consumed per day normally, and about one-half of this amount is sodium chloride. We will discuss briefly the metabolism of the principal inorganic elements.

Sodium. The sodium ion occurs principally in the extracellular fluids of the body.

The sodium ion acts as a stimulant in promoting the irritability of muscle and nerve tissues. It also functions powerfully as a basic ion in the acid-base balance of the body, making up about 92% of the fixed bases of the plasma and extracellular fluids.

The sodium ion has a specific effect on the flow of water through the body. When sodium ions are held back in the tissues, water is likewise retained, and when sodium ions are excreted through the kidneys, water is also eliminated by way of the kidneys.

The level of sodium in the blood plasma is maintained by the action of hormones of the adrenal cortex, of which aldosterone is the most powerful. These hormones promote the reabsorption of the sodium ion from the lumen of the kidney tubules, the plasma level being determined by the amount of hormone secreted. Thus adrenal cortical activity establishes the level of sodium in the blood plasma which, in turn, influences greatly the reabsorption of water from the kidney tubules. The latter mechanism participates prominently in establishing the volume of fluid in the blood vessels, but there are other factors (plasma proteins, antidiuretic hormone, and other ions) that influence the movement of water through the body.

The sodium ion in the blood plasma has an important role in respiration; this ion serves as a carrier of carbon dioxide in the form of sodium bicarbonate ($NaHCO_3$).

In the maintenance of the osmotic pressure of extracellular fluids, the sodium ion participates the same as the ions of other elements or compounds, the quantitative effect of the sodium ion being the same as that of any other ion.

The sodium ion is rapidly absorbed from the intestinal tract. About 95% of the sodium entering into metabolism is excreted in the urine. Small amounts of sodium are excreted in the feces and the sweat.

Potassium. Potassium ions occur in all tissues and predominantly in the intracellular fluids. There is about 20 times as much potassium in the red cells of the blood as in the plasma.

The potassium ion is necessary to maintain the irritability of muscle and nerve tissue. This ion also participates in the maintenance of the osmotic pressure of body fluids. The potassium of the red cells participates in the carbon dioxide–carrying function of the blood (see p. 351). *The potassium ion also takes part in the neutralization, transportation, and elimination of acid radicals by way of the kidneys.*

A distinctive function of the potassium ion is to serve as a cofactor for certain enzymes in carbohydrate metabolism.

The effect of potassium on water balance is opposite to that of sodium; *the potassium ion has a diuretic effect;* that is, it stimulates the excretion of water by the kidneys.

The potassium ion is rapidly absorbed from the intestinal tract. More than 90% of potassium excretion is by way of the urine.

The maintenance of a normal level of potassium in the blood plasma (16 to 22 mg/100 ml) is extremely important. Hypopotassemia leads to paralysis, and hyperpotassemia, if the level is high enough, will cause cardiac arrest.

Calcium. Certain salts of calcium—the phosphate, the carbonate, and the fluoride—are highly insoluble in a neutral or alkaline medium. These compounds are therefore readily adaptable to the formation of structural and supporting tissues in the animal body. Of the 50% to 60% of mineral matter in bone, calcium phosphate makes up about 85%, calcium carbonate about 10%, and calcium fluoride about 1%. The cement and dentin of the teeth have essentially the same calcium composition as bone, and about 90% of the enamel of the teeth consists of calcium compounds, a composition that accounts for the unusual hardness of this substance. *Calcium thus has an important role in supplying material for the skeletal growth of young persons and for the maintenance of the composition of the hard tissues of both young and adult persons.*

Calcium ions are also essential for the clotting of blood and the coagulation of casein following the digestive action of rennin.

Calcium ions have a sedative influence on the motor nervous system. The normal calcium concentration of the blood is 9 to 11 mg/100 ml of serum. If a condition develops that reduces the serum calcium concentration below normal, varying degrees of hyperexcitability results; tetany usually occurs when the level of blood calcium gets below 7 mg/100 ml of serum.

Calcium ions are also concerned in the regulation of heart action. The effect of calcium is to promote systole of the heart; this action is balanced by the antagonistic effect of potassium, which is to reduce contractility of the heart muscle and thus favor diastole. The sodium ion also participates with calcium and potassium ions in maintaining the heartbeat.

The blood calcium concentration is regulated by the secretion of the parathyroid gland; hypersecretion by this gland raises and hyposecretion lowers the blood calcium level. A normally functioning parathyroid gland keeps the blood calcium level normal when the dietary intake of calcium and vitamin D is adequate. Vitamin D promotes the absorption of calcium from the intestinal tract and thus participates in the maintenance of a normal blood calcium level.

The body does not conserve its calcium in periods of calcium deprivation or calcium starvation but apparently continues to excrete this substance at the usual rate. If the calcium of the diet is reduced below minimum requirements, the blood calcium will remain at a normal level for a considerable time because of a withdrawal of calcium from the bones by the action of the secretion of the parathyroid glands; continued deprivation of calcium, however, will bring about a decalcification of the bones and a lowered calcium content of the blood. The outcome of such a dietary deficiency is rickets in children and osteomalacia in adults.

Magnesium. *Magnesium compounds make up about 1% of the mineral constituents of the bones and teeth. This element thus has some importance as a structural substance, being an essential constituent of the hard tissues of the animal body.* Magnesium also exists in small quantities in the soft tissues of the body and in the body fluids and has certain important physiological activities.

The magnesium ion has a sedative effect on nerve and on smooth and skeletal muscle. It is a depressant of the central nervous system, and because of this effect, it has been used as an anesthetic.

The magnesium ion is an activator of a number of enzymes in animal and plant tissues.

Magnesium ions are slowly absorbed from the intestinal tract. For this reason magnesium compounds such as magnesium sulfate (Epsom salt), magnesium citrate, and magnesium hydroxide (milk of magnesia) are used as laxatives. The magnesium ions of these compounds are not absorbed fast enough to do any harm in the body, and they are retained in the intestinal tract in high concentrations. These ions, along with the anions (sulfate, citrate, etc.) of the compound administered, exert a marked osmotic pressure that causes water to pass from the tissues into the lumen of the intestine, producing a laxative effect. The laxative effect of magnesium compounds is illustrative of the action of saline cathartics in general.

Iron. Iron is a constituent of the hemoglobin of blood. The iron in hemoglobin is in the ferrous state. *Ferrous hemoglobin has the unique ability to bind reversibly molecular oxygen in a loose combination called oxyhemoglobin.* In the lungs, where there is a high pressure from gaseous oxygen, hemoglobin and oxygen readily combine, forming oxyhemoglobin, and in the tissues, where there is a lowered oxygen pressure and the pH has shifted to a less alkaline value, oxyhemoglobin gives up its loosely bound oxygen. *Thus hemoglobin serves as a carrier of oxygen from the lungs to the tissues, and the part played by iron is an important function of this element.*

Another function of iron is based on its ability to undergo valence changes readily. *Ferrous iron* (Fe^{++}) *will readily give up an electron and become ferric iron* (Fe^{+++}), *and ferric iron easily accepts an electron to become ferrous iron. The readiness with which iron can accept or release electrons makes this element exceptionally well adapted to serve in the electron transport systems in living tissues.* It is this property of iron that enables this element to function in the iron-protein complexes, such as the cytochromes, cytochrome oxidase, catalase, and peroxidase, which participate in tissue respiration.

Iron is absorbed from the intestinal tract in the form of the ferrous ion (Fe^{++}). Iron in the ferric (Fe^{+++}) state is probably absorbed after it has been reduced to the ferrous form by intestinal bacteria, ascorbic acid, or other reducing agents.

Iron is stored in the body as a constituent of the compound *ferritin,* a colloidal protein-iron complex with a molecular weight of 465,000. The iron makes up about 23% of this complex. *The functions of ferritin are to participate as an intermediate in the absorption of iron and to serve as a storage reservoir of iron in the tissues.* Ferritin is found in significantly large amounts in the intestinal mucosa, liver, spleen, and bone marrow.

The theory concerning iron absorption is that ferrous iron enters the mucosal cells and combines with apoferritin to form ferritin. When the tissues of an animal are well supplied with iron, there is a relatively large amount of ferritin in the intestinal mucosa, and under these conditions only a small amount of iron is absorbed, most of the iron in the food being excreted in the feces. On the other hand, if there is a low supply of ferritin in the intestinal mucosa, as in an animal with an anemia, the absorption of iron from the intestinal tract by way of the ferritin mechanism is much greater than in an animal with a good supply of ferritin in the tissues.

The transport of iron from ferritin in the intestinal mucosa to the tissues of the body is accomplished by a β-globulin in blood plasma called transferrin. This is an iron-binding protein with a molecular weight of 90,000. Iron is released from ferritin in the intestinal mucosal cells into the blood plasma, where it combines with transferrin. In combination with transferrin, iron is transported to various tissues in the body.

The amount of iron absorbed daily from the intestinal tract of a normal individual with an adequate iron storage is quite small. An unusual economy in the use of iron exists in the body because of the functioning of the highly efficient reticuloendothelial cells, which remove iron from the hemoglobin of brokendown red blood cells and make it available for the synthesis of hemoglobin in newly formed red cells or for the formation of other iron-containing compounds.

The iron in foods in excess of the body's needs is excreted in the feces. Only traces of iron are excreted in the urine. Overloading of the body with iron does not occur except as a result of multiple blood transfusions or prolonged excessive iron therapy. When overloading does occur, the excess iron is incorporated into another iron-storage protein complex called *hemosiderin,* which is deposited principally in the liver. Hemosiderin is thought to be a complex formed by the polymerization of ferritin molecules. When extensive hemosiderin deposits are present in the body, the condition is called *hemosiderosis.*

Chloride. Chlorides are present in all tissues, chiefly in the form of sodium or potassium chloride.

Chlorides are highly soluble in water, and they diffuse rapidly either into or out of cells. The property of rapid diffusibility through membranes makes the chloride ion well suited to participate in keeping positively and negatively changed ions balanced in the tissues. If a shift of negative ions from an area takes place in metabolism, chloride ions diffuse into that area to balance the positive ions and thus maintain electrical neutrality. An example of this is the chloride exchange with bicarbonate (chloride shift) between red cells and plasma in respiration (see p. 351). This ability of the chloride ion to diffuse rapidly explains why *it functions effectively in the maintenance of a normal electrolyte pattern in tissues.*

The ability of the chloride ion to diffuse rapidly through tissues also enables this ion *to assist effectively in maintaining a balance in the osmotic pressure in the tissues.*

In conjuction with the sodium ion, the chloride ion *assists in maintaining water balance in the animal body.* If the blood level of sodium and chloride ions of a patient is low, the patient is dehydrated. For this condition, sodium chloride in isotonic solution is administered. *The sodium chloride helps to retain water in the tissues.*

The chloride ion is also used in the production of the hydrochloric acid of gastric juice.

Phosphorus. Inorganic phosphate participates in metabolism as *(1) it enters into the formation of bones and teeth in combination with calcium and magnesium, (2) it provides the phosphate ion for various phosphorylations, (3) it enters into the formation of nucleic acids and phosphatides, and (4) it serves in the trapping, storage, and utilization of energy in the high-energy phosphate bond (such as ATP and creatine PO_4).*

Questions for study

1. What is metabolism?
2. Define anabolism. Catabolism.
3. Describe the absorption of carbohydrates.
4. What is the normal amount of glucose in the blood? How high does the blood sugar rise following a carbohydrate meal? What does the body do with this increased blood sugar normally?
5. How much glycogen is in the whole body when well nourished?
6. What is the role of the liver in carbohydrate metabolism?
7. What is the effect of insulin in carbohydrate metabolism? What is the effect of epinephrine?
8. What is glycolysis? What advantages are achieved by glycolysis?
9. What is the citric acid cycle?
10. Discuss the importance of the citric acid cycle as a common pathway in metabolism.
11. How is the energy obtained from the oxidation of carbohydrates in the body trapped, stored, and later utilized?
12. What is the caloric value of carbohydrates?
13. What is the importance of ATP?
14. What changes occur in the chemistry of the blood and urine of a patient with diabetes mellitus?
15. Mention causes of glucosuria other than diabetes mellitus.
16. Describe the absorption of fats from the intestinal tract.
17. What is the pathway of metabolism of the glycerol part of the fat molecule?
18. Discuss β-oxidation of fatty acids.
19. Discuss the pathway of metabolism of fatty acids after β-oxidation.
20. What is the caloric value of fats when oxidized in the body?
21. Describe the metabolism of the ketone bodies.
22. What ratio of carbohydrate to fat in the diet will prevent the production of excess ketone bodies?
23. What is acidosis?
24. Describe the condition known as fatty liver.

25. Mention several causes of fatty liver in an animal.
26. What dietary constituents will prevent fatty liver?
27. What is the cause of obesity?
28. What abnormal conditions may lead to obesity?
29. What center in the brain has a control over appetite?
30. Describe the anabolism of proteins.
31. What are the functions of amino acids in the body?
32. Are proteins stored in the body?
33. In calculating a diet for a diabetic patient, what percent of protein is considered as carbohydrate?
34. What is oxidative deamination?
35. What is transamination?
36. What is the importance of the determination of serum glutamic oxalacetic transaminase (SGOT)? Serum glutamic pyruvic transaminase (SGPT)?
37. What are the functions of the sodium ion in the animal body?
38. What is the relation of the sodium in extracellular fluids to water balance in the body?
39. How is the level of sodium maintained in blood plasma?
40. What are the functions of the potassium ion in the animal body?
41. What condition results from an abnormally low potassium level in blood plasma? An abnormally high potassium level in blood plasma?
42. What are the effects of sodium and potassium ions on the irritability of skeletal muscle?
43. What are the functions of calcium in the animal body?
44. What hormone regulates the level of calcium in blood plasma?
45. What are the functions of magnesium in the animal body?
46. What are the effects of calcium and magnesium on the irritability of skeletal muscle and nerve tissue?
47. Describe two functions of iron in the animal body.
48. Discuss the mechanism of the absorption of iron from the alimentary tract.
49. What is ferritin? Transferrin? Hemosiderin?
50. What are the functions of the chloride ion in the animal body?
51. What are the functions of inorganic phosphate in the animal body?

22 Nutrition

Nutrition is a science that deals with the relation of food substances to the well-being of the organism using these substances. In this chapter we will discuss animal nutrition only, with principal emphasis on the nutrition of human subjects. *In animal nutrition the following are considered: (1) the need of the body for various constituents of the diet, (2) the function of each dietary constituent, (3) the amount of each dietary component that is required for optimum health in the adult and for growth and optimum health in the young, and (4) the level of intake of food substances below which poor health results.*

A complete diet for the human being includes the following constituents in adequate amounts: proteins that include ten essential amino acids, glucose or carbohydrates that yield glucose by digestion or metabolism, fats that include three essential fatty acids, not less than 18 inorganic constituents, at least 17 vitamins, and water. The diet must contain enough foodstuffs to provide sufficient energy (calories) for the metabolism and activities of the body, and in the young person, additional calories to provide for growth. A correct diet also has a certain balance with respect to the amounts of protein, carbohydrate, and fat.

The specific functions and requirements of the various constituents of the diet in the human subject will now be discussed.

PROTEINS

Nitrogen equilibrium. The determination of the amount of proteins that should be in the diet is based on several considerations. In the first place there must be a balance between the intake of proteins and the loss of proteins from the body as a result of metabolic activities. Since approximately 16% of protein is nitrogen and nitrogen can be readily determined chemically, the protein content of a substance is measured by means of a chemical estimation of its nitrogen content. The intake of proteins may therefore be measured by a chemical determination of the nitrogen content of the food that is ingested, and the loss of proteins from the body may be determined by estimating the nitrogen content of the urine and feces.

Measurement of the nitrogen intake and output of the body is thus a means of studying the adequacy of a diet with respect to its protein content. *When the nitrogen intake of the body is equal to its nitrogen output, the condition is called nitrogen equilibrium. If the nitrogen intake is greater than the nitrogen output, the individual is said to be in positive nitrogen balance.* Such a condition may exist in a growing young person, in persons recovering from a wasting disease or from starvation, and to a very limited extent as a result of vigorous muscular activity. *Negative nitrogen balance exists when the nitrogen intake is less than the nitrogen output.*

A consideration of nitrogen equilibrium only is not a complete criterion for the establishment of a proper protein intake. The degree of protein catabolism, which determines the losses of nitrogen from the body, varies according to the amount of carbohydrate and fat in the diet. When the diet does not contain sufficient carbohydrate, the liver makes glucose from available proteins, a process that increases the losses of nitrogen from the body. The nitrogen losses of the body are high when there is little or no carbohydrate and fat in the diet, and the losses are low when the diet contains an abundant amount of these food constituents. It obviously would not be satisfactory to set the protein requirement at the amount the body loses during starvation or at the quantity an individual metabolizes when on a diet containing only proteins because these situations compel an abnormal protein catabolism to supply the body's need for glucose. It also would not be desirable to establish the protein requirement as the minimum amount that will maintain nitrogen equilibrium when the diet contains an abundant amount of carbohydrate and fat. Such an amount is too close to the borderline requirement for practical maintenance and does not take into consideration the fact that the proteins ingested vary widely in their capacity to satisfy the nutritional requirements of the human body.

Biological value. Some proteins are qualitatively deficient as constituents of the diet of man because they do not contain certain amino acids that the human body needs and is unable to synthesize. Such proteins are said to have a poor biological value. The protein foods that most readily satisfy the requirements of man are those having an amino acid composition most similar to that of his own body. Such foods have a high biological value. Proteins of animal origin have the highest biological value to man because they contain a greater amount of the amino acids that the human body needs but cannot synthesize. Based on rat-feeding experiments, the biological value of the proteins of various foods arranged in order of diminishing value are whole eggs, milk, soybeans, meats, vegetables, and grains.

Essential amino acids. The requirement of essential amino acids varies in different species.

In the rat. The importance of the qualitative nature of the proteins of the diet has been revealed by investigators who studied the effects of synthetic diets on the growth and well-being of rats. The early experiments consisted of feeding a diet containing carbohydrate, fat, salt mixture, vitamins, water, and some purified protein. From such experiments, Osborne and Mendel found that rats would grow on a diet in which the protein constituent was either casein, lactalbumin, egg albumin, glutenin of wheat, glutelin of corn, or globulin of cottonseed. In other experiments these investigators found that when gelatin, gliadin of wheat, zein of corn, hordein of barley, or phaseolin of the white kidney bean was used as the sole source of protein, the rats would not grow. Osborne and Mendel found by further experimentation that the proteins that would not bring about growth of rats were deficient in the amino acids lysine and tryptophan. When lysine and tryptophan were added to a deficient protein (zein) diet, the rats grew normally. Other investigators have confirmed Osborne and Mendel's results and have found further that histidine, methionine, leucine, isoleucine, phenylalanine, valine, threonine, and arginine must be present in the proteins of the diet in adequate amounts to obtain growth of rats. *Thus it is now known that in the proteins of the food there must be present at least ten amino acids (lysine, tryptophan, histidine, methionine, leucine, isoleucine, phenylalanine, valine, threonine, and arginine) to bring about normal growth of the rat. These are known as essential or indispensable amino acids.* A diet may contain considerable protein, but it will not be adequate nutritionally for the rat unless these essential amino acids are present in the protein in sufficient amounts (Figs. 22-1 and 22-2).

Fig. 22-1. Upper photograph shows the effect of a dietary deficiency of the amino acid valine. Lower photograph shows the same animal after valine had been administered for 25 days. (From Rose and Epstein: J. Biol. Chem. **127**:683, 1939.)

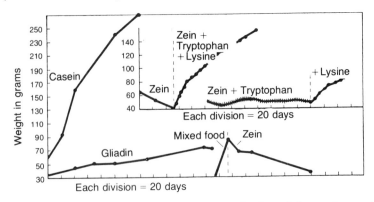

Fig. 22-2. Curves illustrating increase or decrease in body weight (ordinates in grams) of young rats fed on food indicated on curves. Abscissae represent days. The growth with casein is to be regarded as normal. (After Mendel and Osborne; from Tuttle and Schottelius: Textbook of physiology, The C. V. Mosby Co.)

In man. Studies of the amino acid requirement of man based on growth response could not be carried out because of the slow growth rate of the human subject and the expense involved. Hence another method had to be used. The method selected was the nitrogen balance procedure. It is a well-established nutritional principle that if an essential amino acid is left out of the diet of an animal, a negative nitrogen balance results. *Experiments on human subjects have shown that only eight amino acids (lysine, tryptophan, methionine, leucine, isoleucine, phenylalanine, valine, and threonine) must be present in the diet to maintain nitrogen equilibrium.* However, in these experiments some unfavorable clinical observations were made when diets lacking in histidine and arginine were eaten. It may be that the nitrogen equilibrium method is not sensitive enough to show whether histidine and arginine are essential in the diet of man; hence a satisfactory answer concerning the need of these two amino acids in the diet of the human subject cannot be made from available evidence. *Thus it has been shown by nitrogen equilibrium studies that eight amino acids are essential in the diet of man, and there is clinical evidence that two other amino acids (histidine and arginine), found essential in the rat, may also be required in the human diet.*

Human protein requirement. Based on a statistical study of the protein content of the diet of the German people, Voit proposed that the diet should contain about 118 g of protein per day. Chittenden, on the other hand, from an extensive series of experiments with soldiers and laboratory workers on various diets found that a man of average size could be maintained in nitrogen equilibrium on a protein intake of 44 to 53 g per day.

Taking into consideration the conditions necessary for maintaining nitrogen equilibrium and for providing the body with essential amino acids, it is now generally believed that the diet should contain about 1 g of protein per kilogram of body weight per day (60 to 90 g for an adult). This amount is considered ample to maintain an individual in nitrogen equilibrium and to provide a margin of safety for practically all conditions of food intake.

The protein requirement of the growing child is relatively higher than that of the adult because of the need of additional protein for tissue building. The recommended daily allowance for growing children per kilogram of body weight ranges from 3.5 g for infants to 1.5 g for boys and girls 16 years of age.

CARBOHYDRATES

The tissues of the body need glucose at all times. Either glucose or substances that yield glucose by digestion or metabolism must be in the diet. If the blood sugar of an animal is decreased by experimental procedure, the animal becomes sick, and if the concentration of sugar in the blood is lowered sufficiently, the animal will die.

Considerable glucose is eaten as such and thus serves to maintain the stores of glucose in the body. Other assimilable carbohydrates either yield glucose by digestion in the alimentary tract or form glucose by metabolic conversion in the liver. The body thus obtains glucose or glucose precursors from the food that is ingested.

In addition to the glucose absorbed from the alimentary tract, there are three other sources of blood glucose. (1) The glycogen of the liver may be broken down to form glucose, or (2) certain amino acids or (3) fatty acids may be converted into glucose by the liver. *The conversion of amino acids or fatty acids into glucose is called gluconeogenesis.* It is especially important when the glycogen stores are exhausted. This conversion mechanism serves as a safety device to provide the tissues with a constant supply of glucose when alimentary sources of the sugar are not available. It is thus seen that the body has certain mechanisms that operate successfully in case there is an inadequate intake of carbohydrate; however, the diet should contain an abundant amount of carbohydrate to provide directly for the glucose requirements of the body. When carbohydrate is present in the diet in adequate amounts, there is less conversion of amino acids to glucose. This means that there is less nitrogen turnover, hence less work by the liver and kidneys to maintain the normal level of glucose in the blood and tissues. Thus carbohydrates serve as "protein sparers," a desirable nutritional result.

Since the body uses carbohydrates for the production of energy, the amount of carbohydrate in the diet is dependent on the activity of the individual. This amount varies from 300 to 700 g per 24 hours.

FATS

Fats are of nutritive importance as sources of energy and as food substances that are readily stored. They are also of value as structural materials, giving form to the body and physical support to certain internal organs.

The amount of fat in the diet is subject to variations due to activity and must be regulated according to the carbohydrate content, since fats will give rise to excess ketone body formation unless adequate amounts of carbohydrates are being utilized. The quantity of fat desirable in the diet varies for 50 to 150 g per 24 hours.

Essential fatty acids. Three unsaturated fatty acids (linoleic, linolenic, and arachidonic acids, see p. 224) are indispensable for correct nutrition of man, the rat, and other animals. Rats placed on a fat-free but otherwise adequate diet develop dermatitis, alopecia, necrosis of the tail, and body sores. Growth failure occurs, and the animals do not survive long (Fig. 22-3). Addition of linoleic, linolenic, or arachidonic acid to such a diet will correct this condition or prevent its onset. It appears from experiments of this kind that *linoleic, linolenic, and arachidonic acids are essential constituents of the diet.*

Fig. 22-3. These rats are litter mates. The one on the left received a diet containing 20% lard; and the one on the right received no fat. (Courtesy Dr. George Burr; from Hawley and Maurer-Mast: The fundamentals of nutrition, Charles C Thomas, Publisher.)

Table 22-1. Distribution of calorie-producing dietary components in a balanced diet

Foodstuff	Percent of calories	Grams/kg of body weight
Protein	10 to 15	1
Fat	25 to 35	1 to 2
Carbohydrate	50 to 60	5 to 7

THE BALANCED DIET

The amounts of protein, fat, and carbohydrate in the diet should be balanced with respect to each component. In a well-balanced diet the distribution of foodstuffs with respect to total caloric content and grams per kilogram of body weight is approximately as shown in Table 22-1.

When the diet contains adequate calories, the balance shown in Table 22-1 allows enough protein to maintain nitrogen equilibrium with a good margin of safety. The amounts of carbohydrate and fat recommended are in a ratio that prevents ketosis. The fat allowance also provides adequate amounts of essential fatty acids. The total caloric content of the diet depends on the activity of the individual.

Sodium and chlorine. The dietary intake of sodium chloride ranges from 5 to 15 g per day. This is probably a superfluous intake, but the metabolic turnover of a large amount of this salt apparently has no harmful effect in the absence of disease.

Considerable sodium is present in the usual foods, and this source, together with the sodium chloride added voluntarily to the food for its flavoring effect, supplies the needs of the body except in very hot weather, when excessive sweating may make it necessary to ingest additional salt. In hot weather it is the practice in industry to provide sodium chloride in tablets or in the drinking water for workers in occupations that cause severe sweating.

It often becomes necessary in the hospital to administer sodium chloride solution to a patient to maintain normal electrolyte and water balance. This is accomplished by parenteral administration of physiological saline solution.

In certain disease states such as edema and hypertension the restriction of sodium intake is very important to decrease the blood pressure and also to reduce the work of the heart. Low-salt diets are planned for this purpose.

Potassium. The amount of potassium metabolized and excreted by the adult ranges from 2 to 4 g per day. Potassium compounds are found in all food supplies in amounts adequate to satisfy the nutritional requirements of the body. For this reason special attention to the potassium intake is not necessary and a dietary allowance of potassium has not been recommended by nutrition authorities.

Calcium. The Food and Nutrition Board of the National Research Council has recommended that the daily allowance of calcium should be 0.8 g for adults and 0.8 to 1.4 g for children and that the daily intake should be 1.3 g during pregnancy (second and third trimesters) and lactation. The amounts recommended are considered optimal for good nutrition.

The increased allowance for pregnancy is intended to apply especially to the last trimester and is to provide for the needs of the fetus for additional calcium for bone formation. The greater allowance during lactation is to take care of the additional need for this element in milk production. The greater allowance for children is to supply extra calcium for the formation of bone during growth. To provide for the unusual demands for calcium during pregnancy and lactation, many physicians prescribe a calcium compound as a supplement to the recommended food intake.

Magnesium. The daily dietary requirement of magnesium for the adult human subject is 0.2 to 0.4 g. Apparently the foods of all peoples of the world contain an amount of this element adequate for the needs of the human body. A dietary deficiency of magnesium in human subjects has not been reported, but magnesium deficiency states caused by disease are encountered. A dietary deficiency of magnesium in cattle called "grass tetany," caused by the ingestion of grass with a low magnesium content, has been observed.

Phosphorus. According to Sherman, the dietary phosphorus requirement of the adult human is 0.88 g per day. Sherman recommended that 1.4 g of phosphorus be included in the daily food intake, an amount that contains an additional 50% as a factor of safety.

It has been considered safe to assume that the phosphorus content of the diet is adequate when the needs for protein and calcium have been satisfied. This follows from the fact that ingested proteins are combined with phosphate, which is released by digestion and metabolism and thus becomes available for nutritional purposes, and also from the fact that phosphate is in common foods in the form of calcium phosphate. Calcium in this way serves as a carrier of dietary phosphate. Thus foods rich in proteins and calcium are good sources of phosphorus. Therefore special attention to the dietary phosphorus requirement is not necessary because this requirement is taken care of indirectly by satisfying the needs for protein and calcium.

Iron and copper. Iron must be present in the diet for the synthesis of hemoglobin of blood and myoglobin of muscle. Iron is also essential for the formation of the cytochromes —cytochrome oxidase, catalase, and peroxidase—cofactors in respiration.

Copper is essential for the incorporation of iron into hemoglobin; without copper in the diet an animal will develop a hypochromic anemia even though abundant amounts of iron are present in the blood-forming tissues.

The daily dietary allowance of iron recommended by the Food and Nutrition Board of the National Research Council is 10 mg for the adult male, 15 mg for the adult female, and an additional 5 mg per day for women during pregnancy and lactation. The recommended daily allowance for children increases with age, from 8 mg for infants to 15 mg for boys and girls over 9 years of age. The body has a remarkable capacity to conserve its iron; hence the dietary requirement is small.

The dietary requirement of copper is 2 mg per day.

Copper and iron are widely distributed in foods. The usual diet does not require supplementation with these two elements, but iron supplementation may become desirable in disease.

Manganese. Manganese is indispensable for the development of plants, and it is essential in the diet of at least several species of animals. *In animals it has been found necessary for growth, reproduction, lactation, the proper development of bone, and the activity of many enzymes.*

Mice on a low-maganese diet failed to grow as well as control animals receiving adequate amounts of manganese.

A high mortality was observed in the young of female rats on a low-manganese diet. Apparently manganese is necessary for the development of the fetus during gestation. Male rats on a manganese-free diet develop a testicular degeneration that results in sterility.

In fowls the need for manganese is as great as in mammals. Eggs from hens on a low-manganese diet showed a hatchability of less than 10%, and the embryos of those eggs that hatched were poorly developed. On a low-manganese diet chickens develop perosis or "leg weakness." In this condition the tendon of the knee joint slips from its normal position, producing a crippling deformity.

In view of the evidence from animal experimentation, it is assumed that manganese is essential in the diet of the human subject.

Zinc. Zinc is necessary for the growth of both plant and animals. Many foods contain small amounts of zinc.

Zinc is a constituent of carbonic anhydrase, an enzyme that stimulates the formation of carbonic acid (H_2CO_3) from carbon dioxide and water. Zinc is also a constituent of alcohol and lactic dehydrogenases and certain peptidases, and it is essential for the action of these enzymes. *In view of the activity of zinc in several enzyme systems, it is clear that this element is an essential constituent of the diet of man.*

Zinc is poisonous if taken into the body in unusual amounts. Poisoning may result from the oral ingestion of zinc salts or from the inhalation of particles of zinc oxide or zinc dust.

Iodine. *Iodine is an essential constituent of thyroxin and other active hormones of the thyroid gland. If iodine is not presented in the diet in adequate amounts, simple goiter results.* This condition is discussed on p. 254. The Food and Nutrition Board of the National Research Council has recommended that the diet should contain about 0.15 to 0.30 mg of iodine per day to meet the needs of the adult person.

Fluorine. In the human body, fluorine in the form of fluoride is found in very small amounts in the soft tissues and in slightly higher concentrations in the bones and teeth. Bone contains about 0.01% to 0.02% and tooth enamel about 0.03% of this element.

Fluoride is toxic to the body if ingested in appreciable quantities. The fluoride ion precipitates the calcium of the blood, and it inhibits the activity of a number of the body's enzymes. Cases of fatal poisoning from the ingestion of fluoride (a common ingredient of certain roach exterminator powders) have occurred. Obviously this element could not be tolerated in the food or drinking water if present in appreciable quantities.

Fluoride occurs in many foods and in small amounts in the drinking water of many communities. The content in the food and water is usually related to the amount in the soil of the community. When the food contains more than 10 mg/100 g or when the drinking water has in it more than 2 parts per million (0.2 mg/100 ml) of this element, a defect in the enamel of the teeth called mottled enamel develops in children who ingest such food or water at the time the permanent teeth are being formed. Mottled enamel, also called *fluorosis,* is a condition characterized by chalky-white patches in the enamel of the teeth. It has been produced experimentally in rats by placing fluorides in the drinking water.

Investigations have shown that *small amounts of fluorides in the food and drinking water are beneficial in preventing dental caries.* Dean and associates in the United States Public Health Service observed relatively few cases of dental caries in communities where the drinking water contains more than 0.5 part of fluorides per million parts of water. These authors also found a low incidence of mottled enamel in regions where the drinking water did not contain more than 1.2 parts of fluorides per million parts of water. *Thus, with regard to the well-being of the teeth, it appears ideal to have drinking water of a fluoride concentration of 1 part per million and foods with a comparable fluoride content.* This concentration of fluoride would be expected to produce a minimum of dental caries and at the same time would not result in mottled enamel.

Cobalt. Cobalt is a constituent of vitamin B_{12} (cyanocobalamin), the vitamin that prevents pernicious anemia and that probably has other vital functions. Vitamin B_{12} occurs in foods of animal origin, such as meats, milk, and eggs. The richest source of this vitamin is liver. Severe deficiency disease attributed to a lack of this vitamin in the diet has been reported in people using an exclusively vegetable diet. The daily administration of 1 μg (one millionth of a gram) of vitamin B_{12} will prevent the development of pernicious anemia.

Molybdenum. Molybdenum is a constituent of the enzyme xanthine oxidase. Trace amounts of molybdenum are essential in the diet, but higher levels of this element in the food are toxic.

Selenium. Selenium, when present in the food in very small amounts, prevents liver necrosis in the rat and pig. Selenium appears to be in its most beneficial dietary form in an organic compound, temporarily designated as factor 3. Selenium is toxic when present in the foods in amounts greater than trace levels. Poisoning of cattle allowed to feed on grass of a high selenium content has been observed extensively in certain parts of the United States.

Vitamins. The specific action of vitamins is known in only a relatively few instances. When the action was discovered, it was almost always that of serving as a cofactor for an enzyme.

Unlike proteins, carbohydrates, and fats, the vitamins do not supply materials from which tissues are constructed or energy is produced. Functionally, they resemble certain inorganic constituents of the diet that assist in the coordination of chemical processes and are effective in small amounts.

In regard to dietary intake, attention is given to six vitamins: vitamin A, thiamine, riboflavin, niacin, ascorbic acid, and vitamin D. The daily allowances of these nutrients recommended by the Food and Nutrition Board of the National Research Council are given in Table 22-2. It is generally agreed that diets adequate in these six vitamins will almost certainly contain all the other vitamins in sufficient quantities.

Table 22-2. Recommended daily dietary allowances designed for the maintenance of good nutrition of practically all healthy persons in the United States (allowances[1] intended for persons normally active in a temperate climate)*

	Age[2] years from to	Weight kg. (lb)	Height cm (in)	Calories	Protein gm	Calcium gm	Iron mg	Vitamin A value I.U.	Thiamine mg	Ribo-flavin mg	Niacin equiv.[3] mg	Ascorbic Acid mg	Vitamin D I.U.
Men	18–35	70 (154)	175 (69)	2900	70	0.8	10	5000†	1.2	1.7	19	70	
	35–55	70 (154)	175 (69)	2600	70	0.8	10	5000	1.0	1.6	17	70	
	55–75	70 (154)	175 (69)	2200	70	0.8	10	5000	0.9	1.3	15	70	
Women	18–35	58 (128)	163 (64)	2100	58	0.8	15	5000	0.8	1.3	14	70	
	35–55	58 (128)	163 (64)	1900	58	0.8	15	5000	0.8	1.2	13	70	
	55–75	58 (128)	163 (64)	1600	58	0.8	10	5000	0.8	1.2	13	70	
	Pregnant (2nd and 3rd trimesters)			+200	+20	+0.5	+5	+1000	+0.2	+0.3	+3	+30	400
	Lactating			+1000	+40	+0.5	+5	+3000	+0.4	+0.6	+7	+30	400
Infants[4]	0– 1	8 (18)		kg × 115 ±15	kg × 2.5 ±0.5	0.7	kg × 1.0	1500	0.4	0.6	6	30	400
Children	1– 3	13 (29)	87 (34)	1300	32	0.8	8	2000	0.5	0.8	9	40	400
	3– 6	18 (40)	107 (42)	1600	40	0.8	10	2500	0.6	1.0	11	50	400
	6– 9	24 (53)	124 (49)	2100	52	0.8	12	3500	0.8	1.3	14	60	400
Boys	9–12	33 (72)	140 (55)	2400	60	1.1	15	4500	1.0	1.4	16	70	400
	12–15	45 (98)	156 (61)	3000	75	1.4	15	5000	1.2	1.8	20	80	400
	15–18	61 (134)	172 (68)	3400	85	1.4	15	5000	1.4	2.0	22	80	400
Girls	9–12	33 (72)	140 (55)	2200	55	1.1	15	4500	0.9	1.3	15	80	400
	12–15	47 (103)	158 (62)	2500	62	1.3	15	5000	1.0	1.5	17	80	400
	15–18	53 (117)	163 (64)	2300	58	1.3	15	5000	0.9	1.3	15	70	400

*Food and Nutrition Board, National Academy of Science, National Research Council: Recommended daily dietary allowances, revised 1963, Publication No. 1146.
† 1000 I. U. from preformed vitamin A and 4000 I. U. from beta-carotene.
[1] The allowance levels are intended to cover individual variations among most normal persons that live in the United States under usual environmental stresses. The recommended allowances can be attained with a variety of common foods providing other nutrients for which human requirements have been less well defined.
[2] Entries on lines for age range 18 to 35 years represent the 25-year age. All other entries represent allowances for the midpoint of the specified age periods; that is, line for children 1 to 3 is for age 2 years (24 months); 3 to 6 is for age 4½ years (54 months), and so on.
[3] Niacin equivalents include dietary sources of the preformed vitamin and the precursor, tryptophan; 60 mg tryptophan represent 1 mg niacin.
[4] The calorie and protein allowances per kilogram for infants are considered to decrease progressively from birth. Allowances for calcium, thiamine, riboflavin, and niacin increase proportionately with calories to the maximum values shown.

Nutrition authorities agree that all the vitamins are present in adequate amounts in a well-selected diet, consisting of meats, eggs, milk, green and yellow vegetables, and fruits, ingested in amounts that satisfy the individual's appetite.

The chemistry and physiological effects of the vitamins will be considered in Chapter 24.

MILK

Milk, the secretion of the mammary glands of mammals, is the food designed by nature for the nourishment of the newborn; hence it is an excellent nutritive substance. It has all the constituents of a correct diet in practically balanced quantities, with the exception of being deficient in iron, copper, and vitamins C and D. The average energy value of cow's milk containing 4% butterfat is about 675 Calories per quart.

The milks of all species of animals have essentially the same constituents, but they vary considerably in the relative amount of these constituents. This variation has an important influence on the rate of growth of each species. In general, the speed of growth of the young varies in direct proportion to the richness of the milk in proteins and inorganic salts: the higher the content of proteins and inorganic salts in the milk of any species, the more rapid the rate of growth of that species.

Principal constituents. The gross constituents of milk are shown in Table 22-2. It will be of interest to discuss these substances in a little more detail.

The proteins of milk are casein, lactalbumin, and lactoglobulin. Casein is a phosphoprotein. It makes up about 80% of the total protein content. These three proteins are entirely adequate for the nutrition of an animal insofar as protein requirement is concerned.

The carbohydrate of milk is lactose. This sugar does not occur anywhere except in milk. It is synthesized by the mammary gland from the glucose of the blood.

The fats of milk consist of glycerol esters of all the fatty acids mentioned in Chapter 17. These are tributyrin, tricaproin, tricaprylin, tricaprin, trilaurin, trimyristin, tripalmitin, tristearin, and triolein.

The principal inorganic constituents of milk are salts of calcium, potassium, sodium, and magnesium, in the form of phosphate and chloride. Iron and copper are present although not in amounts adequate for the nutrition of the young. Milk is a rich dietary source of calcium and phosphate.

Milk contains all the vitamins necessary for the nutrition of the species producing it. Cow's milk, however, contains only one-third as much vitamin C as human milk, and, as a result of standing and pasteurization, its vitamin C content is reduced. Cow's milk is thus inadequate in vitamin C for human nutrition, and during the months of the year when there is reduced exposure to sunshine, it does not contain enough vitamin D for the human infant.

The average composition of human milk and of cow's milk is shown in Table 22-3.

Supplying the food deficiencies. As just mentioned, cow's milk is deficient in vitamin C and does not contain enough vitamin D to prevent rickets in children during the time of the year when there is limited exposure to sunlight. Milk is also deficient in iron and copper. Milk does not contain enough iron and copper to bring about normal hemo-

Table 22-3. Comparison of human and cow's milk

Constituent	Human milk (%)	Cow's milk (%)
Water	84 to 88	84 to 88
Proteins	1 to 1.5	3 to 4
Carbohydrate (lactose)	6.5 to 8	3.5 to 5
Fat	3 to 5	3 to 5
Inorganic salts	0.15 to 0.3	0.6 to 0.7

globin regeneration. Young rats placed on a diet of only milk develop an anemia referred to in the literature as a "nutritional anemia." The addition of only iron to the milk diet will not cure this anemia, but the feeding of small amounts of copper in addition to iron supplements brings about a rapid restoration of the normal red cell and hemoglobin content of the blood.

At birth the infant has a reserve supply of iron and copper stored in its body, which is utilized for the formation of red blood cells during the early periods of life while he is on an exclusive milk diet. If continued on a diet of only milk, the infant will develop a nutritional anemia. It is therefore necessary, beginning about one month after birth, to supplement the milk diet of the infant with food containing iron and copper to prevent nutritional anemia. To provide the necessary vitamin C, it is the practice to give orange juice, tomato juice, or crystalline vitamin C to the infant. Vitamin D is supplied by giving a vitamin D concentrate. The necessity for additional iron and copper is met by supplementing the milk diet with purées of green vegetables.

Souring of milk. Milk becomes sour on standing because of the production of lactic acid by the action of certain becteria or lactose. When milk sours, the casein separates from solution as an insoluble curd, a change caused by the lactic acid formed. Casein may be precipitated from milk by very small amounts of other acids and by the action of rennin. When casein is removed from milk by acids or by rennin, it carries with it considerable fat and traces of other constituents of milk. This combined mixture of casein, fat, and other substances is known as cottage cheese. The casein mixture prepared from milk is the basis of the various cheeses that are widely used as foods.

Pasteurization. In principle, pasteurization of milk consists of heating the milk to a temperature that will kill most bacteria and then of cooling it rapidly. Pasteurization is carried out most successfully by keeping the milk at 65° C for 30 to 40 minutes and then chilling it promptly to low temperatures. *Satisfactory milk should not contain more than 200,000 bacteria per milliliter before pasteurization, and after pasteurization it should contain less than 10,000 bacteria per milliliter.*

ENERGY METABOLISM

Calorimetry. The human body is capable of doing work, for which it requires fuel. The fuels used are proteins, carbohydrates, and fats. In the body these substances undergo chemical degradation through the activity of enzymes, as shown in previous chapters, and release their energy slowly. The energy that the body obtains from the chemical breakdown of food substances is used to carry on its living processes such as respiration, heartbeat, muscle tonus, and the muscular movements producing work.

The energy value of foods is determined with an instrument called the "bomb calorimeter," which gets its name from the fact that it is shaped like a bomb. This instrument is a steel cylinder, which can be sealed tightly and immersed in a vessel of water whose temperature can be measured by delicate thermometers. When a test is to be made, food and oxygen are introduced into the apparatus, and it is closed tightly. An electric current is then passed through a platinum wire in contact with the food, which heats the wire to redness and ignites the food. The food is oxidized completely, and heat liberated is estimated by the temperature change in the water surrounding the bomb.

The heat liberated by foods burned in the human body or in the body of an animal may be determined by an interesting application of calorimetry. There are two general methods of calorimetric measurement, the direct and the indirect, used with animals or human subjects.

Direct calorimeters involve the same principle of heat measurement as the bomb calorimeter. One of the best direct calorimeters consists of a large double-walled copper chamber surrounded by cooling water pipes in which the temperature of the water can be taken as it enters and leaves the system. Appropriate devices for supplying measured amounts of oxygen for respiration and for removing and measuring carbon dioxide and water vapor are connected with the chamber. An individual can be placed in such an instrument and his heat output can be determined when he is fasting or has been fed any type of food, while lying down or in a sitting position, or while doing work of graded types of severity.

Indirect calorimetry involves the use of instruments that measure the oxygen consumption and carbon dioxide output of an individual for a given time. From such data, together with the urinary nitrogen and facts obtained by laboratory experimentation, the heat production in the body can be calculated. The Atwater-Rosa-Benedict calorimeter is a combination of both the direct and the indirect systems of measurement. The inventors secured remarkably close agreement in the results obtained by direct and indirect methods of calorimetry with this calorimeter (Table 22-4).

It has been found by investigators that carbohydrates and fats when burned in the human body have the same caloric value as when burned in a bomb calorimeter. Proteins, however, show a higher caloric value when burned in a bomb calorimeter than when oxidized in the body. This is explained by the fact that the nitrogen of proteins is not oxidized in the human body but is oxidized in the bomb calorimeter, and the oxidation of nitrogen liberates heat. If we deduct the amount of heat due to the oxidation of nitrogen when proteins are burned in a bomb calorimeter, we get the same value as obtained from the burning of proteins in the body. The law of the conservation of energy therefore holds good for the energy exchanges that take place in the human body.

The rate of metabolism may be expressed as the number of Calories per kilogram of body weight or as the number of Calories per square meter of body surface. The relationship to body surface is generally used, since it gives more satisfactory information in com-

Table 22-4. Caloric value of some common foods

Energy value	Food	Calories per pound
High	Lard	4080
	Butter	3491
	English walnuts	3199
	Brazil nuts	3162
	Almonds	2940
	Bacon	2840
	Chocolate	2768
	Peanuts	2490
	American cheese	2102
Intermediate	Sugar	1815
	Oatmeal	1811
	Wheat	1635
	Macaroni	1625
	Corn meal	1620
	Rice	1591
	Beans, dried	1565
	Mutton	1543
	Pork chops	1530
	Honey	1481
	Fresh ham	1458
	Olives	1357
	Lamb, breast	1311
	Apple pie	1233
	Beef, porterhouse steak	1230
	White bread	1182
Low	Veal, breast	817
	Bluefish	744
	Liver, beef	583
	Chicken	564
	Potatoes, white	378
	Milk, whole	314
	Pears	288
	Apples	285
	Grapefruit	235
	Oranges	233
	Oysters	228
	Carrots	204
	Peaches	188
	String beans	184
	Beets	180
	Muskmelon	180
	Cabbage	143
	Cauliflower	135
	Pumpkin	117
	Spinach	109
	Tomatoes	104
	Lettuce	87
	Celery	84
	Cucumbers	79

paring individuals of widely differing sizes. The total metabolism of the body is expressed in terms of the number of Calories of heat produced in the body as a whole. The total metabolism of the human being ranges from 2200 to 10,000 Calories per 24 hours. Metabolism is influenced by many factors, the most important of which will be discussed.

Factors affecting the rate of metabolism. Factors affecting the rate of metabolism are activity, age, size, food, and glandular secretions.

Activity. The factor that affects the rate of metabolism the most is activity. The more work an individual does, the greater will be the oxidation of foods in his body to supply the needed energy. Persons engaged in sedentary work have a metabolism of 2200 to 2800 Calories per 24 hours. In the same period of time a carpenter will average around 3400 Calories, and very hard laborers may have a metabolic activity as high as 5000 to 6000 Calories.

Age. Metabolism is much higher in the young than it is in older persons. The metabolic rate shows a gradual decline after the second year until old age is reached. The greater rate in the young is due to greater activity, to the requirement of growth, and to difference in size.

Size. Metabolism varies according to size. The more surface an individual has, the greater will be the amount of heat radiated from his body. Hence the metabolic rate is roughly proportional to the body surface. The more surface per unit of weight, the higher the rate of metabolism will be. The metabolic rate is therefore higher in a small individual than in a large one, other factors being equal.

Food. The ingestion of foods brings about an increase in metabolism. In this respect the three classes of food substances vary widely. When ingested, carbohydrates will increase the rate of metabolism an average value of about 6%, fats will increase the metabolic rate about 4%, and proteins stimulate the rate of metabolism about 30%. The acceleration of metabolism by foods is known as their *specific dynamic action* or the *calorigenic action*. A high-protein diet will reduce body weight because proteins stimulate the rate of metabolism (specific dynamic action), bringing about an increased burning of food and tissues in the body.

Glandular secretions. The secretions of the endocrine organs have a regulatory influence on metabolism. The secretion of the thyroid, for instance, has a stimulating influence. A person with hyperthyroidism shows a high metabolic rate, and a person with hypothyroidism has a subnormal metabolism. For this reason the determination of an individual's basal metabolism is an important aid in the diagnosis of thyroid conditions.

Basal metabolism. *The basal metabolism is the metabolic activity of the body 12 to 15 hours after the last meal while the individual is awake, comfortably warm, and at complete rest.* It represents the lowest amount of chemical changes that take place in the body in a relaxed position, the subject being awake, in a comfortable environment, at complete rest, and free from the influence of food. It is the energy production necessary for the maintenance of a state of tonus in the muscles, for the movements of the chest and the diaphragm in respiration, for the heartbeat and muscular contractions involving other internal organs—in short, for all the life processes that take place when the subject is living under the conditions just mentioned. *The normal basal metabolic rate of the*

Nutrition 319

Fig. 22-4. Basal metabolism determination. (Courtesy Warren E. Collins, Inc.)

male adult between the ages of 20 and 40 years is 39 Calories per square meter of body surface per hour. In the case of the adult female at the same ages, the rate is lower, being about 36 Calories per square meter of body surface per hour. For a 24-hour period the total heat production of the normal male adult under basal conditions (postabsorptive and resting) ranges from 1500 to 1800 Calories.

A basal metabolism determination is made with the individual lying at rest about 15 hours after ingestion of food (Fig. 22-4). The process consists of measuring the number of liters of pure oxygen consumed by the individual in a definite time and deriving from this obtained value the amount that would be consumed in 1 hour. Since the oxidation brought about by 1 liter of oxygen in a normal individual at complete rest produces 4.83 Calories of heat, the calculation consists of multiplying the number of liters of oxygen consumed per hour by 4.83. This value represents the total number of Calories of heat produced per hour. When this value is divided by the number of square meters of body surface of the individual (obtained from height-weight tables), the basal metabolic rate, which is the number of Calories per square meter of body surface per hour, is obtained.

Questions for study

1. What factors with respect to human subjects and animals are considered in the science of nutrition?
2. What is an adequate diet?
3. What is meant by nitrogen equilibrium in the body? Positive nitrogen balance? Negative nitrogen balance?
4. What proteins have the highest nutritional value to man? Arrange groups of foods in a descending order of biological value.
5. What is an essential amino acid? How many amino acids are essential for man? For the rat?
6. What is the recommended daily allowance of proteins for the adult? For the growing young person?
7. What factors determine the amount of carbohydrate that should be in the diet?
8. What is the recommended allowance of carbohydrate in the diet in grams per 24 hours?
9. What is the nutritive importance of fats in the diet?
10. What is the importance of essential fatty acids in the diet? Name the three essential fatty acids.
11. What is the recommended allowance of fat in the diet in grams per 24 hours?
12. Discuss the balanced diet.
13. What is the range of intake of sodium chloride per day?
14. What conditions call for sodium chloride intake in addition to what is in the food?
15. For what conditions does it become desirable to administer sodium chloride solution parenterally?
16. What conditions call for restricted sodium intake? Why is such restriction helpful to the patient?
17. Why is it not necessary to make special provisions for supplying the potassium intake?
18. What is the recommended dietary allowance of calcium per day for the normal adult? The child? For a woman during pregnancy and lactation?
19. Have magnesium deficiency states been observed in man that are due to a dietary deficiency? To disease?
20. How are the body's needs for inorganic phosphate supplied by the diet?
21. What is the nutritive value of iron in the diet? Copper?
22. Why does anemia result from the intake of a diet consisting exclusively of milk?
23. What is the recommended daily allowance of iron for the adult male? The adult female? During pregnancy and lactation?
24. What is the dietary copper requirment per day?
25. What conditions will develop in an animal on a diet that is deficient in manganese?
26. What is the nutritive value of zinc in the diet?
27. What is the physiological importance of a small amount of iodine in the diet?
28. What is the importance of fluoride in very small amounts in the drinking water or the food?
29. What is the optimum concentration of fluoride in the drinking water?
30. What is the result of the intake of fluoride in excess of the optimum amount?
31. Is fluoride toxic when excess is ingested? If so, why?
32. What is the nutritive value of cobalt?
33. In general, how do the vitamins participate in the metabolism of animals?
34. What vitamins require special attention in the planning of adequate diets?
35. Are vitamin supplements necessary when a well-selected diet is used?

36. Discuss the value of milk as a food.
37. What is the relation of composition of milk to the rate of growth of different species of animals?
38. Compare the composition of human milk with that of cow's milk.
39. How is the milk diet of the infant made adequate in vitamins C and D?
40. How are copper and iron supplied in the diet of the infant?
41. State the changes that occur in the souring of milk.
42. What is cottage cheese?
43. How is milk pasteurized? What is the permissible number of bacteria in milk that is considered safe for consumption?
44. How does the human body obtain energy?
45. What unit is used to measure metabolism?
46. What is a bomb calorimeter? For what is it used?
47. Describe a direct calorimeter used for measuring the metabolism of the body.
48. What are the procedures used in indirect calorimetry?
49. Do carbohydrates and fats yield the same heat when oxidized in the body as when burned in a bomb calorimeter?
50. Explain why proteins yield more heat when oxidized in a bomb calorimeter than when oxidized in the body.
51. What are the factors that change the rate of metabolism?
52. Why will a high-protein diet tend to reduce body weight?
53. What is basal metabolism?
54. What is the average basal metabolism per 24 hours in terms of Calories?
55. Describe a basal metabolism determination.
56. What is the normal basal metabolic rate in Calories per square meter of body surface per hour for the male? For the female?

23 Vitamins

The science of nutrition had its beginning in systematic efforts to learn the effect of different food substances when fed to small animals. In such experiments rats, mice, and guinea pigs have been most uueful because such animals have a short span of life and consume realtively small amounts of food substances, and because usually, although not always, the results obtained with these animals have been found to apply similarly to the human being. The general procedure has been to prepare a diet of known purified constituents in known amounts. Such an artificially prepared diet will be called a synthetic diet in this discussion. This diet is fed to these small animals, and its effects on growth and general well-being are noted. It is interesting that the early attempts to maintain mice and rats on an artifically prepared diet resulted in complete failure; animals fed such diets did not live. We know now that their failure to survive was due to the absence from the diet of the substances we call vitamins.

The term vitamin arose from a suggestion made in 1912 by Funk, who intended to incorporate in a name for these unknown dietary constituents the ideas that they are vital or essential for life and that they are amines; hence, the name "vitamine" was proposed. Later, when the chemistry of the vitamins became better know, it was found that the meaning of this term did not fit the chemical nature of these substances, but the name was retained for its convenience, the terminal letter "e" of Funk's designation being dropped to form the term vitamin.

The earliest significant experiments in the search for vitamins were those made in Switzerland in 1888 by Lunin, who fed mice a synthetic diet of foodstuffs prepared in the laboratory by chemical methods. On this diet consisting of proteins, carbohydrates, fats, and inorganic salts, the animals did not survive. When milk was added, the animals lived. In interpreting his experiments the investigator expressed the view that there are unknown substances in milk that must be present in the diet of an animal to maintain life. These highly suggestive laboratory experiments received little attention until Pekelharing in the Netherlands in 1905 repeated them and confirmed the correctness of Lunin's observations; Pekelharing also found that he could maintain life in the mice receiving synthetic diets by giving them milk to drink instead of water.

Hopkins in England in 1906 published the results of his work showing that rats would not survive when fed a highly purified food mixture consisting of casein, starch, cane sugar, lard, and inorganic salts. When milk was added to this diet, the rats lived and grew well. Thus it was demonstrated again by feeding experiments with synthetic diets prepared from purified ingredients that there are substances in milk—small in amounts and presumably not proteins, carbohydrates, fats, or inorganic salts—that must be present in the diet to maintain life. These observations, together with others to be mentioned later, led to the discovery of the highly important dietary constituents called vitamins.

Some vitamins are soluble in fats and oils, and other are soluble in water. It has been found useful to classify these substances in two groups based on their solubility—the fat-soluble vitamins and the water-soluble vitamins.

FAT-SOLUBLE VITAMINS
Vitamin A (retinol)

In 1913, McCollum and Davis and Osborne and Mendel independently announced results of experiments showing that there is a substance in butterfat that promotes the growth and well-being of rats. These investigators found that rats would not grow on a synthetic diet in which lard was used as the source of fat, but when butter was substituted for lard, the animals grew well and remained in good health. Thus it was revealed for the first time that there is a fat-soluble dietary substance essential for satisfactory nutrition of animals. This substance was first called "fat-soluble A" factor and is now known as vitamin A.

Vitamin A is found in butter, cod-liver oil, milk, egg yolk, liver, and animal fats in general. Substances giving rise to vitamin A (precursors) also occur in these sources and are especially prevalent in green and yellow vegetables.

Vitamin A has several functions in the body. In the first place *it must be present in the diet to obtain growth of the young*. This function need not be thought of as characteristic since probably any food deficiency resulting in disordered health will retard growth.

A more characteristic function of vitamin A is the maintenance of a normal condition of the eyes. A dietary deficiency of this substance produces an early disturbance of the eyes known as *night blindness* or *nyctalopia*. An individual with night blindness cannot see well in dim light. This is due to a subnormal amount of visual purple in the retina. *Vitamin A enters into the formation of the visual purple of the retina. When there is a dietary deficiency of vitamin A, an inadequate amount of this substance is available for the formation of visual purple; hence limitations in vision at night arise.*

Nyctalopia has been observed for a long time, and it is more widespread throughout the world than might be suspected. Studies in the United States have shown an incidence of mild nyctalopia of 10% to 20% in children of lower economic classes. The early physicians of Egypt and Greece recognized the condition and treated it by feeding liver, a practice that was an interesting forerunner of the modern usage of cod-liver oil as a source of vitamin A.

Further deprivation of vitamin A results in a more severely diseased condition of the eyes known as *xerophthalmia. In xerophthalmia the eyes become dry because the cells of the lacrimal glands undergo a degenerative change called keratinization, which decreases the secretion of tears; consequently infection develops in the eyes, and blindness results if the diet is not corrected.*

Xerophthalmia has been observed in Japan, in certain regions of Africa, in India, and in Denmark during World War I. It is found in people on a diet of rice, cereals, and nonleafy vegetables, with limited amounts of meats, milk, animal fats, and green and yellow vegetables. The disease is cured by giving foods containing vitamin A if treatment is instituted in time.

When the vitamin A content of the diet is inadequate, the mucous membranes of the mouth, nose, throat, respiratory passages, and genitourinary tract become thickened and horny in character, a degenerative change called keratinization. Such changes in epithelial tissues lower the resistance of the structures involved and lead to infection in these tissues. Animals that have become markedly depleted in vitamin A have a lowered resistance to respiratory infections. The vitamin has no direct antibacterial action; its effect is the restoration of the epithelium to the normal state that actively resists bacterial entrance.

A dietary deficiency of vitamin A may produce an unhealthy condition of the skin. Dryness and scaliness may develop, and sometimes small pustules containing pigment are formed around the hair follicles. The latter condition is called *follicular hyperkeratosis* (Fig. 23-1).

The origin of vitamin A is in plants. In certain fruits and vegetables there occurs a yellowish compound called carotene, of which there are several isomeric forms. It has

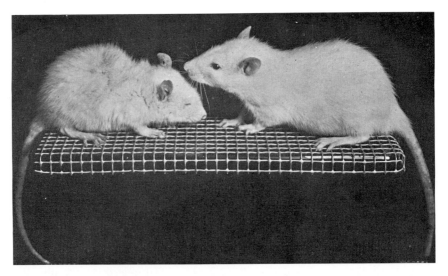

Fig. 23-1. Inhibition of growth in rat produced by restriction of vitamin A in diet. Animals, litter mates, were 21 days old at the start of the experiment, which was continued for 33 days. Animal at left received diet containing all nutritive substances except vitamin A; animal at right received adequate diet. Note the growth failure and xerophthalmia in vitamin A–deprived rat. (Courtesy The Upjohn Co., Kalamazoo, Mich.)

been found that carotene when present in the diet has a protective influence similar to that of vitamin A. Experimental work has shown that carotene is converted into vitamin A in the body by an enzyme called carotenase. Carotene is thus a provitamin or a precursor of vitamin A. The animal obtains vitamin A either by ingesting foods containing this vitamin or by feeding on carotene-containing foods that form vitamin A in the body.

The reaction by which one molecule of β-carotene is split in the liver by the enzyme carotenase, forming two molecules of vitamin A, is shown below:

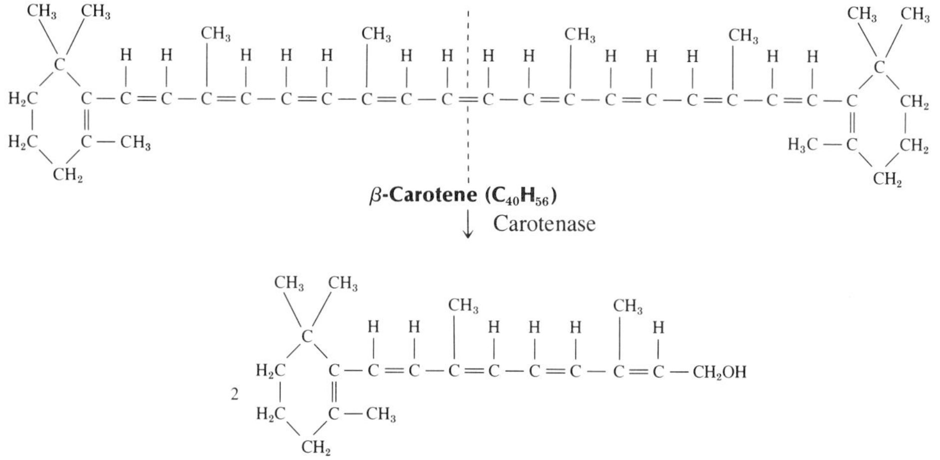

$$\beta\text{-Carotene } (C_{40}H_{56})$$
$$\downarrow \text{Carotenase}$$
$$2 \text{ Vitamin A}$$

Vitamin A is destroyed by prolonged heating in the presence of oxygen. The usual cooking and preserving processes used for foods do not impair appreciably their content of vitamin A or vitamin A precursors.

In summary, the functions of vitamin A may be stated briefly as follows: *A mild dietary deficiency of vitamin A results in early changes in the eyes known as night blindness or nyctalopia. A more marked deprivation of this vitamin produces xerophthalmia, a disease of the eyes characterized by dryness, keratinization, and ulceration. A dietary deficiency of vitamin A lowers resistance to infection and may give rise to a skin lesion known as follicular hyperkeratosis.*

Vitamin D (calciferol)

If an adequate amount of vitamin D is not present in the diet, the condition known as rickets results. Rickets is essentially a disease of infancy and childhood. It is characterized by a failure in the deposition of calcium phosphate in the bones, which results in various bone malformations such as bowlegs, knock-knees, distorted spine, twisted pelvis, and enlargement of the osteochondral junctions of the bones. In addition to faulty calcification of the bones, rachitic children show general symptoms such as nervousness, irritability, loss of appetite, loss of weight, and anemia. Deficiency of this vitamin in the diet has probably contributed more than any other cause to faulty structural formations in the human race in the past.

Vitamin D occurs in fish liver oils, egg yolk, butter, milk, liver, oysters, and clams.

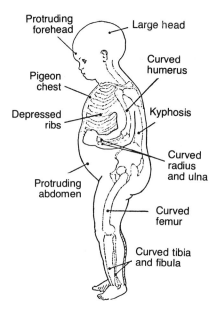

Fig. 23-2. Symptoms of rickets. (From Harris: Vitamins in theory and practice, The Macmillan Co.)

We are now in possession of considerable chemical knowledge of this vitamin. Of special interest is the fact that ergosterol, a sterol, is the parent substance of one antirachitic compound. When ergosterol is exposed appropriately to ultraviolet light, several isomers of this substance are formed. One of these isomers, called calciferol, has been obtained in a pure crystalline form and has been found to have an extremely high antirachitic potency.

Calciferol
(vitamin D_2)

Recent experimental work has shown that there is more than one antirachitic substance in nature. Antirachitic activity has also been imparted to a number of substances by physical and chemical treatment. All compounds with antirachitic activity that have been observed are derivatives of either ergosterol or 7-dehydrocholesterol, two well-known compounds in nature that belong to the class of sterols.

In 1919, Huldschinsky showed that infantile rickets could be cured by exposure of a rachitic child to the ultraviolet rays of a quartz-mercury vapor lamp. In 1921, Hess and Unger observed that a similar result is obtained by exposure of the rachitic child to direct sunlight. In 1924, Steenbock and Black found that certain foods incapable of preventing rickets may be made antirachitic by appropriate irradiation with ultraviolet light.

The imparting of antirachitic potency to certain foods by irradiation with ultraviolet light is due to the transformation of ergosterol or 7-dehydrocholesterol into antirachitic compounds. The development of antirachitic substance in the body by exposure to sunlight is the result of a similar photochemical effect on 7-dehydrocholesterol in the skin. This effect is the opening of one of the rings in the steroid nucleus. (See the formula for calciferol.)

Calciferol is prepared by irradiation of ergosterol with ultraviolet light and subsequent isolation by chemical procedures. *Calciferol is known as vitamin D_2: Vitamin D_3 is 7-dehydrocholesterol activated by irradiation with ultraviolet light.* The term vitamin D_1 is obsolete; it was once used for an impure mixture of products formed by the irradiation of ergosterol. Other vitamin D preparations (for example, D_4 and D_5) have been made, but they are of no special interest. Solutions of calciferol in oil have the official name of *viosterol*.

From the foregoing discussion it is clear that the human race obtains its vitamin D both from food intake and from exposure of the body to direct sunlight.

The irradiation of foods to give them antirachitic potency has assumed industrial importance. A considerable amount of the milk now consumed is fortified in antirachitic potency by the addition of vitamin D concentrates.

Vitamin D is not affected by boiling at 100° C for several hours, and it is fairly resistant to oxidation. The vitamin D content of foods is not impaired by the usual cooking and preserving processes.

Vitamin E (α-tocopherol)

The dietary substance known as vitamin E has been shown to prevent sterility in animals. Animals in which sterility is known to occur as a result of vitamin E deficiency are rats, cattle, chickens, and rabbits. Whether this dietary substance has an influence on reproduction in human beings has not been proved conclusively, but the evidence from studies on human subjects thus far has favored the idea that this vitamin is essential for reproduction and perhaps other bodily functions.

The effects of this vitamin deficiency on the male and the female are different. Sterility in the male is caused by a pathological degeneration of the germinal cells of the testes that form spermatozoa. The inclusion in the diet of foods containing vitamin E prevents this reproductive failure, but after definite sterility has developed, it cannot be cured since the pathological changes in the gonadal cells are permanent.

The female animal on a diet deficient in vitamin E experiences a normal estrus cycle. In the presence of this dietary deficiency, apparently normal ovulation takes place and pregnancy also occurs. During the early stages of pregnancy pathological changes occur in the placenta, the developing embryo dies, and the fetus is resorbed. The degenerative changes are in the placenta and embryo and are not in the ovarian tissue or ova; consequently this form of sterility may be cured as well as prevented by the administration of foods containing vitamin E.

In chickens reproductive failure is exhibited by a low hatchability of eggs laid by hens on a vitamin E–deficient diet. Young chicks on a diet deficient in vitamin E also develop extensive capillary damage and an encephalomalacia.

Rabbits, guinea pigs, and rats on a vitamin-deficient diet develop a severe muscular dystrophy leading to paralysis; these animals also survive poorly on such a diet. For survival, a dietary source of vitamin E has been found essential for the rabbit, rat, mouse, guinea pig, chicken, and duck.

Vitamin E occurs in the embryo of grains such as wheat and corn, in leafy vegetables such as lettuce, spinach, or water cress, and in small amounts of meats.

Vitamin E is a fat-soluble vitamin. It is not destroyed by heating to 250° C. It is stable to drying and to ordinary light. Rapid destruction of vitamin E is caused by oxidizing agents and by the "rancidity" substances in rancid fats. The vitamin E content of foods is not diminished by the usual cooking processes.

Vitamin E has been isolated, and its chemistry has been worked out. The chemical name of this vitamin is tocopherol. There are six forms of this compound. The formula of the most potent form, α-tocopherol, is given below. The other tocopherols differ from the α form in the number and position of the methyl groups attached to the benzene ring.

α-Tocopherol

Vitamin K (menadione)

While studying the metabolism of sterols at the University of Copenhagen, Dam, in 1929, observed a hemorrhagic condition in chickens having a certain type of diet. The chicks, when placed on this diet, developed subcutaneous and deep intramuscular hemorrages. This hemorrhagic condition was soon found not to be related to scurvy since the administration of ascorbic acid did not improve the lesion. Subsequent work by Dam showed that a vitaminlike substance in the diet prevents this condition. Dam named this factor vitamin K.

Vitamin K is a fat-soluble thermostable dietary principle. It is synthesized by plants, its occurrence being most abundant in green plants. Alfalfa, kale, spinach, cabbage, and grass contain an abundance of vitamin K, whereas potatoes, mushrooms, and tomatoes—foodstuffs with little or no chlorophyll—have a low vitamin K content.

The effect of vitamin K is to stimulate the production of the prothrombin of the blood by the liver. When there is a deficiency of this vitamin, the blood has a prolonged blood-clotting time because of its reduced content of prothrombin, one of the factors entering into the formation of the blood clot. There must be bile in the intestinal tract to bring about the absorption of vitamin K.

A diet containing the usual portions of green vegetables is apparently adequate in its content of vitamin K. Human subjects and animals also receive protection against vitamin K deficiency through the synthesis of this vitamin by intestinal bacteria. An avitaminosis due to an inadequate supply of vitamin K to the tissues may occur in the following conditions or situations:

1. When bile is absent from the intestinal tract, as occurs in obstructive jaundice
2. In ulcerative colitis or sprue, conditions in which the absorption of vitamin K is poor
3. When the bacterial flora of the intestinal tract have been suppressed by medication with sulfa drugs
4. In the newborn infant when the mother's diet has been poor in vitamin K

Vitamin K is used clinically with success in the conditions listed. It is especially important in the control of the bleeding tendency in obstructive jaundice and in the prevention of hemorrhage in the newborn infant. About 1% of babies develop the bleeding tendency when attention has not been given to the mother's diet before parturition. Hemorhage in the newborn infant is prevented either by giving the mother a liberal supply of vitamin K before delivery or by administering vitamin K to the baby immediately after birth.

The chemical formula for a synthetically prepared vitamin K, *menadione* or vitamin K_3, is given below.

Menadione
(synthetic vitamin K)

Naturally occurring vitamin K contains the chemical group shown in the formula for menadione (naphthoquinone group), which is the physiologically active part of the molecule, together with a hydrocarbon side chain attached to the ring at the carbon atom adjacent to the one holding the methyl group. There are two naturally occurring forms of vitamin K—K_1 and K_2—differing from each other only in the length and structure of the hydrocarbon side chain attached to the naphthoquinone group. Vitamin K_1 is synthesized by green plants; it was first isolated from alfalfa. Vitamin K_2 was first prepared from fish meal; it is synthesized by bacteria.

WATER-SOLUBLE VITAMINS
Ascorbic acid (vitamin C)

Ascorbic acid is the chemical name for the compound known also as vitamin C. The terms ascorbic acid and vitamin C are used interchangeably to designate the substance that must be present in adequate amounts in the tissues of animals to prevent the condition known as scurvy. *Scurvy is a disease that occurs in man, monkeys, and guinea pigs.* This deficiency syndrome has also been produced in the bulbul, a songbird native to India. Other animals are able to synthesize their own ascorbic acid through the biosynthetic activity of their liver or kidney tissues and are therefore not dependent on a dietary source of this substance.

Scurvy is a disease that has long been known to be caused by a lack of fresh fruits and vegetables in the diet. Sailors, explorers, invading armies, and all such groups of people who have had to exist for a time on salted meats and preserved foods, have been victims of this disease in the past. As early as 1795, the British Navy adopted the practice of giving lime juice to all sailors to prevent the onset of scurvy. As a result, all British sailors were called limeys.

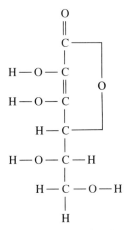

Ascorbic acid or vitamin C

The earliest indication of an ascorbic acid deficiency in the guinea pig is a failure in the formation of the dentin of the teeth by the odontoblast cells. This is followed by a failure in the activity of the fibroblasts, which form the collagen fibers of connective tissue. As a result of inadequate connective tissue formation in the endothelium of the blood capillaries, the walls of these vessels rupture and hemorrhage occurs. There is also a failure in the maintenance of intercellular substances in the matrices of bone and cartilage, and poor healing of wounds results when the ascorbic acid deficiency is marked. As the condition advances there is a retardation of the activity of the osteoblasts, resulting in poor calcification of bone (Fig. 23-3).

The chief symptoms of scurvy in man are swollen, bleeding gums, loosening and demineralization of the teeth, demineralization of the bones, hemorrhage in the mucous membranes beneath the skin and in the joints, and loss of weight. This disease is prevented

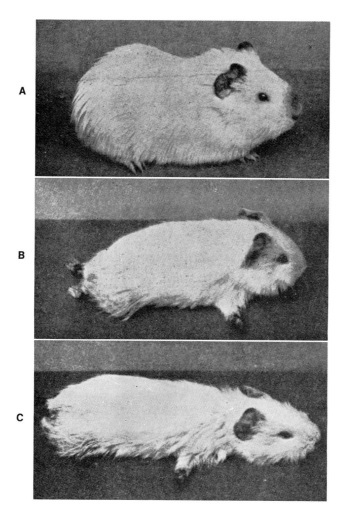

Fig. 23-3. Guinea pigs with scurvy. **A,** Normal guinea pig. **B** and **C,** Guinea pigs with scurvy, showing characteristic position. (From Rose: Foundations of nutrition, The Macmillan Co.)

or cured by eating foods that contain ascorbic acid in adequate amounts or by administering synthetically prepared ascorbic acid.

Ascorbic acid is found in the greatest quantity in peppers, lemons, oranges, grapefruit, limes, strawberries, cantaloupe, pineapple, cabbage, kale, broccoli, spinach, potatoes, turnips, and tomatoes. Some ascorbic acid occurs in practically all fresh fruits and vegetables.

Ascorbic acid is destroyed by oxidation. When oxidizing agents are present, its destruction takes place slowly in an acid medium and rapidly in a neutral or alkaline medium. The canning industries protect the ascorbic acid of their products with considerable success by avoiding the introduction of alkali into their preparations and by removing the oxygen before sealing the cans. Cooking decreases considerably the ascorbic acid

332 Roe's principles of chemistry

content of foods. Losses of ascorbic acid in cooking arise from two factors: oxidation of the vitamin and passage of this highly soluble substance into the cooking water.

Ascorbic acid has been isolated in pure form from adrenal glands, lemon juice, and peppers; it has also been made synthetically in the laboratory. As indicated by its formula, ascorbic acid has a structure like a monosaccharide.

Thiamine (vitamin B_1)

Thiamine is a water-soluble vitamin, which originally was called vitamin B_1. This vitamin is essential in the diet to bring about normal functioning of the nervous and cardiovascular systems. (Fig. 23-4). The disease resulting from thiamine deficiency in man is called beriberi. *Early symptoms of thiamine deficiency are loss of appetite, fatigue, weakness, muscular cramps, and nervous irritability. As the condition advances, muscular incoordination, paralysis, and tachycardia or bradycardia develop. Edema may also occur.* The pathological changes present are lesions in the central nervous system, muscular atrophy, and cardiac enlargement.

Thiamine, when linked with pyrophosphoric acid, serves as a coenzyme (cocarboxylase) for pyruvic oxidase, an enzyme that functions in the metabolic breakdown of pyruvic acid, one of the compounds formed in the intermediary metabolism of carbohydrate. *Thus thiamine is a factor that aids in bringing about an essential step in the metabolism of*

Fig. 23-4. A, Rat, 24 weeks old, received practically no thiamine; notice lack of muscle coordination. **B,** Same rat 24 hours later after having received sufficient thiamine. (From Bureau of Human Nutrition and Home Economics, U. S. Department of Agriculture; in Tuttle and Schottelius: Textbook of physiology, The C. V. Mosby Co.)

carbohydrates. This behavior is of special interest because it is an instance in which a vitamin supports the action of an enzyme.

Beriberi has been observed in Japan, China, the Malay Peninsula, the Dutch East Indies, the Philippine Islands, and in regions of South America, Central America, and Africa. Occasional cases are observed in the United States in individuals who have subsisted for a time on a restricted diet.

Beriberi is prevented, or successfully treated if the disease has already developed, by by the administration of foods containing thiamine.

The first suggestion that beriberi is a disease caused by faulty diet was made by Eijkman, a Dutch physician, in 1897. Eijkman proved his contention by producing a polyneuritis with paralysis in chickens with a diet of polished rice and then curing the condition by feeding rice polishings to the fowls. Thus it was demonstrated that there is a nutritive principle in the hulls of rice that prevented a condition in chickens that resembeled beriberi in man. Subsequent work has shown that this principle is thiamine.

Thiamine occurs in yeast, whole grain cereals, unpolished rice, meats (especially pork), eggs, and milk, and in vegetables and fruits in general. Yeast is the richest source of thiamine. It is also found in large amounts in the pericarp of grains. The modern milling practice of removing the hulls from the grains, such as in the preparation of white flour or polished rice, removes a considerable part of the thiamine.

Thiamine hydrochloride

When materials containing an abundance of thiamine, such as yeast or rice polishings, are shaken with fuller's earth, an active absorbing agent, at a certain acidity, the vitamin is absorbed on the fuller's earth. Later the vitamin is removed from the fuller's earth by the addition of alkali in appropriate amounts. In this way thiamine has been separated from natural sources in amounts that have permitted active investigation. Methods of purification have been developed, and thiamine has been isolated in a pure crystalline state. It is now prepared synthetically for therapeutic use as the hydrochloride salt.

Thiamine is fairly stable chemically. Modern methods of cooking and preserving, except pressure cooking, do not affect the thiamine content of foods to an appreciable extent.

VITAMIN B COMPLEX

In the early days of vitamin discoveries it was thought that the water-soluble substances in yeast, rice bran, whole wheat, and so on, which prevents polyneuritis, was a single compound—vitamin B. Later it was shown that a second factor was present in the same water extract, which produced a gain in weight in rats on certain diets. Hence, the factor preventing polyneuritis was called vitamin B_1, and the substance that caused a weight gain was called vitamin B_2. The chemistry of these two substances was worked out, and vitamin B_1 was given the name thiamine; vitamin B_2 was found to be riboflavin. In the course of investigations of the water-soluble extracts of materials containing vitamins B_1 and B_2, other nutritive factors were discovered. Since the source of these new factors was somewhat the same as that of the original vitamin B, it became the custom to designate the new factors by the letter B with a subscript (B_3, B_4, B_5, B_6, and so on), and this whole group of vitamins was considered as belonging to the "vitamin B complex." We will discuss those members of the B complex whose chemistry and physiological significance are well known. These include niacin, riboflavin, pyridoxine, choline, pantothenic acid, biotin, folic acid, and vitamin B_{12}.

Niacin—the antipellagra vitamin

The word pellegra is derived from two Italian words, *pelle agra,* that mean rough skin. *One of the outstanding characteristics of pellagra is a skin lesion. In addition to dermatitis, pellagra is characterized by glossitis, diarrhea, anemia, and lesions in the central nervous system producing mental confusion, dementia, and mania, As expressed by some writers, pellagra is characterized by the three D's—dermatitis, diarrhea, and dementia.* Pellagra is caused by a diet deficient in niacin.

The scientific solution of the problem of pellagra is due largely to the investigations made by Goldberger and associates. Working in state-controlled institutions in the southern United States, Goldberger came to the conclusion that pellagra is a disease caused by a dietary deficiency. He proved his assumption to be correct by a series of dramatic experiments in which he cured persons with pellagra in certain institutions by placing them on an adequate diet. He also produced the disease in a group of volunteer subjects (convicts from the Mississippi state penitentiary who were given their freedom for submitting to experimentation) by placing them on a diet such as that eaten by pellagra victims and later cured those affected by giving them adequate diet.

Many investigators have attempted to produce pellagra in animals by dietary means, and out of their efforts have come the prolific results leading to the discovery of other factors of the vitamin B complex. The lesion in dogs known as "black tongue," first produced by Goldberger with an experimental diet, is the analogue of pellagra in man. In 1937, Elvehjem obtained an extract from liver, highly active in curing black tongue, from which he separated niacinamide. He found that this substance and also synthetically prepared niacin would cure black tongue in dogs. In a few months clinical trials at several hospitals showed that *niacin is effective in curing human pellagra.*

Niacin

Niacinamide

The antipellagra vitamin is a water-soluble substance found in lean meats, eggs, vegetables, fruits, and most grains. This vitamin is quite stable chemically and is not destroyed by cooking.

Niacinamide is a constituent of diphosphopyridine nucleotide (DPN) and triphosphopyridine nucleotide (TPN). These are instances in which a vitamin serves as the functioning group in the molecule of a coenzyme.

The amino acid tryptophan is the precursor of niacin. From this amino acid, by a series of metabolic transformations, niacin is formed in yeast, bacteria, plants, and most animals, including man.

The dietary requirement of niacin is related to the amount and quality of the proteins in the foods ingested. Milk and eggs have a high antipellagra value because their proteins have a high tryptophan content. Corn, on the other hand, can give rise by biogenesis to only small amounts of niacin because its proteins have a very low tryptophan content. Therefore there was a real basis for the once maintained theory that ingestion of diet rich in corn would cause pellegra. The dietary requirement is also influenced by the bacterial flora of the intestinal tract, which have an unknown capacity to synthesize niacin.

Riboflavin (vitamin B$_2$)

As previously mentioned, from the efforts of investigators to learn the nutritive significance of the vitamin B complex came the discovery of unknown nutritive substances of importance. One of the substances was riboflavin.

Riboflavin is a water-soluble substance that occurs rather widely in foodstuffs. Rich sources of this vitamin are liver, kidney, eggs, yeast, and milk.

In the rat a riboflavin deficiency results in dermatitis, loss of hair, corneal vascularity (an abnormal extension of blood vessels into the cornea), and corneal opacity (Fig. 23-5).

In man a dietary riboflavin deficiency produces eye disturbances. These are photophobia, lacrimation, conjunctivitis, corneal vascularity, and corneal opacity. *In the human subject a dietary deficiency of riboflavin also produces lesions in the angles of the mouth, or the lips, and around the nose and eyes. Reddened macerated lesions form in the corners of the mouth and progress to transverse linear fissures, a condition called*

Fig. 23-5. Photographs showing effects of riboflavin deficiency in the diet. **A,** Rat that received a riboflavin-deficient diet. **B,** Rat that received a riboflavin-deficient diet supplemented with 90 μg of riboflavin weekly; in this animal there was no evidence of alopecia or cataract. The animals were photographed on the seventy-fifth day of experiment, at which time they weighed 35 and 150 g, respectively. (From Day, Darby, and Langston: J. Nutrition **13:**392, 1937.)

cheilosis. In addition to the cheilosis, a scaly, slightly greasy dermatitis appears around the nasolabial folds and in some cases around the eyes and ears.

Riboflavin is a fairly stable substance chemically but is affected by light. The formula for riboflavin is as follows:

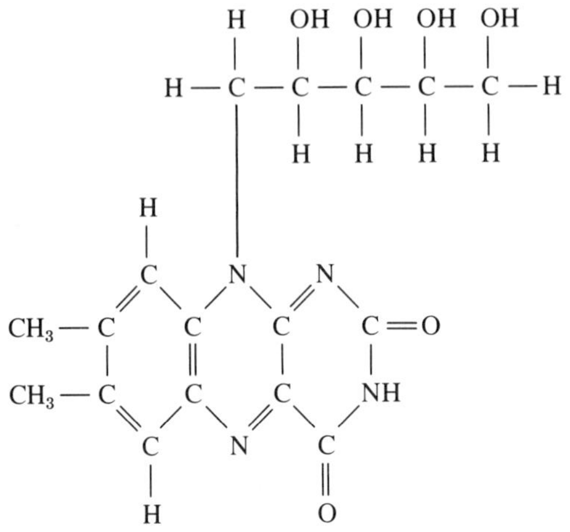

Riboflavin

It is not destroyed by cooking. Some of the effects of a riboflavin dietary deficiency are shown in Fig. 23-5.

Riboflavin is the functioning group in the molecule of a number of enzymes that participate in the biological oxidations that take place in plants and animals. In biological oxidations in living tissues, hydrogen is removed from foodstuffs by the action of certain enzymes, and by further enzyme action it is made to combine with oxygen, which is brought to the tissues by respiration. Thus by a series of reactions catalyzed by enzymes, hydrogen from foodstuffs is made to combine with oxygen from the air, a process that liberates energy. A number of enzymes participate in this transfer of hydrogen from foodstuffs to oxygen. Some of these enzymes are the flavoproteins in which riboflavin is the active functioning group. Riboflavin combines with hydrogen at the site of the two double bonds attached to 2 of its nitrogen atoms (see formula for riboflavin). In this way the hydrogen from foodstuffs is taken up, held, or "carried" until another enzyme acts to remove it from its loose combination with riboflavin and to continue its transportation toward final combination with oxygen. Thus riboflavin participates in a vital step in biological oxidations.

Pyridoxine

Pyridoxine is a component of the vitamin B complex. Before its chemical structure was known, it was called vitamin B_6.

A dietary deficiency of pyridoxine produces a dermatitis in rats. This dermatitis is characterized by its location on the paws, nose, and ears of the rat.

Another manifestation of pyridoxine deficiency is a convulsive state, of the nature of epileptiform fits. This condition has been produced experimentally in rats, pigs, dogs, and chicks.

In the dog, pyridoxine deficiency produces a microcytic, hypochromic anemia. This anemia appears in three to five months after the withdrawal of pyridoxine from the diet.

In certain microorganisms the metabolic removal of the carboxyl group (decarboxylation) of certain amino acids by enzymes called decarboxylases has been shown to be stimulated by pyridoxine and two other forms of this vitamin, *pyridoxal* and *pyridoxamine*. The three forms of the vitamin are equally effective in animal nutrition, but pyridoxal is much more effective than pyridoxine in stimulating growth of bacteria.

Pyridoxine serves as a coenzyme in a considerable number of metabolic transformations of amino acids. Outstanding instances of pyridoxine function as a cofactor are decarboxylations and transminations of amino acids.

Pyridoxine is a water-soluble vitamin. It was first isolated in 1938, and it is now prepared by chemical synthesis. The formulas for the three forms of this vitamin are as follows:

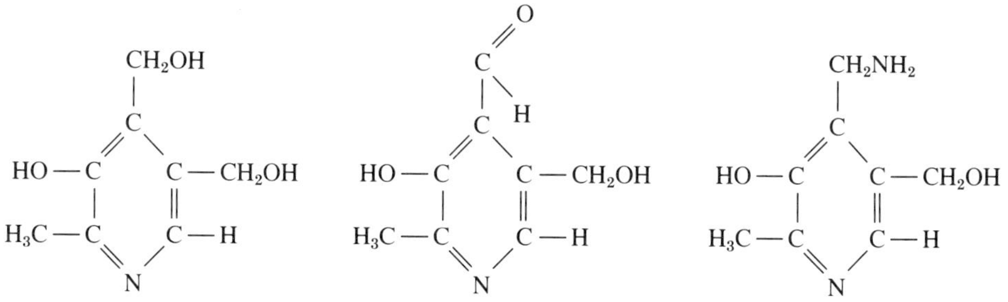

Pyridoxine   **Pyridoxal**   **Pyridoxamine**

Pyridoxine activity in the rat seems to be related to the metabolism of the essential fatty acids. These fatty acids apparently are necessary in the diet to obtain maximum response from pyridoxine administration.

From the effects observed in animals and microorganisms it would be expected that pyridoxine has some function in man. Some interesting probable functions in man have been suggested by clinical trials.

Rich dietary sources of pyridoxine are liver, yeast, rice bran, egg yolk, legumes, cereals, seeds, and milk. It is a fairly stable substance chemically.

Choline

Choline is effective in the animal body in relatively small amounts; hence it is classed as a vitamin. It is considered a member of the vitamin B complex. The formula for choline is as follows:

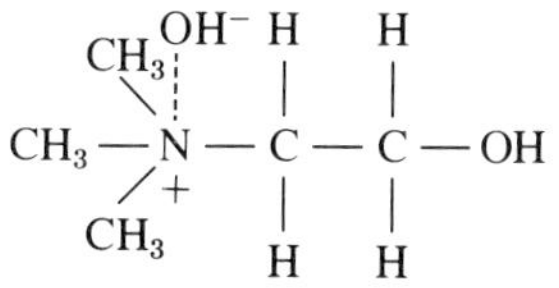

Choline

Choline is a constituent of lecithin, an important compound in nerve, brain, and other body tissues (see p. 230). Its acetic acid derivative, acetylcholine, is a stimulatory substance liberated at the parasympathetic nerve endings and at the end plates of the motor nerves of skeletal muscle when nerve action occurs.

Choline has regulatory influence on the deposition of fat in the liver. *If the diet of an animal contains adequate choline, the fat content of the liver remains normal, but in choline deficiency there is a fatty infiltration of the liver, a condition called "fatty liver."* The normal fat content of the liver of experimental animals, which ranges from 4% to 8%, may increase to 15% to 25% in a marked choline deficiency (see p. 294).

Besides opposing fatty infiltration of the liver, choline is effective in preventing a hemorrhagic degeneration of the kidneys in young rats—a lesion that can be produced only in young rats during a period of a few days after weaning.

Choline deficiency will not develop in animals on a diet high in protein. This is because proteins supply the amino acids methionine and serine, from which the animal can synthesize choline sufficient for its needs.

Pantothenic acid

The biological importance of pantothenic acid was first suggested in 1930 by studies that indicated the existence of an unknown dietary factor *that prevented a dermatitis in chicks.* The chemical identity of the substance remained unknown until 1939 when it was

Fig. 23-6. The rat on the left, whose hair was originally black, shows graying of the hair as the result of a diet deficient in pantothenic acid. Its litter mate on the right received a normal diet. (Reproduced from Therapeutic Notes; courtesy Parke, Davis & Co.)

shown that the "chick antidermatitis factor" is pantothenic acid, the formula of which is as follows.

$$HO-CH_2-\underset{\underset{CH_3}{|}}{\overset{\overset{CH_3}{|}}{C}}-\underset{H}{\overset{OH}{\underset{|}{C}}}-\overset{O}{\overset{\|}{C}}-N-\underset{\underset{H}{|}}{\overset{\overset{H}{|}}{C}}-\underset{\underset{H}{|}}{\overset{\overset{H}{|}}{C}}-\overset{O}{\overset{\|}{C}}-OH$$

Pantothenic acid

An interesting nutritional effect of pantothenic acid is its prevention of the graying of hair in fur-bearing animals. On a pantothenic acid–deficient diet, black rats will turn gray, and dogs and silver foxes show a loss of dark pigment from the hair (Fig. 23-6). The addition of pantothenic acid to the deficient diet will restore the natural pigment to the hair, and on a diet adequate in pantothenic acid, depigmentation of hair does not occur. The absence of pigment from hair is called *achromotrichia*. The prevention of achromotrichia in certain animals by pantothenic acid has been well established. However, studies of the ability of this vitamin to restore pigment to gray hair in human subjects have not yielded positive results.

In addition to the prevention of dermatitis in chicks and achromotrichia in fur-bearing animals, pantothenic acid has other effects in animals. Absence of this vitamin from the diet causes degenerative changes in the adrenal cortex, with resulting hypoadrenal function in rats; fatty liver and gastrointestinal pathology in the dog; and poor feathering, fatty liver, and nerve tissue degeneration in chicks.

Pantothenic acid is a part of the molecule of coenzyme A. As noted earlier in the section on metabolism, coenzyme A is an essential factor in the metabolism of pyruvic and fatty acids.

Good natural sources of pantothenic acid are yeast, liver, meats, grains, egg yolk, and milk. It is now prepared by chemical synthesis.

Biotin

Biotin is one of the more recently discovered members of the vitamin B complex. It is essential in the diet of animals and man. *The most general symptom of biotin deficiency is a dermatitis.* In rats, besides causing a dermatitis, biotin deficiency results in retarded growth, fatty liver, and gangrene of the tail. In human subjects, biotin deficiency produces such symptoms as scaly dermatitis, grayish pallor of the skin, anorexia, nausea, muscle pains, depression, and lassitude. These pathological effects disappear promptly when the subject receives an adequate amount of biotin.

The discovery of biotin came in a roundabout way. For a number of years it was known that a characteristic dermatitis could be produced in rats when they were fed a diet containing a rather high amount of raw egg white. This effect could be prevented by feeding liver, kidney, yeast, and certain other foods. The substance in the foods that prevented the undesirable effects of a diet having a high amount of raw egg white was called vitamin H and also the "anti-egg-white injury factor." In 1940, it was shown that biotin and vitamin H are the same substance. The part played by the raw egg white has been explained as follows. Raw egg white contains a protein called *avidin* that combines with biotin and makes it inactive. The feeding of raw egg white results in a loss of biotin to the animal because this compound becomes bound to avidin in the intestinal tract and is excreted in the feces. Heated egg white has no biotin-binding effect because heating transforms avidin to a form that will not combine with biotin.

Rich sources of biotin are liver, kidney, yeast, and egg yolk. Experimental work has indicated that biotin is synthesized by bacteria in the intestinal tract of animals. It is probable that there is sufficient biotin synthesis by bacteria in the intestine of the rat to meet the biotin needs of this animal; but the chick does not receive enough biotin in this way to meet its requirement. The biotin-deficiency syndrome has been produced experimentally in human subjects by feeding a diet containing one-third of the calories in the form of raw egg white.

Synthetically prepared biotin is now available for use. The structural formula for this vitamin as follows:

$$\underset{\text{Biotin}}{\begin{array}{c} O \\ \parallel \\ C \\ HN \diagup \quad \diagdown NH \\ | \qquad | \\ HC \text{——} CH \\ | \qquad | \\ H_2C \diagdown \quad \diagup CH\text{—}CH_2\text{—}CH_2\text{—}CH_2\text{—}CH_2\text{—}COOH \\ S \end{array}}$$

Recent experimental work has shown that biotin is essential in animals for the formation of aspartic acid, a compound needed in the bioxynthesis of purines and in other metabolic processes. Biotin has also been found to be a component of an enzyme that is involved in the synthesis of fatty acids. Since this vitamin is widely distributed in foodstuffs and is synthesized by intestinal bacteria, the human requirement does not need attention.

Folic acid

Folic acid is essential for the formation of the red and white cells of the blood and the blood platelets. When the diet is deficient in folic acid, animals develop macrocytic anemia, leukopenia, granulocytopenia, and thrombocytopenia. The animals requiring this nutritional factor are man, monkeys, chickens, turkeys, rats, guinea pigs, swine, and probably many others. This factor also promotes the growth of young animals and certain bacteria.

Folic acid has been shown to be effective in the treatment of macrocytic anemias. Pernicious anemia, nutritional anemia, macrocytic anemia of pregnancy, and sprue respond dramatically to the administration of this compound. The red and white cell counts of the blood of patients suffering with any of the macrocytic anemias are rapidly restored to normal when folic acid is given orally or intravenously. This vitamin has no effect on aplastic and hypochromic anemias.

Folic acid is present in liver extract, but it is only one of the factors in liver extract that is effective in preventing pernicious anemia. Another antipernicious anemia factor in liver extract is vitamin B_{12}. Qualitatively the hematological response to folic acid and to vitamin B_{12} is apparently the same, but quantitatively the effect of vitamin B_{12} is much greater. Furthermore, folic acid will not prevent the neurological symptoms (paresthesias) of patients suffering with pernicious anemia, whereas vitamin B_{12} is very effective in preventing or relieving these symptoms.

In the past, folic acid has been designated by the names "vitamin M," "vitamin B_c," "norite eluate factor," and "lactobacillus casei factor." These names were applied to this nutritional factor before its chemistry was known. Its effects were first reported in 1935 in studies on monkeys. Its chemical structure and synthesis were announted in 1946; at that time enough studies had been completed to show that the substances referred to by these several names were the chemical compound. Pteroylglutamic acid (PGA) is the chemical name for folic acid. The molecule of this compound contains glutamic acid, paraaminobenzoic acid, and an organic nucleus belonging to the class of pigments known as pterins. The chemical structure of synthetic folic acid is shown at the top of the following page.

Folic acid has a widespread occurrence in the biological world. It is found in unusual amounts in green leaves, liver, kidney, yeast, and mushrooms. It appears that certain bacteria, notably the coliform organisms of the intestinal tract, have the power to synthesize this vitamin.

Another compound having the biological effects of folic acid is *folinic acid*.

Folic acid

Folinic acid is a derivative of folic acid. It is a formylated, reduced folic acid; that is, it is folic acid containing a formyl ($-$CHO) group and three additional hydrogen atoms attached to its pterin nucleus. Folinic acid has a much greater growth-promoting action on certain bacteria than does folic acid.

Experimental work has shown that folic acid and folinic acid are required for the biosynthesis of purines and pyrimidines.

Vitamin B_{12} (cyanocobalamin)

In 1948, a crystalline compound having a positive hematological effect in pernicious anemia was isolated from liver extract. This substance was first designated vitamin B_{12}. Later it was given the name cyanocobalamin. This compound has remarkable potency: A positive hematological response was observed in three patients with permicious anemia after single intramuscular injections of 3, 6, and 150 μg of the crystalline compound. (A microgram [μg] is 1 millionth of a gram.) *The responses to these small doses show that this compound is far more potent than folic acid in stimulating blood cell formation.* Clinical tests have shown that this substance also brings about a rapid regression of the neurological manifestations (paresthesias) of the pernicious anemia syndrome.

Analyses of crystalline vitamin B_{12} have shown that it is an organic compound containing cobalt, phosphorus, and nitrogen. The finding of cobalt in this compound is of great interest because it has been known for some time that cobalt in trace amounts has a stimulating effect on the formation of red blood cells in animals.

It has been suggested that the daily requirement of cyanocobalamin for the human subject is around 1 μg per day.

Pernicious anemia results from a lack of intrinsic factor in gastric juice, which is essential for the absorption of vitamin B_{12}. In pernicious anemia the failure is in the gastric juice. When given with a source of intrinsic factor or when the gastric juice is normal, vitamin B_{12} is absorbed and is effective in preventing pernicious anemia. Thus pernicious anemia is usually due to defective gastric secretion and not to inadequate amounts of vitamin B_{12} in the diet.

BACTERIAL SYNTHESIS OF VITAMINS IN THE ALIMENTARY TRACT

It has been shown that the bacterial flora inhabiting the alimentary tract of man and animals synthesize certain vitamins that are essential for the host. The evidence indicates that the following vitamins are synthesized by bacteria: thiamine, biotin, riboflavin, niacin, pantothenic acid, vitamin K, and folic acid.

The feeding of diets containing 1% or 2% of sulfa drugs to rats was the basis of experiments used to demonstrate the bacterial synthesis of vitamins. Rats fed on such diets developed lesions that resembled those observed in certain vitamin-deficiency diseases. These lesions could have been due to the direct effect of the sulfa drug on the tissues; however, the addition of the vitamins, whose deficiency was suspected, to the diets containing sulfa drug resulted in a restoration of the animals to normal. These results showed that the absence of the vitamins from the animal tissues was the cause of the disease and that the sulfa drug was not producing the lesion by direct action on the tissues. The most logical explanation of these results is that the sulfa drugs sterilized the animal's intestinal tract and thus removed the bacteria that had been synthesizing certain vitamins.

Questions for study

1. Give the origin of the name vitamin.
2. Discuss the earliest significant experiments leading to the discovery of vitamins.
3. What is nyctalopia? Xerophthalmia?
4. Summarize the functions of vitamin A.
5. What condition is prevented by the presence of vitamin D in the diet?
6. What chemical substance may be naturally occurring vitamin D?
7. What is the relationship of ultraviolet light to the prevention of rickets?
8. Name some foods that are fortified in vitamin D.
9. What is vitamin D_2? Vitamin D_3? Viosterol?
10. What is the effect of a dietary deficiency of vitamin E in the male? In the female?
11. What is the physiological effect of vitamin K? What substance must be present in the alimentary tract to bring about the absorption of vitamin K?
12. What clinical applications of vitamin K have been made?

13. What condition does vitamin C prevent? How is the vitamin C of canned foods preserved by modern methods of canning?
14. What is vitamin C chemically?
15. What conditions are prevented by thiamine?
16. What is the vitamin B complex?
17. Describe the lesion known as pellagra.
18. What dietary substance will prevent pellagra?
19. What is the biological precursor of niacin?
20. Why does a diet rich in corn favor the development of pellegra?
21. What are the effects of a dietary deficiency of riboflavin?
22. How does riboflavin participate in biological oxidations?
23. What are the effects of a dietary deficiency of pyridoxine in the rat? In the dog?
24. Name the other two members of the pyridoxine group. What does this group of vitamins do in metabolism?
25. Of what important biological compounds is choline a constituent?
26. What pathological conditions arise in animals as a result of choline deficiency?
27. What are the principal pathological effects of a dietary deficiency of pantothenic acid in chicks? In rats? In dogs and silver foxes?
28. What is the outstanding importance of pantothenic acid in metabolism?
29. What are the effects of a dietary deficiency of biotin in the rat? In man?
30. What is avidin and how may it bring about a dietary deficiency of biotin?
31. What are the effects of a dietary deficiency of folic acid?
32. What is folinic acid? What are its physiological effects?
33. What are the physiological effects of vitamin B_{12}? From what source material was this vitamin isolated?
34. What is the chemical name for vitamin B_{12}?
35. What two factors are involved in preventing pernicious anemia?
36. What is the failure that causes pernicious anemia?
37. What vitamins are probably synthesized in the alimentary tract by bacteria?
38. Describe the experiments that seem to prove that bacteria synthesize certain vitamins in the intestinal tract.

24 The blood

The blood is a circulating tissue. It consists of a fluid part called the *plasma* and of certain cellular or formed elements uniformly distributed in the plasma. The formed elements of the blood are the *red cells* or *erythrocytes,* the *white cells* or *leukocytes,* and the *platelets.* In human blood the average number of red cells per cubic millimeter is from 4.5 to 5 million; of white cells, from 5000 to 10,000; and of platelets, from 300,000 to 800,000. The total amount of blood in the body is about 8.8% of the body weight. A man weighing 154 pounds (70 kg) would therefore have approximately 6 liters (6000 ml) of blood.

The principal chemical constituents of the blood are water, proteins, amino acids, glucose, lipids, inorganic salts, waste products of metabolism, hormones, vitamins, and enzymes. A detailed composition of the blood is given in Table 24-1.

Functions. Five important functions of blood are listed below.
1. The blood carries food materials from the intestinal tract to the tissues and oxygen from the lungs to the tissues. The food materials are carried in solution. Of the oxygen transported, about 1% is carried in solution and 99% is carried in combination with the hemoglobin of the red cells as oxyhemoglobin.
2. The blood carries waste products of metabolism from the tissues to the excretory organs of the body (kidneys, lungs, skin, and intestine).
3. The blood distributes internal secretions (hormones) to their appropriate sites of action.
4. The blood defends the body against infective agents. The function is carried out by the antibodies and by the leukocytes, which engulf and digest pathogenic organisms and cellular debris.
5. The blood in circulating brings about a uniform distribution of heat. Hence it aids in the regulation of body temperature.

Coagulation. The coagulation, or clotting, of blood is a mechanism designed to prevent the loss of blood from the body when some point in the closed system of vessels in which it circulates is ruptured. The components that form the clot exist in the blood and in the tissues surrounding the blood vessels. In the blood there are calcium salts, fibrinogen, and prothrombin, as well as antithrombin, a substance that prevents clotting. In the

Table 24-1. Composition of blood

Water, 75% to 80%	Waste products, about 0.5%
Proteins, about 20%	Carbon dioxide
Hemoglobin	Urea
Fibrinogen	Uric acid
Serum albumin	Creatine
Serum globulin	Creatinine
	Ammonia
Carbohydrate, about 0.1%	
Glucose	Enzymes
	Catalase
Lipids, 1% to 2%	Amylase
Fats	Maltase
Cholesterol	Protease
Phospholipid	Lipase
Inorganic salts, about 2%	Antibodies
Sodium, Potassium, Calcium, Magnesium, Iron	
Chlorides, Sulfates, Phosphates, Bicarbonates	

blood platelets and the tissues there is a substance called *thromboplastin* that facilitates clotting. When a blood vessel is cut, blood is shed and becomes mixed with the surrounding tissues. Under these conditions thromboplastin is liberated from the platelets and the cut tissues. Thromboplastin in the presence of calcium ions reacts with prothrombin to produce thrombin. Thrombin finally acts on the fibrinogen of the blood, forming fibrin, the insoluble threadlike mass of protein that enmeshes the blood cells and forms the clot. The mechanism is summarized by the following reactions:

$$\text{Prothrombin} \xrightarrow{\text{Thromboplastin, Ca}^{++}} \text{Thrombin}$$

$$\text{Fibrinogen} \xrightarrow{\text{Thrombin}} \text{Fibrin}$$

As just noted, calcium ions are essential for the coagulation of blood. We may therefore prevent the coagulation of drawn samples of blood by adding some substance that will precipitate the calcium ions from solution. If it is desired to have a sample of blood in the uncoagulated state, it is only necessary to add some calcium precipitant such as potassium oxalate or sodium fluoride to prevent coagulation. This is why a little powdered potassium oxalate is placed in the bottles or tubes in which blood is collected for chemical analysis.

Sodium citrate prevents the coagulation of blood because the citrate ion combines with the calcium ion to form un-ionized calcium citrate, and thus the calcium is converted into a form that prevents its entering into the clotting process. Sodium citrate is used to prevent coagulation of blood in blood transfusions or when blood is stored in blood banks.

Heparin. Coagulation of blood is also prevented by heparin, a mucopolysaccharide with a molecular weight of 17,000. Heparin functions by interfering with the normal conversion of prothrombin to thrombin. Very small amounts of heparin are present in the body tissues, and this compound has been isolated in crystalline form from liver and lung. The small amounts of heparin in the body assist in maintaining the blood in the uncoagulated state. Heparin is used in medicine to reduce the coagulability of the blood. Heparin may be used in the syringe and collecting vial, when a blood sample is collected for analysis, to maintain the sample in the uncoagulated state.

Blood buffers. *A buffer is a substance that resists a change in the pH of a solution. The buffers of the blood are proteins, amino acids, bicarbonates, and phosphates.* Amino acids are effective buffers, but there is a relatively small amount of these substances in blood. The proteins are far more effective because the blood contains a large amount of them. Proteins behave as buffers because of their ability to neutralize either acid or base. We learned in Chapter 15 that proteins consist of amino acids and that amino acids contain amine groups, which will neutralize acids, and carboxyl groups, which will neutralize bases. Thus letting

$$P\begin{cases} COOH \\ NH_2 \end{cases}$$

stand for the protein molecule, either of the following two reactions may take place:

$$P\begin{cases} COOH \\ NH_2 \end{cases} + HCl \rightarrow P\begin{cases} COOH \\ NH_3^+ \, Cl^- \end{cases} \quad (1)$$

$$P\begin{cases} COOH \\ NH_2 \end{cases} + NaOH \rightarrow P\begin{cases} COONa \\ NH_2 \end{cases} + H_2O \quad (2)$$

In reaction 1 acid is neutralized, and in reaction 2 base is neutralized.

Bicarbonates act as buffers against acidity by reacting with a strong acid to produce carbonic acid (H_2CO_3), which is a weak acid. For example, if a strong acid (HCl) is added to a solution containing sodium bicarbonate, the following change occurs:

$$NaHCO_3 + HCl \rightarrow NaCl + H_2CO_3$$

In this reaction the strong acid (HCl) is replaced by H_2CO_3, which is so weak an acid that there is very little change in the pH of the solution. The carbonic acid thus formed acts as a buffer against strong bases, neutralizing them and producing water and the poorly ionized compound, sodium bicarbonate, which changes the pH of the solution but slightly.

$$H_2CO_3 + NaOH \rightarrow NaHCO_3 + H_2O$$

A mixture of sodium bicarbonate and carbonic acid acts as a buffer preventing changes in the pH of the solution, which would occur if it were not present.

Phosphates are effective buffers against either acid or base. For examples, let us consider the following two reactions:

$$Na_2HPO_4 + HCl \rightarrow NaCl + NaH_2PO_4 \qquad (3)$$

Disodium   **Monosodium**
phosphate   **phosphate**

$$NaH_2PO_4 + NaOH \rightarrow H_2O + Na_2HPO_4 \qquad (4)$$

Monosodium   **Disodium**
phosphate   **phosphate**

In reaction 3 the strong acid (HCl) is replaced by an acid salt (NaH_2PO_4), which is an acid it is true, but a much weaker acid than HCl. Hence disodium phosphate serves to resist the increase in the acidity of the solution that would result from the addition of the HCl in the absence of the phosphate. In reaction 4 the phosphate behaves in an opposite way. The strong base (NaOH) is replaced by the weakly basic disodium phosphate, and thus a marked shift toward alkalinity in the solution is avoided by the presence of the phosphate.

The story of buffers in connection with the blood is the same for all the tissues of the body. Without its buffers the human body (or the body of an animal) could never withstand the acids produced in normal metabolism, the excesses of acids or bases that are sometimes encountered as a result of accidental intake of extra acid or base, or the abnormal amounts of acid or base sometimes resulting from an unbalanced diet or disease.

The kidneys are able to produce ammonia, which may serve to neutralize acids. The renal production of ammonia is thus an additional buffering defense of the body against acidity.

Application of blood chemistry to the diagnosis of disease. With the development of simple methods of analysis, blood chemistry has become an invaluable aid to the physician in the diagnosis of disease. This application of chemistry is based on the fact that the blood in many diseases undergoes changes in composition that are characteristic of the disease present (Table 24-2).

To obtain chemical blood findings that may be used in diagnosis it is necessary to collect the blood after the patient has fasted for about 15 hours. This is done by taking the sample of blood in the morning before the patient has eaten breakfast. Blood collected under these conditions has been found to give values on chemical analysis that show very little variation in normal subjects of the same species. By analyzing the blood of a large number of normal human subjects, collected after a 15-hour fast, we are able to establish a value for each of the constituents of the blood that would be considered normal for human blood.

Having obtained normal values for each of the constituents of human blood, we find it an easy matter to interpret variations from the normal in terms of pathology. For example, the blood sugar of a normal individual has been found to be within the limits of 65 to 100 mg/100 ml of blood under fasting conditions. If a sample of blood is collected from a patient in the morning before breakfast and the chemical analysis of this sample shows a blood sugar higher than 130 mg/100 ml, a diabetic condition is usually indicated.

Table 24-2. Blood constituents of importance in the diagnosis of disease (values for whole blood unless otherwise stated)

Constituent	Normal range (mg/100 ml)	Comment
Nonprotein nitrogen (NPN)	25–40	Increased in nephritis
Blood urea nitrogen (BUN)	10–20	Increased in nephritis
Amino acid nitrogen	4–8	Increased in severe liver disease
Ammonia nitrogen	0.14–0.19	
Uric acid	2–6	Elevated in gout and nephritis
Creatinine	1–2	Elevated in advanced nephritis
Bilirubin (serum)	0.2–1.0	High in biliary obstruction and hemolytic anemia
Glucose	65–100	Increased in diabetes mellitus; low in hyperinsulinism
Cholesterol (serum)	130–250	High in diabetes, biliary obstruction, and overintake of calories
Triglycerides (serum)	40–175	High in diabetes, idiopathic hyperlipemia, and over-intake of calories
Chlorides as NaCl	450–530	High in congestive heart failure; low in penumonia
Chlorides, mEq/l	77–90	
Calcium (serum)	9–11	Low in tetany; high in hyperparathyroidism
Phosphorus as PO_4 (serum)	2–5	Increased in nephritis; low in rickets
Sodium (serum)	310–350	Low in diabetic acidosis and alkali deficit
Sodium (serum), mEq/l	135–150	
Potassium (serum)	16–20	Low: trend toward paralysis; high: trend toward cardiac arrest
Potassium (serum), mEq/l	3.5–5.0	
Bicarbonate (plasma), mEq/l	24–30	Low in diabetic acidosis; high in alkali excess

Gross constituents	Normal range (grams/100 ml)	Comment
Total proteins (serum)	6.0–8.0	Low in nephrosis, poor protein intake, and severe liver disease
Albumin (serum)	4.0–6.0	
Globulins (serum)	2.0–3.0	High in infections
α-1-globulin (serum)	0.3–0.6	
α-2-globulin (serum)	0.4–0.9	
β-globulin (serum)	0.6–1.1	
γ-globulin (serum)	0.7–1.5	
Hemoglobin	12–16	Low in anemia; high in polycythemia

Micro constituent	Normal range (μg/100 ml)	Comment
Protein-bound iodine (PBI) (serum)	3.8–8.0	Low in hypothyroidism; high in hyperthyroidism
Iron	50–150	Low in anemia; high in polycythemia
Copper	8–16	

Enzyme (serum)	Normal range (units*)	Comment
Amylase	60–180[1]	High in acute pancreatitis
Lipase	1–10[2]	High in acute pancreatitis
Phosphatase, alkaline	2–8[3]	High in biliary obstruction, rickets, and hyperparathyroidism
Phosphatase, acid	1–4[4]	Increased in carcinoma of prostate
Transaminase (SGOT)	10–40[5]	Increased in myocardial infarction
Transaminase (SGPT)	5–35[5]	Increased in liver disease

*Explanation of units: [1]Somogyi, [2]Roe-Byler, [3]Bodansky, [4]King-Armstrong, [5]Wroblewski.

Likewise it is known from chemical analysis of normal bloods that the nonprotein nitrogen constituents of the blood (waste products of protein and purine metabolism) make a total of 25 to 35 mg of nitrogen per 100 ml. If the blood chemistry of a patient shows that the nonprotein nitrogen is above 40 mg/100 ml of blood under fasting conditions, this patient is not eliminating nitrogenous waste products properly, and kidney disease or other eliminating nitrogenous waste products properly, and kidney disease or other elimination disorders are suspected. These are illustrations of the numerous instances in which, knowing the normal composition of blood under fasting conditions, we may be able to make a diagnosis of a diseased condition from the blood chemistry when a variation from the normal values is obtained.

The applications of blood chemistry in the diagnosis of disease are suggested by the data of Table 24-2.

RESPIRATION

Pure air consists of about 20% oxygen, 79% nitrogen, and 0.03% to 0.04% carbon dioxide. When air is breathed into the lungs and retained there for a fraction of a minute, a considerable change in composition occurs. The air that is expired from the lungs of a human subject contains about 16% oxygen, 79% nitrogen, and 4% carbon dioxide. In other words, as a result of its passage into and out of the lungs, the air loses approximately 4% oxygen and gains about 4% carbon dioxide. The nitrogen content of the air remains constant during breathing because nitrogen does not enter into chemical reactions in the body.

The absorption of oxygen from the air and its conveyance to the tissues in the case of the human body are interesting processes and are typical of all higher animals. When air is breathed into the lungs, oxygen passes through the thin layers of cells of the lung alveoli into the blood. The blood contains a protein called hemoglobin, which combines with oxygen to form oxyhemoglobin. Hemoglobin is dark in color, and oxyhemoglobin is bright red. For this reason venous blood, which comes from the tissues and contains considerable hemoglobin, appears dark in color, and arterial blood, which comes from the lungs and contains this protein largely in the form of oxyhemoglobin, has a bright red appearance. After the oxygen has combined with hemoglobin in the lung capillaries, the oxyhemoglobin formed is carried by the blood to the tissues. In the tissues oxygen is released from the oxyhemoglobin combination and used to oxidize or "burn" food substances that have been carried to the tissues from the alimentary tract by the blood and lymph. The oxidation of foods in the tissues produces heat that is used to keep the body warm and to provide energy for work and for the living processes of the body. Oxygen also oxidizes the worn-out tissues of the body and certain poisonous substances, and in this manner it aids in the removal from the body of undesirable products, at the same time producing more heat and energy.

Further consideration of respiration involves a discussion of the mechanisms that enable the blood to carry oxygen from the lungs to the tissues and carbon dioxide from the tissues to the lungs.

Transportation of oxygen by the blood. As already stated, oxygen passes from the air in the lungs into the blood of the alveolar capillaries and combines with hemoglobin to form oxyhemoglobin. About 99% of the oxygen used by the body is carried to the tissues by hemoglobin, the remaining 1% being carried in solution in the blood plasma.

Three factors operate to facilitate the combination of oxygen with hemoglobin. In the first place the pressure under which oxygen exists in the lungs is greater than in the blood and tissues. In the lung alveoli oxygen has a pressure that will raise the mercury in a manometer to a height of about 100 mm. The oxygen in arterial blood is held under a tension equal to about 80 mm Hg, and that in the fluids surrounding the capillaries in the tissues is under a pressure of 20 to 40 mm Hg. The greater pressure of oxygen in the lungs physically promotes both the passage of this gas from the lung alveoli into the blood and its combination therein with hemoglobin. On the other hand, the low oxygen pressure in the tissues is favorable for the dissociation of oxyhemoglobin into hemoglobin and oxygen. Thus the high oxygen pressure in the lungs favors the uptake of oxygen by hemoglobin in the pulmonary blood, and the low oxygen pressure in the tissues is favorable to the release of oxygen from oxyhemoglobin in the sites where oxygen is needed.

When a person is given pure oxygen, the pressure of oxygen as a gas in the lungs is several times greater than when air is breathed. The increased oxygen pressure does two things: it promotes the uptake of oxygen by hemoglobin, and it brings a greater amount of oxygen into solution in the blood plasma. Both these factors promote a better supply of oxygen in the tissues.

A second factor influencing the uptake of oxygen by hemoglobin is the pH of the blood. A shift of pH toward the alkaline side favors combination of oxygen with hemoglobin, and a shift of pH toward the acid side favors the dissociation of oxyhemoglobin into hemoglobin and molecular oxygen. In the lungs the blood pH shifts toward the alkaline side because carbon dioxide is given off; hence the combination of oxygen with hemoglobin is promoted in the lungs. In the tissues the pH of the blood changes toward the acid side because carbon dioxide is taken up by the blood; this favors the release of oxygen from oxyhemoglobin. Thus the shift of the pH of the blood toward alkalinity in the lung capillaries favors the uptake of oxygen from the pulmonary alveoli, and the shift of pH toward acidity in the tissues promotes the release of oxygen from oxyhemoglobin in those regions where oxygen is needed.

A third factor of importance in respiration is temperature. The lowering of the temperature of the blood as it passes through the lung capillaries promotes the combination of oxygen with hemoglobin. In the tissues the temperature of the blood becomes slightly warmer, an influence that favors the dissociation of oxyhemoglobin. Thus the cooling of the blood in the lungs promotes the uptake of oxygen by hemoglobin, and the warming of the blood in the tissues favors the release of oxygen from oxyhemoglobin in sites where this element is needed.

Transportation of carbon dioxide by the blood. Carbon dioxide is one of the end products of the oxidation of foodstuffs in the body. It is carried away from the tissues by the blood. The substances that assist in this process in blood are bicarbonates, carbonates, proteins, and an enzyme, carbonic anhydrase.

As the blood passes through the tissue capillaries, CO_2 in higher concentration diffuses into the red cells and then combines with H_2O to form H_2CO_3. The latter process is facilitated by the enzyme carbonic anhydrase. The H_2CO_3 dissociates and forms bicarbonate. These reactions are as follows:

$$CO_2 H_2O \xrightleftharpoons{\text{Carbonic anhydrase}} H_2CO_3 \rightleftharpoons H^+ + HCO_3^-$$

When the second reaction takes place to the right, the H^+ ion is buffered by hemoglobin, and the HCO_3^- ion associates with the K^+ ion (released from oxyhemoglobin) to form potassium bicarbonate, $KHCO_3$. The HCO_3^- ion of $KHCO_3$ diffuses through the cell wall into the plasma where it associates with the Na^+ ion, forming sodium bicarbonate, $NaHCO_3$, the compound that serves as the principal carrier of CO_2 from the tissues to the lungs. The red cell wall is impermeable to the K^+ ion but is highly permeable to the Cl^- ion; hence Cl^- ions pass into the red cells as HCO_3^- ions pass out of the latter and associate with K^+ ions, forming KCl. Thus the HCO_3^- and Cl^- ions exchange places across the red cell wall in the tissues. The movement of the Cl^- ion is called the *chloride shift,* the result of which is to maintain electrical neutrality on each side of the cell wall by balancing the positive and negative ions.

Thus, while the blood passes through the tissue capillaries, CO_2 is taken up by the red cells and converted into HCO_3^-; the latter diffuses back into the plasma, where it becomes associated with Na^+ ions, forming $NaHCO_3$. The blood in the tissue capillaries moves into the venous system. The venous blood passes to the lungs where the $NaHCO_3$ gives up part of its CO_2. The CO_2 passes into the lung alveoli and is removed by respiration.

The ionic shift, previously described as occurring in the tissue capillaries, also takes place in the lungs but in the opposite direction. Some bicarbonate (HCO_3^-) in the plasma of the lung capillaries diffuses back into the red cells, and an equivalent of chloride (Cl^-) ions passes back into the plasma to maintain a balance of oppositely charged ions. The bicarbonate (HCO_3^-) dissociates and forms carbonic acid (H_2CO_3). The latter decomposes into CO_2 and H_2O under the catalyzing influences of carbonic anhydrase; the CO_2 diffuses from the red cells into the plasma and then into the lung alveoli.

Proteins, particularly hemoglobin, assist in the transportation of carbon dioxide in two ways: (1) they yield basic ions (chiefly sodium and potassium) to combine with carbon dioxide and form carbonates and bicarbonates, as just shown, and (2) carbon dioxide will combine with the amino group of the protein molecule to form a protein carbamino compound. This compound releases its carbon dioxide as it passes through the lung capillaries and thus serves as a carrier of carbon dioxide from the tissues to the lungs. The amount carried by this process, illustrated below, is approximately 30% of the total amount transported.

In the blood of peripheral tissues

$$\underset{\text{Protein}}{P \cdot NH_2} + CO_2 \rightarrow \underset{\substack{\text{Protein carbamino}\\\text{compound}}}{P \cdot NH\,COOH}$$

In the blood of the lung capillaries

$$P \cdot NH\,COOH \rightarrow P \cdot NH_2 + CO_2$$

A small amount of carbon dioxide is physically dissolved in blood plasma. The amount carried by the blood in this way is about 5% to 10% of the total transportation. A small fraction of the physically dissolved CO_2 is hydrated and exists as carbonic acid, H_2CO_3.

Summary. About 1% of the total oxygen carried from the lungs to the tissues is in physical solution, and the remaining 99% is transported by the compound oxyhemoglobin in the red cells. Factors that affect favorably the combination of oxygen with hemoglobin in the pulmonary blood to form oxyhemoglobin are (1) the pressure of oxygen in the arterial blood in the lungs being greater than in the blood of other tissues (head of oxygen pressure in the lungs), (2) a shift in pH toward alkalinity as the blood passes through the lung capillaries, which results from the output of carbon dioxide (shift toward alkalinity), and (3) a slight lowering of temperature as the blood passes through the lung capillaries (cooling effect). Factors that favor the dissociation of oxyhemoglobin in the tissues (releasing free molecular oxygen) are the reverse of the factors that favor the combination of hemoglobin with oxygen in the lungs. Placing a patient in an oxygen tent facilitates respiration by increasing the pressure of oxygen in the lungs. This increases the amount of oxygen physically dissolved in the plasma and promotes the uptake of oxygen by hemoglobin.

Of the total carbon dioxide carried by the blood from the tissues to the lungs, 5% to 10% is physically dissolved in the plasma; approximately 60% is transported in the form of bicarbonate, $BHCO_3$ (B represents a basic ion that in this case is almost wholly sodium); and around 30% is carried by the carbon dioxide–transporting compound, carbaminohemoglobin.

The enzyme carbonic anhydrase has an important role in catalyzing reversibly the formation of H_2CO_3 from H_2O and CO_2.

Questions for study

1. What is blood plasma? What are the formed elements of the blood?
2. What is the average number of red cells per cubic millimeter of human blood? White cells? Platelets?
3. Name the principal chemical constituents of blood.
4. What are the functions of blood?
5. What are the components that enter into the formation of a blood clot? Describe the coagulation of blood.
6. What is a buffer? Name the buffers of the blood.
7. How does the body resist the introduction of excesses of acids or bases?
8. Why is potassium or sodium oxalate placed in the containers in which blood is collected for a blood chemistry?
9. Why is sodium citrate solution mixed with blood for blood transfusions?
10. How are normal values for the composition of the blood obtained?
11. Under what conditions are samples of blood collected for blood analysis?
12. Explain how blood chemistry is used in the practice of medicine.
13. What would be the interpretation of a blood chemistry report showing a blood sugar of 200 mg/100 ml of blood? A blood nonprotein nitrogen of 80 mg/100 ml?
14. What changes occur in the composition of the air that is breathed into and out of the lungs in a normal manner?
15. What causes the decrease in oxygen in the lung alveoli? The increase in carbon dioxide?

16. How is oxygen transported from the lungs to the tissues?
17. Explain how the differences in the pressure of oxygen as a gas favor the uptake of oxygen in the lungs and its release from oxyhemoglobin in the tissues.
18. Give two reasons why the administration of pure oxygen enables the patient to get a better supply of oxygen in the tissues.
19. How do changes in the pH of the blood influence the combination of oxygen with hemoglobin?
20. How do changes in the temperature of the blood affect oxygen uptake in the lungs and oxygen release from oxyhemoglobin in the tissues?
21. What are the carriers of carbon dioxide in the blood?
22. Explain how bicarbonates serve as carriers of carbon dioxide from the tissues to the lungs.
23. How do proteins function as carriers of carbon dioxide?
24. Is any carbon dioxide carried by the blood in solution as carbonic acid?

25 The urine, electrolyte balance, water balance, and acid-base balance

Urine

COMPOSITION

The urine is a solution of inorganic salts and organic waste products of metabolism, consisting of about 96% water and 4% solids. It is an important excretion. It contains practically all the nitrogenous waste products of the body, the nonvolatile acids produced in metabolism, the excess inorganic salts not needed by the body, and many other metabolic end products that would be toxic to the body if retained in the tissues.

The principal constituents of the urine of normal human subjects are outlined in Table 25-1.

Table 25-1. Composition of normal urine

Water, 96%	*Inorganic ions*
Organic constituents	Sodium
Urea	Potassium
Uric acid	Calcium
Creatinine	Magnesium
Creatine	Ammonium
Amino acids	Iron (trace)
Glucose (0.01% to 0.04%)	Copper (trace)
Fructose (very small amount)	Chloride
Pentoses (very small amount)	Sulfate
Purines: adenine, guanine, xanthine, hypoxanthine	Phosphate
Glucuronic acid	Bicarbonate
Hippuric acid	Other elements (traces)
Indican	
Lactic, pyruvic, citric, oxalic acids	
Pigments: urobilin, urochrome, riboflavin	
Ketone bodies (very small amount)	
Vitamins	
Enzymes	
Hormones	

The urine, electrolyte balance, water balance, and acid-base balance **355**

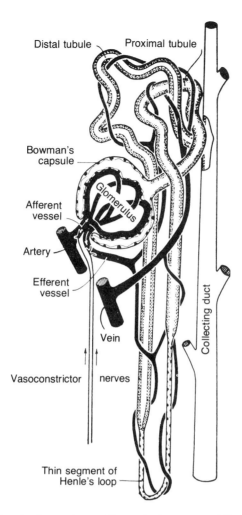

Fig. 25-1. Diagram of a single nephron. (From Amberson and Smith: Outline of physiology, The William & Wilkins Co.)

SECRETION OF URINE

To understand a discussion of the secretion of urine, it is necessary to first review the microscopical anatomy of the kidney.

Structure of the nephron

The functional unit of the kidney that brings about the secretion of urine is called the nephron. Each kidney in a healthy person contains about 1 million nephrons. The nephron consists of two parts: (1) a spherical unit about 0.2 mm in diameter called the *glomerulus* and (2) a winding tubule 40 to 60 mm long that arises in the glomerulus. A network of blood capillaries makes up a large part of the glomerulus, and blood capillaries surround the tubules throughout its length. (The student should study and frequently refer to Fig. 25-1 in this discussion.)

The nephron begins with the glomerulus. A single afferent arteriole enters the glomerulus and subdivides into many fine capillaries, which extend for a short distance and then anastomose into a single vessel, the efferent arteriole. The network of capillary blood vessels in the glomerulus is surrounded by a membranous chamber, the glomerular capsule. The network of blood vessels is known as the malpighian corpuscle, and the membranous pouchlike chamber is called Bowman's capsule. The entire network of capillaries and its surrounding capsule make up the glomerulus.

Bowman's capsule is the beginning of the uriniferous tubule. The tubule consists of three parts: (1) the proximal tubule, the first part, which arises from the capsule, (2) the midportion, including Henle's loop (a part of which has thin walls and is called the thin segment), and (3) the distal tubule, the part between the midportion and a collecting tubule. Many collecting tubules join together to form a collecting duct, which passes through the medulla of the kidney to the renal pelvis where urine is discharged.

Mechanism of secretion

The secretion of urine begins in the glomerulus. Blood coming from the renal artery enters the afferent arteriole, passes through the capillaries of the glomerulus, emerges by way of the efferent arteriole, and continues its passage through many capillaries that surround the tubules. Water and small-molecule substances (smaller than colloids) in the blood in the capillaries of the glomerulus diffuse into the capsule of the glomerulus. The force that drives these substances through the capillary walls is the blood pressure, which is produced by the heartbeat. If the blood pressure falls too low (below 40 to 50 mm Hg), urine will not be secreted and anuria will result. The fluid that diffuses in this way through the glomerular capillaries into the capsule is called glomerular filtrate. This filtrate has approximately the same composition and pH as the blood plasma minus its colloids.

A very large amount of water is used in transferring excretory products from the blood into the uriniferous tubules. The amount of water that diffuses from the glomerular capillaries into the proximal tubules in the adult human subject is about 180,000 ml per day. About 99% of this water is reabsorbed in passing through the tubules, and 1% is excreted in the urine.

When glomerular filtrate appears in the uriniferous tubules, a remarkable reabsorbing and reconstituting process takes place. In the proximal tubule, water, glucose, amino acids, inorganic ions, and some end products of metabolism are reabsorbed. This process continues in Henle's loop and the distal tubule. In the distal tubule, the process is more selective. A hormone secreted by the posterior pituitary gland, vasopressin, controls the reabsorption of water. If secretion of this hormone is deficient, the urine volume is greatly increased, a condition called diabetes insipidus.

Certain hormones secreted by the adrenal cortex (deoxycorticosterone and aldosterone) exert a regulatory influence on the reabsorption of sodium ions in the distal tubule. If these hormones become deficient, the condition known as Addison's disease will develop. In Addison's disease there is excessive passage of sodium from the blood into the urine, with a consequently increased passage of water from the blood into the urine. This lowers the blood volume and the blood pressure.

In the distal tubules another important process occurs, the acidification of urine. In the cells of the tubular walls carbon dioxide combines with water to form carbonic acid. This process is catalyzed by the enzyme carbonic anhydrase. The H_2CO_3 formed dissociates into hydrogen ions and bicarbonate ions. The processes are as follows:

$$CO_2 + H_2O \xrightleftharpoons{\text{Carbonic anhydrase}} H_2CO_3$$
$$H_2CO_3 \rightleftharpoons H^+ + HCO_3^-$$

The H^+ ions formed diffuse into the lumen of the tubule, and Na^+ ions diffuse from the lumen back into the tubular cells, an exchange of hydrogen ions for sodium ions. This mechanism accomplishes a saving of sodium ions to the body and an excretion of hydrogen ions into the urine. Finally, the fluid (which had a pH around 7.4—the pH of plasma—in the proximal tubule) has its pH lowered to the acid range of 5 to 7 by the time it is excreted as urine.

In the uriniferous tubules the reabsorption that occurs is an adjustment to maintain the normal volume, electrolyte pattern, and osmotic pressure of the blood and extracellular fluids of the body. From the tubules there is taken back into the bloodstream the amount of each substance that is needed to maintain a normal blood volume and to keep a normal level of plasma electrolytes and other noncolloidal constituents; the amount of each constituent of glomerular filtrate not needed in the body is excreted in the urine.

Summary

1. Water and noncolloidal constituents of the blood plasma from the capillaries of the kidney glomerulus diffuse into the capsule (Bowman's) that surrounds them. The concentration of the substances in glomerular filtrate is approximately the same as in blood plasma.

2. Water, glucose, amino acids, inorganic ions, and other metabolites, also small amounts of end products of metabolism are reabsorbed from the kidney tubules. This reabsorption is selective. The tubules function to maintain the normal electrolyte pattern of the blood and body fluids and thereby the normal blood volume, and at the same time they excrete substances that are not needed or are undesirable in the body.

3. Acidification of urine takes place in the distal kidney tubules by a mechanism involving exchange of hydrogen ions for sodium ions.

4. Foreign substances are excreted through the glomerulus and directly across the tubular walls into the lumen of the tubule.

QUALITATIVE ANALYSIS

Laboratory examination of the urine reveals important information concerning the patient. A detailed discussion of items involved in a urine analysis report will be presented.

Amount. The amount of urine excreted by a normal adult in 24 hours varies from 800 to 1800 ml. The volume excreted is influenced greatly by climate. In warm weather the volume of urine is low because of the increased elimination of water through the skin in the sweat. In cold weather the elimination of water by sweating is reduced, and as a consequence there is a marked increase in the volume of urine excreted.

Color. The color of urine varies from light yellow to dark brown. This color is due to the presence of pigments, the principal ones being urobilin, urochrome, and riboflavin. In general, the depth of color of urine is proportional to the concentration. Dilute urines are more or less pale in color, whereas concentrated urines are highly colored.

Odor. The odor of urine is faintly aromatic when freshly voided. On standing, it tends to become ammoniacal because of decomposition of nitrogenous substances by bacterial action. The odor may be influenced somewhat by food or drug ingestion and by disease. Thus the ingestion of asparagus imparts the peculiar odor of methyl mercaptan to urine, and the urine of a patient with diabetes mellitus sometimes has a fruity odor because of the presence of acetone bodies.

Sediment. The urine normally contains a few white blood cells, an occasional red blood cell, and a small amount of epithelial cells because of desquamation of tissue from the urinary tract. The amount of cellular material in urine may be great enough to give the urine a cloudy appearance. Whether the cloudy appearance has any significance is revealed by a microscopical examination.

The urine may be turbid because of the presence of certain substances that have crystallized or precipitated from solution. The substances appear if conditions that lower their solubility develop after the sample is voided, such as a decrease in the temperature of the specimen or an ammoniacal decomposition, which makes the urine more alkaline. Of these substances, phosphates, amorphous urates, crystals of uric acid, and calcium oxalate occur most commonly. Phosphates may occur either in the crystalline or the amorphous state. They are found in the form of sediment principally in alkaline or neutral urines and sometimes in a very faintly acid urine. Amorphous urates may be seen in acid urines, appearing as "brick-dust" deposits. Uric acid crystals will often form in acid urines on standing. Calcium oxalate crystals are sometimes found in neutral or very faintly acid urines. These substances do not have any pathological significance. Turbidity in a sample of urine because of the presence of phosphate is considered normal, except that phosphates may possibly indicate an abnormally alkaline urine.

Specific gravity. The specific gravity of urine is related to the amount voided. In general, the greater the volume excreted, the lower will be the specific gravity. Taking into account the variations in amount excreted, a specific gravity of 1.008 to 1.025 is generally considered normal.

pH. The pH of urine is determined by the diet. A diet of proteins will make the urine acid, whereas a diet of fruits and vegetables tends to make the urine alkaline. The pH of the urine of man is normally slightly acid. This is because man is naturally a carnivorous animal, eating a considerable amount of meats, which are largely proteins. In general, the urine of carnivorous animals is acid, and that of herbivorous animals is alkaline.

The pH of urine is usually in the range of 5 to 7. However, the kidneys may secrete urine with a pH as low as 4.6 or as high as 8.2. A very low or very high pH may indicate a marked dietary imbalance, or it may be the result of pathology.

Proteins. Normally the urine does not contain proteins in amounts greater than trace quantities. If proteins are found in the urine in amounts sufficient to give the usual qualitative chemical tests, pathology in the urinary tract may be present. The proteins most often encountered in urine under pathological conditions are albumin and globulin. These proteins are the serum albumin and serum globulin of the blood, which pass through the kidneys when they are not functioning normally. The globulin occurs in urine less frequently and in smaller amounts than the albumin. Proteinuria usually indicates the existence of a nephritis or a nephrosis, diseased conditions of the kidneys due to inflammatory or degenerative changes. Proteinuria may also be found in patients with inflammation in the urinary tract below the kidneys, such as a cystitis or inflammation of the bladder, in which there is a discharge of protein material into the urine. Renal proteinuria is practically always accompanied by the occurrence of tube casts in the urine.

Proteinuria that is not pathological may sometimes occur. Thus proteinuria may be due to severe muscular exercise; it may also be the form known as orthostatic or postural proteinuria, which appears after the subject has stood on his feet for long periods of time and which disappears after he rests in bed. Orthostatic proteinuria is believed to be due to a stasis of blood in the kidneys that occurs during standing.

Sugar. The urine normally contains about 0.01% to 0.04% sugar. This amount will not give the usual qualitative tests for sugar such as Benedict's or Fehling's tests. In abnormal conditions of carbohydrate metabolism, sugar is excreted in the urine in amounts readily detected by the usual tests. The sugar found in the urine under abnormal conditions is nearly always glucose, and this condition is called glucosuria. Glucosuria is in nearly all cases the result of hyperglycemia or increased blood sugar. Occasionally subjects are observed, however, who excrete sugar in the urine when the blood sugar is within normal limits. Such subjects are said to have a low sugar threshold, and the condition is known as *renal diabetes*.

Glucosuria is a characteristic finding in diabetes mellitus. Moderate glucosuria may also occur from the ingestion of a very large quantity of carbohydrate and from fright or nervous excitement.

Lactose is sometimes excreted in the urine during late pregnancy and lactation. Lactosuria is not a pathological condition and is of clinical interest principally because it may lead to a mistaken diagnosis of glucosuria by the usual urine tests for sugar.

Ketone bodies. The ketone bodies are acetone, acetoacetic acid, and β-hydroxybutyric acid. They appear in the urine when not enough carbohydrate is metabolized to prevent their production in excess by the liver. They are found in the urine in diabetes mellitus, in carbohydrate starvation, and when the fat content of the diet is too high—that is, when the diet contains more than 1.5 g of fat to 1 g of carbohydrate. Ketone bodies may also be found in the urine after general anesthesia.

The ketone bodies are acid in character. When formed in excess by the liver, they require bases to neutralize them before they are excreted in the urine. Hence the overproduction of ketone bodies reduces the fixed bases of the body and thus produces an acidosis. The finding of ketone bodies in the urine may indicate that a metabolic acidosis of greater or less degree exists.

Indican. Indican is a metabolic product formed in the liver by a combination of the sulfate radical with indole, a substance produced in the intestinal tract by bacterial action on the amino acid tryptophan. The quantity of this substance excreted in the urine is considered a general measure of the amount of bacterial action on amino acids in the intestinal tract. A little indican in the urine is considered normal. When found in the urine in increased quantities, it indicates that excessive putrefaction is taking place in the intestinal tract.

Bile. Bile is normally excreted through the intestine along with the feces. When found in the urine, it indicates the existence of an obstruction to the flow of bile into the intestine. Such obstruction may be due to a calculus, a neoplasm, or a catarrhal inflammation, which would interfere with the flow of bile through the bile ducts, or it may be due to severe liver damage, which produces obstruction to the flow of bile in the bile canaliculi. Examination of the urine for bile has less clinical value than the more sensitive tests for biliary dysfunction such as examination of the blood serum for bilirubin, which gives earlier and more reliable indications of existing pathology.

Blood. The most common causes of blood appearing in the urine are acute nephritis, stone in the kidneys or bladder, renal tuberculosis, and tumor in the urinary passages. The chemical tests for blood are exceedingly sensitive and are used largely to corroborate the finding of red blood cells in the microscopical examination.

Electrolyte balance
ELECTROLYTE COMPOSITION

A typical electrolyte balance in the human body is shown in Table 25-2. The data are for blood plasma at a pH of 7.4. They show that there are 155 milliequivalents (mEq) of cations and 155 mEq of anions in 1 liter (l) of plasma. These values reveal that there is an exact balance of oppositely charged ions and, hence, an electrolyte balance in the plasma. It is noteworthy that in plasma the sodium ions constitute 92% of the total cations and that the chloride ions and the bicarbonate (HCO_3) ions make up 66% and 17%, respectively, of the total anions.

The concentrations of ions of electrolytes in the interstitial fluids and the blood plasma are essentially the same, except that the plasma contains considerably more protein anions. The large protein particles do not pass through the walls of the capillaries and into the interstitial fluids that surround them. The smaller electrolyte ions can make this passage, however, and these ions exist in almost similar concentrations on each side of the capillary wall.

There is a characteristically high concentration of large protein anions inside the cells because of the necessity of proteins for vital physiological functions within the cell (hemoglobin for respiration, enzymes that promote metabolism, etc.). This functional need for protein requires a different pattern of ionic components within the cell, and there is a considerable difference between the electrolyte compositions of extracellular and intracellular fluids.

There is an overall ionic balance between the extracellular and intracellular fluids. Hence the osmotic pressure of extracellular and intracellular fluids is the same. Hypotonicity or hypertonicity cannot be tolerated in any of the body fluids.

Table 25-2. The electrolyte content of blood plasma with a pH of 7.4

	mEq/l
Cations	
Sodium	143
Potassium	5
Calcium	5
Magnesium	2
Total	155
Anions	
Bicarbonate	27
Chloride	103
Phosphate	2
Sulfate	1
Organic acids	6
Proteinate	16
Total	155

BALANCE BETWEEN ANIONS AND CATIONS IN EXTRACELLULAR FLUID

The columns of data in Table 25-2 show a normal electrolyte pattern. These data reveal both the relative quantity of each of the ions in blood plasma and the balance between the total anions and the total cations. The milliequivalent values—143, 5, 5, and 2, for Na, K, Ca, and Mg, respectively—indicate the quantitative relationship of cations to each other and their significance as a group were added; similarly, the ionic concentrations of bicarbonate, chloride, phosphate, sulfate, organic acids, and proteinate represented by these milliequivalents per liter values—27, 103, 2, 1, 6, and 16, respectively—show the quantitative relationships of the anions and, when added, their significance as a group. The sum of all the cations is 155 mEq/l and that of all of the anions is 155 mEq/l. These data show that a balance exists between the total anions and cations in plasma. This balance is representative of the electrolyte pattern in the extracellular fluids. An overall balance between anions and cations exists in intracellular fluids also, but the quantitative relationships differ from those in extracellular fluid.

Water balance

Amount of water in the body. The water content of the body ranges from 42% to 70% of the body weight. It varies widely, being dependent on age, sex, and—most prominently—the amount of body fat. In general, the less the fat content, the greater the percentage of water in the body. Mean values for the adult male and female are 60% and 54% of the body weight, respectively.

Fluid compartments. Body water may be said to exist in compartments as follows:
1. Intracellular compartment
2. Extracellular compartment
   a. Interstitial subcompartment
   b. Intravascular subcompartment

The intracellular water makes up 35% to 50% of the body weight. The extracellular water is distributed as follows: interstitial water, the water in the tissues outside the cells, accounts for about 15% of the body weight; and intravascular water, the water within the blood vessels, makes up about 5% of the body weight.

Intake. Water is supplied to the body through the fluids drunk and the foods ingested. Foods contain considerable water as such (35% to 95%), and they yield water when oxidized. One gram of food, when oxidized in the body, yields an average 0.9 g of water.

Absorption. Some absorption of water may occur as a result of fluid pressure in the lumen of the intestine, produced by peristaltic movements, but the major uptake is due to osmotic pressure. As shown previously in our discussion of osmosis, the movement of water across membranes is always from the higher concentration of water to the lower concentration of water. Therefore active absorption of water occurs when the fluids in the intestinal tract are hypotonic to the solutions in the cells and blood capillaries of the intestinal epithelium. If the fluid concentration on each side of the intestinal epithelium is the same, there is little, if any, water absorption. (This is why some water is always in the lumen of the intestine.) If, on the other hand, the water concentration within the lumen of the intestine is lower than that in the epithelial cells and adjacent tissues, the reverse of absorption occurs; that is, water passes from the tissues back into the lumen of the intestine, a passage which, if marked, produces diarrhea. The latter occurs when a concentrated sugar solution or a saline cathartic (solution of $MgSO_4$ or Na_2HPO_4) is ingested.

Distribution—hydrostatic versus osmotic pressures. The distribution of water to the tissues after its absorption into the blood is determined by the operation of the hydrostatic and osmotic pressures of the blood. The hydrostatic pressure of the blood is the fluid pressure imparted to the latter by heart action (blood pressure). Osmotic pressure is the force that causes water movement through membranes; in the blood it is due to the presence of plasma proteins, glucose, amino acids, ions of electrolytes, and waste products, the major effect being exerted by the plasma proteins. In the circulatory system hydrostatic pressure and osmotic pressure oppose each other and water movement goes in the direction determined by the greater force.

After water is absorbed, it is moved through the major arterial vessels into the capillaries of the tissues by the predominant force of hydrostatic pressure. In the arterial side of the capillary circulation the hydrostatic pressure of the blood emerging from the heart is around 110 and 120 mm Hg; when the blood reaches the peripheral tissues the hydrostatic pressure has been diminished to around 60 to 70 mm Hg. This is much higher than the opposing osmotic pressure of the blood, which is about 25 mm Hg; consequently, water is forced into the tissues by the greater force of hydrostatic pressure. Further along in the capillaries the hydrostatic pressure of the blood is gradually diminished to a point where it becomes less than the osmotic pressure. When this situation arises, the force of osmotic pressure in the blood takes over and causes water to pass from the tissues into the venous capillaries. This water is returned to the vascular system by osmotic pressure, and its circulation is continued.

In the tissues of the body, water moves freely and rapidly under its own diffusion energy, performing its function of carrying nutriment and oxygen, and removing carbon dioxide and other waste products. Some water passes to the epithelial cells of the skin and escapes as invisible perspiration, or as visible sweat; some passes to the lungs and escapes as invisible vapor; a small amount is excreted in the feces; and a considerable amount is excreted through the kidneys, in which special mechanisms (described on p. 356) operate to eliminate the water not needed.

If an imbalance in tissue fluid concentration occurs anywhere in the body, due to the introduction of substances that produce a temporarily higher osmotic pressure, water rapidly moves to the point of greater osmotic pressure, making equal the concentrations and restoring an osmotic equilibrium.

Hormonal control. Water movement through the body is influenced by two hormones: (1) aldosterone, a secretion of the adrenal cortex, which promotes reabsorption of sodium ions in the kidney tubules and thereby aids in the retention of water in the body, and (2) vasopressin, the antidiuretic hormone of the posterior pituitary gland, which has a regulatory effect on the reabsorption of water from the kidney tubules (p. 261).

Excretion. Water in the liquid state is excreted from the body in the urine, the feces, and the sweat. It is also eliminated from the outer surfaces of the body and through the lungs in the form of water vapor, which is not ordinarily perceptible.

There is a daily *obligatory* loss of water from the body of about 1500 ml. This loss is distributed as follows: 400 ml through the skin as insensible perspiration, 400 ml in the expired air, 200 ml in the feces, and 500 ml in the urine. Additional losses occur in the urine, and from sweating, especially when the individual consumes extra water as a result of increased exercise or being in a warm environment. The total water intake varies widely in proportion to bodily activity and the temperature of the environment. The intake obviously must be in considerable excess of the obligatory output. An average daily intake is about 3000 ml.

Importance of water balance. Normally a balance between intake and output of water is automatically maintained when a person is in health and has free access to water. In many illnesses, however, water balance becomes a problem, and the maintenance of a normal water balance in the patient is an important responsibility of the health team. The patient is definitely in an unfavorable state (dehydration) when a water loss equal to 6% of the body weight has occurred, and a water loss equivalent to 20% of the body weight is incompatible with life. If water cannot be taken by mouth, an adequate quantity is administered parenterally. A useful guide in the maintenance of water balance is to see that there is a normal excretion of urine. There should be a urinary output of 1000 to 1500 ml each 24 hours to eliminate the 30 to 70 g of waste products that pass through the kidneys.

Laboratory procedures. Laboratory methods are available for measuring (1) the total body water, (2) the plasma volume, or (3) the extracellular fluid—procedures which yield data that are helpful to the physician.

The total body water is measured by injecting intravenously a compound that is highly soluble in water, readily penetrates all cell membranes, and rapidly establishes an equilibrium concentration in all fluid compartments. Substances that have been used are deuterium oxide (heavy water), tritium oxide (radioactive water), and antipyrine.

To measure plasma volume, test substances have been selected that exist in the form of large particles. Examples are (1) a dye, Evans blue, (2) plasma albumin labeled with radioactive iodine, and (3) red cells labeled with radioactive iron or chromium. These substances may be used because their large particles escape very slowly, or not at all, from the blood vessels.

For the measurement of the fluid volume in the extracellular compartment, substances have been selected that have a very slow rate of diffusion into the intracellular compartment. Such substances are inulin, sucrose, mannitol, sodium ions, and thiosulfate. These substances penetrate so slowly into the intracellular area that measurement of the extracellular volume can be made satisfactorily by their use.

In the procedures outlined above, at an appropriate time after injection of the test substance, a sample of blood is collected and analyzed for the substance administered. Knowing the amount of test substance injected, one can calculate the volume of fluid in which it is distributed, from the values obtained by the analysis.

Acid-base balance

The pH of the blood plasma of the normal human subject is 7.4. Extremes of variation that would be considered normal would come within the range of pH 7.35 to 7.45. A plasma pH of 7.3 (or lower) means that an acidosis exists, and a pH of 7.5 (or higher) indicates the presence of an alkalosis.

The pH of fluids in the interstitial and intracellular compartments is slightly lower than the pH of plasma. In the interstitial and intracellular areas the pH is difficult to determine; however, analysis of plasma can yield quite satisfactory information for diagnosis and treatment of patients who have a disturbed acid-base balance.

One important indicator of the acid-base balance of a patient is the pH determination of the blood plasma. It is not a complete answer, however, as there are buffers in the tissues that compensate to some extent for the effects of additional acid or base production, and with mild amounts of extra acid or base introduced into the tissues, the plasma pH may remain within normal limits. To obtain full information on the acid-base balance, it is necessary to determine not only the pH but also the milliequivalents of bicarbonate (HCO_3) and carbonic acid (H_2CO_3) in the plasma, as the following material will explain.

The blood has three powerful buffer systems that protect against the introduction of acid or base. These are (1) proteins in the plasma and hemoglobin in the red cells, (2) the phosphates (B_2HPO_4 and BH_2PO_4) in the plasma, and (3) the bicarbonate ($BHCO_3$) and carbonic acid (H_2CO_3) system in the plasma. The behavior of proteins as buffers is described on p. 346. Phosphates may neutralize acid or base and discharge them through the kidneys into the urine (p. 347). The bicarbonate and carbonic acid system will now be discussed.

THE $BHCO_3/H_2CO_3$ SYSTEM

In blood plasma there is a fairly constant amount of bicarbonate in the form of the sodium and potassium salt. Plasma bicarbonate is designated by the general formula

$BHCO_3$. In this formula the B stands for Na or K, the Na being the predominant ion, as it makes up 92% of the total basic ions. There is also a small amount of CO_2 dissolved in the plasma, which exists in the form of carbonic acid, H_2CO_3. In normal plasma there are 27 mEq of $BHCO_3$ and 1.35 mEq of H_2CO_3 (average normal values). These two components form the $BHCO_3/H_2CO_3$ system, which is sensitive and very efficient in the regulation of acid-base balance. The formation and behavior of carbonic acid and bicarbonate are illustrated in the following equations.

$$H_2O + CO_2 \rightleftarrows H_2CO_3 \quad (1)$$
$$H_2CO_3 \rightleftarrows H^+ + HCO_3^- \quad (2)$$
$$HCO_3^- + B^+ \rightleftarrows BHCO_3 \quad (3)$$

In equation 1, CO_2 combines with water to form H_2CO_3. This reaction is catalyzed by an enzyme, carbonic anhydrase. In equation 2, the H_2CO_3 dissociates into H^+ and HCO_3^- ions. In the presence of base the H^+ ions are neutralized and the bicarbonate ions become associated with basic ions, forming $BHCO_3$. The latter change is illustrated by equation 3. These reactions are indicated as reversible. In the tissues they take place predominantly to the right; that is, the forward reaction prevails. In the lungs the reverse reactions (to the left) predominate; here CO_2 is released from the plasma and passes into the lung alveoli. By these reversible reactions 600 to 1000 g of CO_2 are eliminated from the body in one day. Thus the $BHCO_3/H_2CO_3$ system is a highly efficient mechanism for buffering and removing CO_2 from the tissues and protecting the body against a potential acid (CO_2) that is continuously produced in metabolism.

The reactions in equations 1, 2, and 3 take place in the tissues and lungs at rates that keep the ratio of $BHCO_3$ to H_2CO_3 in the plasma at the fairly constant value of 20:1. Stated in the form of an equation, normally

$$\frac{BHCO_3}{H_2CO_3} = \frac{27 \text{ mEq/l}}{1.35 \text{ mEq/l}} = \frac{20}{1} \text{ at pH} = 7.4$$

A moderate excess of acid or base introduced into the tissues may change the 20:1 ratio of $BHCO_3$ to H_2CO_3 without changing the pH. In such instances there is said to be present an acidosis or an alkalosis that is compensated. If, however, the $BHCO_3/H_2CO_3$ ratio becomes less than about 16:1, or more than about 25:1, the pH of the plasma is changed and an uncompensated acidosis (at the lower ratio) or an uncompensated alkalosis (at the higher ratio) is said to exist. Thus the plasma pH does not show changes in a compensated acidosis or alkalosis, and a determination of the pH alone will not give satisfactory information concerning the acid-base balance of the patient. Complete information is obtained by ascertaining the milliequivalents of $BHCO_3$ and H_2CO_3 in addition to the pH.

CAUSES OF ACID-BASE IMBALANCE

Disturbances in acid-base balance may result from abnormal conditions arising in (1) the alimentary tract, (2) the body tissues, (3) the lungs, or (4) the kidneys.

Disturbances in the alimentary tract

Vomiting. In vomiting, the emesis of gastric fluid brings about a loss of HCl. This means a loss of acid from the body; therefore the immediate change, when vomiting occurs, is the development of an alkalosis. If the condition persists, the balance may change back toward the side of acidosis, due to starvation, which has an acidosis-producing effect. The production of acidosis by starvation is discussed on p. 293.

Diarrhea. The fluids in the lower intestinal tract and the feces are slightly alkaline. Hence, if there is an abnormal loss of water from the lower intestinal tract, as in a diarrhea, the effect on acid-base balance is the production of acidosis. Diarrhea also causes electrolyte depletion, a serious accompaniment of the acidosis. The acidosis and electrolyte losses caused by diarrhea are of a more serious import in infants.

Imbalance arising in body tissues

Ketosis. In diabetes mellitus, in starvation, or in those on a diet with a very high fat-to-carbohydrate ratio the liver produces more ketone bodies (acetone, acetoacetic acid, and β-hydroxybutyric acid) than the other body tissues can oxidize; consequently there is an increase in these substances in the body. Two of the ketone bodies, acetoacetic acid and β-hydroxybutyric acid, are neutralized by tissue bases (Na and K) and excreted in the urine. These changes bring about a loss of basic ions (Na and K) from the tissues and, consequently, a decrease in the bicarbonate content and the pH of the plasma. The result of these abnormal changes is a metabolic acidosis.

Excessive intake of alkaline powders. The ingestion of alkaline preparations such as sodium bicarbonate, magnesium oxide, or potassium lactate or citrate may produce an alkalosis, if the intake is excessive.

Imbalance due to abnormal respiration

In the normal exchange of gasses in the lungs the respiratory rate in the adult is 13 to 18 per minute. If a situation arises in the body that causes an increase in the respiratory rate, there is an increased output of CO_2. This condition is called hyperpnea (hyperventilation). The effect of this condition on the acid-base balance is the production of a respiratory alkalosis, resulting from the abnormal loss of CO_2. If, on the other hand, a condition arises that decreases the respiratory rate, there is a diminished excretion of CO_2 into the lung alveoli (hypoventilation), and consequently a greater amount of CO_2 is retained in the blood and tissues. The result in this condition is the development of a respiratory acidosis.

Imbalance due to failure in the kidneys

The result of poor kidney function is a trend toward acidosis. The failure involves (1) a decreased excretion of H ions in the kidney tubules and (2) a diminished production of ammonia, which normally neutralizes considerable acid. These two functional activities are diminished by kidney disease. If the kidney disease is severe, an acidosis called nephritic acidosis develops.

The urine, electrolyte balance, water balance, and acid-base balance 367

LABORATORY FINDINGS

Aid in the diagnosis of acid-base balance disturbances is obtained by chemical analysis of the blood plasma for the $BHCO_3/H_2CO_3$ ratio and the pH. A complete laboratory report would include, in addition to the $BHCO_3/H_2CO_3$ ratio and the pH, values for the milliequivalents per liter of sodium, potassium, and chloride. The latter are important because electrolyte imbalance accompanies acid-base imbalance.

Questions for study

1. In general, what does urine contain? Name the more important constituents.
2. Give a summary of the mechanisms involved in the secretion of urine.
3. What is the origin of the pressure that forces water and noncolloidal constituents of the plasma through the walls of the glomerular capillaries into the renal (Bowman's) capsule?
4. How much water is filtered through the walls of the glomerular capillaries into the renal capsules in the normal adult in 24 hours?
5. What is the normal amount of urine excreted per 24 hours?
6. What substances may make the urine turbid? Is their presence normal?
7. What is the specific gravity of normal urine?
8. What foods make the urine acid? What foods may make it alkaline?
9. What is the usual pH range of the urine?
10. What does the finding of protein in the urine usually indicate?
11. What is the significance of glucose in the urine?
12. What is the interpretation of lactosuria?
13. What would produce ketone bodies in the urine?
14. What does increased indican in the urine indicate?
15. What pathology would cause bile to appear in the urine?
16. What conditions would cause blood to appear in the urine?
17. Expressed in milliequivalents per liter, what is the approximate overall balance of anions and cations in human plasma at a pH of 7.4?
18. What is the relative quantitative importance of the sodium, chloride, and bicarbonate ions in blood plasma?
19. Compare the concentrations of electrolyte ions in blood plasma and interstitial fluids.
20. What is the relationship between the total osmotic pressures of extracellular and intracellular fluids? What is the importance of this relationship?
21. Compare the concentrations of large protein anions in extracellular and intracellular fluids.
22. What percentage of body weight is water? What factors have an influence in determining the body water content?
23. Explain what is meant by (a) the intracellular water compartment and (b) the extracellular water compartment.
24. What are the two components of the extracellular water compartment?
25. Explain how water is absorbed from the intestinal tract.
26. Explain how water may pass from the tissues back into the lumen of the intestine.
27. Explain how water is made to pass from arterial blood into the tissues of the body.
28. Explain how water is returned from the tissues into the venous side of the blood circulation.
29. What are the channels of excretion of water from the body?
30. What is meant by the term "daily obligatory loss" of water?
31. What is an average daily intake of water (per 24 hours)? What factors bring about a wide variation in the daily intake of water?
32. What is the pH of the blood plasma in a normal subject? At what pH of plasma does uncompensated acidosis exist? Uncompensated alkalosis?

33. What are the principal buffer systems in the body?
34. How is CO_2 taken up in the tissues and carried to the lungs?
35. What is meant by the $BHCO_3/H_2CO_3$ ratio in blood plasma? What is the normal $BHCO_3/H_2CO_3$ ratio? At what ratio of $BHCO_3$ to H_2CO_3 does an acidosis exist? An alkalosis?
36. What is the normal $BHCO_3/H_2CO_3$ ratio, expressed in milliequivalents per liter of plasma?
37. How may acid-base imbalances arise from abnormal function in the alimentary tract? In the body tissues? In the lungs? In the kidneys? Explain each imbalance.
38. What laboratory analyses are essential for diagnosis and treatment of disturbances in acid-base balance?

Appendix

LOGARITHMS

A logarithm is an algebraic exponent applied to a certain mathematical base. One of the common numbers used as a base is ten. In the following mathematical calculations the superscripts 1, 2, 3, 4, and 5 are logarithms and denote the number of times 10 is used as a factor.

$$10 = 10^1 = 10 \text{ (10 is used one time)}$$
$$10 \times 10 = 10^2 = 100 \text{ (10 is used two times)}$$
$$10 \times 10 \times 10 = 10^3 = 1000 \text{ (10 is used three times)}$$
$$10 \times 10 \times 10 \times 10 = 10^4 = 10,000 \text{ (10 is used four times)}$$
$$10 \times 10 \times 10 \times 10 \times 10 = 10^5 = 1000,000 \text{ (10 is used five times)}$$

In mathematics, logarithms are used for their great convenience in calculating and in expressing large numbers, the greater the number, the more convenient the system. A number such as 156,000,000 can conveniently be written 1.56×10^8, which simply means that 1.56 multiplied by 10 eight times would be 156,000,000.

Logarithms 1,2,3, and so on represent numbers of the magnitude of 10 and larger. Smaller number can also be represented by logarithms, for example:

$$1 = 1.00000 = 10^0$$
$$\frac{1}{10} = 0.10000 = 10^{-1}$$
$$\frac{1}{100} = 0.01000 = 10^{-2}$$
$$\frac{1}{1000} = 0.00100 = 10^{-3}$$
$$\frac{1}{10000} = 0.00010 = 10^{-4}$$
$$\frac{1}{100000} = 0.00001 = 10^{-5}$$

The convenience experienced in writing large numbers as exponential values of 10

370 Appendix

can also be found in writing small fractional numbers in the same manner. A fraction such as 1/10000000 can be written 0.0000001 or 10^{-7}.

To multiply exponential figures

When numbers are multiplied, exponents are added.

$$10^5 \times 10^3 = 10^5$$
$$10^{-2} \times 10^3 = 10$$

To divide exponential figures

When numbers are divided, exponents are subtracted.

$$\frac{10^4}{10^2} = 10^2$$

$$\frac{10^{-6}}{10^{-4}} = 10^{-2}$$

To raise exponential figures to a certain power

Numbers raised to a power results in multiplying exponents.

$$(10^2)^3 = 10^6$$
$$(10^{-4})^{-2} = 10^8$$

To take roots of exponential figures

Exponents are divided when roots are taken of numbers.

$$\sqrt{10^4} = (10^4)^{\frac{1}{2}} = 10^2$$
$$\sqrt[4]{10^8} = (10^8)^{\frac{1}{4}} = 10^2$$
$$\sqrt{10^{-2}} = (10^{-2})^{\frac{1}{2}} = 10^{-1}$$

The use of log tables

The convenience of logarithms in expressing very large and very small numbers has been discussed. Numbers can be expressed as logarithms, and these logarithms, or "logs," can be used to simplify mathematical calculations. Then the logs obtained in the answers can be converted back to numbers. To make such conversions easier, log tables have been devised.

As we have seen, for example, the log of 1 is 0 and the log of 10 is 1. This leaves the numbers from 1 to 10 unrepresented in logs. Actually these numbers have logs that are fractional or decimal logs in the range from 0 to 1. A log of any number except 0, 1, and multiples of 10 is a two-part number consisting of a *characteristic* and a *mantissa*. The characteristic is a whole number, zero included, which may be positive or negative. The characteristic shows the number of times 10 can be used as a factor in computing the number the logarithm represents. The mantissa is always positive and a decimal and represents the fractional difference between the number and the next multiple of 10.

To illustrate these principles a simple problem is presented.

$$1743 \div 797$$

The 1743 is larger than 1000 but smaller than 10,000. Its log will have a characteristic that will be the same value as the log of 1000, or 3. The log of 1743 will have a mantissa derived from the log table. The first two figures of the number, 17, will be found eight lines down in the verticle column marked "N" for natural numbers. Following this column horizontally five columns to the right under the column headed "4" (the third figure of the number), one will find a mantissa, 2405. Still following the horizontal line opposite the number 17 to the column headed "Proportional parts" under the number 3 (the fourth figure of the number), one will find the figure 7. This figure is to be added to 2405 to obtain the mantissa of 2412. The log for the number 1743 is 3.2412.

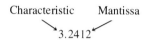

A similar procedure is followed with the number 797. Since the number lies between 100 and 1000, its characteristic is 2. The mantissa will be located in the horizontal column opposite the natural number 79 and under the vertical column headed 7, 9015. The log of 797 is 2.9015.

When numbers are divided exponents (logs) are subtracted. 3.2412 − 2.9015 = 0.3397. Using the log table, one can see that the mantissa closest to 3397 is 3385 located under the number 8 and horizontal to the number 21. The difference between 3385 and 3397 is 12. In the same horizontal column under proportional parts, 12 is found under number 6. The number whose log is 0.3397 is 2.186.

The quotent for the mathematical process 1743 ÷ 797 is 2.186, with a decimal point after the 2 dictated by the characteristic of 0, which indicates that the number lies between 1 and 10.

FOUR-PLACE LOGARITHMS

N	0	1	2	3	4	5	6	7	8	9	Proportional Parts 1 2 3 4 5 6 7 8 9
10	0000	0043	0086	0128	0170	0212	0253	0294	0334	0374	*4 8 12 17 21 25 29 33 37
11	0414	0453	0492	0531	0569	0607	0645	0682	0719	0755	4 8 11 15 19 23 26 30 34
12	0792	0828	0864	0899	0934	0969	1004	1038	1072	1106	3 7 10 14 17 21 24 28 31
13	1139	1173	1206	1239	1271	1303	1335	1367	1399	1430	3 6 10 13 16 19 23 26 29
14	1461	1492	1523	1553	1584	1614	1644	1673	1703	1732	3 6 9 12 15 18 21 24 27
15	1761	1790	1818	1847	1875	1903	1931	1959	1987	2014	*3 6 8 11 14 17 20 22 25
16	2041	2068	2095	2122	2148	2175	2201	2227	2253	2279	3 5 8 11 13 16 18 21 24
17	2304	2330	2355	2380	2405	2430	2455	2480	2504	2529	2 5 7 10 12 15 17 20 22
18	2553	2577	2601	2625	2648	2672	2695	2718	2742	2765	2 5 7 9 12 14 16 19 21
19	2788	2810	2833	2856	2878	2900	2923	2945	2967	2989	2 4 7 9 11 13 16 18 20
20	3010	3032	3054	3075	3096	3118	3139	3160	3181	3201	2 4 6 8 11 13 15 17 19
21	3222	3243	3263	3284	3304	3324	3345	3365	3385	3404	2 4 6 8 10 12 14 16 18
22	3424	3444	3464	3483	3502	3522	3541	3560	3579	3598	2 4 6 8 10 12 14 15 17
23	3617	3636	3655	3674	3692	3711	3729	3747	3766	3784	2 4 6 7 9 11 13 15 17
24	3802	3820	3838	3856	3874	3892	3909	3927	3945	3962	2 4 5 7 9 11 12 14 16
25	3979	3997	4014	4031	4048	4065	4082	4099	4116	4133	2 3 5 7 9 10 12 14 15
26	4150	4166	4183	4200	4216	4232	4249	4265	4281	4298	2 3 5 7 8 10 11 13 15
27	4314	4330	4346	4362	4378	4393	4409	4425	4440	4456	2 3 5 6 8 9 11 13 14
28	4472	4487	4502	4518	4533	4548	4564	4579	4594	4609	2 3 5 6 8 9 11 12 14
29	4624	4639	4654	4669	4683	4698	4713	4728	4742	4757	1 3 4 6 7 9 10 12 13
30	4771	4786	4800	4814	4829	4843	4857	4871	4886	4900	1 3 4 6 7 9 10 11 13
31	4914	4928	4942	4955	4969	4983	4997	5011	5024	5038	1 3 4 6 7 8 10 11 12
32	5051	5065	5079	5092	5105	5119	5132	5145	5159	5172	1 3 4 5 7 8 9 11 12
33	5185	5198	5211	5224	5237	5250	5263	5276	5289	5302	1 3 4 5 6 8 9 10 12
34	5315	5328	5340	5353	5366	5378	5391	5403	5416	5428	1 3 4 5 6 8 9 10 11
35	5441	5453	5465	5478	5490	5502	5514	5527	5539	5551	1 2 4 5 6 7 9 10 11
36	5563	5575	5587	5599	5611	5623	5635	5647	5658	5670	1 2 4 5 6 7 8 10 11
37	5682	5694	5705	5717	5729	5740	5752	5763	5775	5786	1 2 3 5 6 7 8 9 10
38	5798	5809	5821	5832	5843	5855	5866	5877	5888	5899	1 2 3 5 6 7 8 9 10
39	5911	5922	5933	5944	5955	5966	5977	5988	5999	6010	1 2 3 4 5 7 8 9 10
40	6021	6031	6042	6053	6064	6075	6085	6096	6107	6117	1 2 3 4 5 6 8 9 10
41	6128	6138	6149	6160	6170	6180	6191	6201	6212	6222	1 2 3 4 5 6 7 8 9
42	6232	6243	6253	6263	6274	6284	6294	6304	6314	6325	1 2 3 4 5 6 7 8 9
43	6335	6345	6355	6365	6375	6385	6395	6405	6415	6425	1 2 3 4 5 6 7 8 9
44	6435	6444	6454	6464	6474	6484	6493	6503	6513	6522	1 2 3 4 5 6 7 8 9
45	6532	6542	6551	6561	6571	6580	6590	6599	6609	6618	1 2 3 4 5 6 7 8 9
46	6628	6637	6646	6656	6665	6675	6684	6693	6702	6712	1 2 3 4 5 6 7 7 8
47	6721	6730	6739	6749	6758	6767	6776	6785	6794	6803	1 2 3 4 5 5 6 7 8
48	6812	6821	6830	6839	6848	6857	6866	6875	6884	6893	1 2 3 4 4 5 6 7 8
49	6902	6911	6920	6928	6937	6946	6955	6964	6972	6981	1 2 3 4 4 5 6 7 8
50	6990	6998	7007	7016	7024	7033	7042	7050	7059	7067	1 2 3 3 4 5 6 7 8
51	7076	7084	7093	7101	7110	7118	7126	7135	7143	7152	1 2 3 3 4 5 6 7 8
52	7160	7168	7177	7185	7193	7202	7210	7218	7226	7235	1 2 2 3 4 5 6 7 7
53	7243	7251	7259	7267	7275	7284	7292	7300	7308	7316	1 2 2 3 4 5 6 6 7
54	7324	7332	7340	7348	7356	7364	7372	7380	7388	7396	1 2 2 3 4 5 6 6 7
N	0	1	2	3	4	5	6	7	8	9	1 2 3 4 5 6 7 8 9

*Interpolation in this section of the table is inaccurate. From Handbook of chemistry and physics, ed. 46, Weast, R. C., editor, 1965. Used by permission of CRC Press, Inc.

N	0	1	2	3	4	5	6	7	8	9	Proportional Parts								
											1	2	3	4	5	6	7	8	9
55	7404	7412	7419	7427	7435	7443	7451	7459	7466	7474	1	2	2	3	4	5	5	6	7
56	7482	7490	7497	7505	7513	7520	7528	7536	7543	7551	1	2	2	3	4	5	5	6	7
57	7559	7566	7574	7582	7589	7597	7604	7612	7619	7627	1	2	2	3	4	5	5	6	7
58	7634	7642	7649	7657	7664	7672	7679	7686	7694	7701	1	1	2	3	4	4	5	6	7
59	7709	7716	7723	7731	7738	7745	7752	7760	7767	7774	1	1	2	3	4	4	5	6	7
60	7782	7789	7796	7803	7810	7818	7825	7832	7839	7846	1	1	2	3	4	4	5	6	6
61	7853	7860	7868	7875	7882	7889	7896	7903	7910	7917	1	1	2	3	4	4	5	6	6
62	7924	7931	7938	7945	7952	7959	7966	7973	7980	7987	1	1	2	3	3	4	5	6	6
63	7993	8000	8007	8014	8021	8028	8035	8041	8048	8055	1	1	2	3	3	4	5	5	6
64	8062	8069	8075	8082	8089	8096	8102	8109	8116	8122	1	1	2	3	3	4	5	5	6
65	8129	8136	8142	8149	8156	8162	8169	8176	8182	8189	1	1	2	3	3	4	5	5	6
66	8195	8202	8209	8215	8222	8228	8235	8241	8248	8254	1	1	2	3	3	4	5	5	6
67	8261	8267	8274	8280	8287	8293	8299	8306	8312	8319	1	1	2	3	3	4	5	5	6
68	8325	8331	8338	8344	8351	8357	8363	8370	8376	8382	1	1	2	3	3	4	4	5	6
69	8388	8395	8401	8407	8414	8420	8426	8432	8439	8445	1	1	2	2	3	4	4	5	6
70	8451	8457	8463	8470	8476	8482	8488	8494	8500	8506	1	1	2	2	3	4	4	5	6
71	8513	8519	8525	8531	8537	8543	8549	8555	8561	8567	1	1	2	2	3	4	4	5	5
72	8573	8579	8585	8591	8597	8603	8609	8615	8621	8627	1	1	2	2	3	4	4	5	5
73	8633	8639	8645	8651	8657	8663	8669	8675	8681	8686	1	1	2	2	3	4	4	5	5
74	8692	8698	8704	8710	8716	8722	8727	8733	8739	8745	1	1	2	2	3	4	4	5	5
75	8751	8756	8762	8768	8774	8779	8785	8791	8797	8802	1	1	2	2	3	3	4	5	5
76	8808	8814	8820	8825	8831	8837	8842	8848	8854	8859	1	1	2	2	3	3	4	5	5
77	8865	8871	8876	8882	8887	8893	8899	8904	8910	8915	1	1	2	2	3	3	4	4	5
78	8921	8927	8932	8938	8943	8949	8954	8960	8965	8971	1	1	2	2	3	3	4	4	5
79	8976	8982	8987	8993	8998	9004	9009	9015	9020	9025	1	1	2	2	3	3	4	4	5
80	9031	9036	9042	9047	9053	9058	9063	9069	9074	9079	1	1	2	2	3	3	4	4	5
81	9085	9090	9096	9101	9106	9112	9117	9122	9128	9133	1	1	2	2	3	3	4	4	5
82	9138	9143	9149	9154	9159	9165	9170	9175	9180	9186	1	1	2	2	3	3	4	4	5
83	9191	9196	9201	9206	9212	9217	9222	9227	9232	9238	1	1	2	2	3	3	4	4	5
84	9243	9248	9253	9258	9263	9269	9274	9279	9284	9289	1	1	2	2	3	3	4	4	5
85	9294	9299	9304	9309	9315	9320	9325	9330	9335	9340	1	1	2	2	3	3	4	4	5
86	9345	9350	9355	9360	9365	9370	9375	9380	9385	9390	1	1	2	2	3	3	4	4	5
87	9395	9400	9405	9410	9415	9420	9425	9430	9435	9440	0	1	1	2	2	3	3	4	4
88	9445	9450	9455	9460	9465	9469	9474	9479	9484	9489	0	1	1	2	2	3	3	4	4
89	9494	9499	9504	9509	9513	9518	9523	9528	9533	9538	0	1	1	2	2	3	3	4	4
90	9542	9547	9552	9557	9562	9566	9571	9576	9581	9586	0	1	1	2	2	3	3	4	4
91	9590	9595	9600	9605	9609	9614	9619	9624	9628	9633	0	1	1	2	2	3	3	4	4
92	9638	9643	9647	9652	9657	9661	9666	9671	9675	9680	0	1	1	2	2	3	3	4	4
93	9685	9689	9694	9699	9703	9708	9713	9717	9722	9727	0	1	1	2	2	3	3	4	4
94	9731	9736	9741	9745	9750	9754	9759	9763	9768	9773	0	1	1	2	2	3	3	4	4
95	9777	9782	9786	9791	9795	9800	9805	9809	9814	9818	0	1	1	2	2	3	3	4	4
96	9823	9827	9832	9836	9841	9845	9850	9854	9859	9863	0	1	1	2	2	3	3	4	4
97	9868	9872	9877	9881	9886	9890	9894	9899	9903	9908	0	1	1	2	2	3	3	4	4
98	9912	9917	9921	9926	9930	9934	9939	9943	9948	9952	0	1	1	2	2	3	3	4	4
99	9956	9961	9965	9969	9974	9978	9983	9987	9991	9996	0	1	1	2	2	3	3	3	4
N	0	1	2	3	4	5	6	7	8	9	1	2	3	4	5	6	7	8	9

Glossary

acetonuria ketone bodies in the urine; also called ketonuria.

acid a substance that produces hydrogen ions or protons in aqueous solution; also a proton donor and an electron pair acceptor.

acidosis a condition of the body in which there is an increase in the acidity of the tissues. This may result from excessive administration of acids into the body, from the overproduction of acids by faulty metabolism, from respiratory obstruction, or from a diminution in the factors that neutralize acidity.

acceleration the rate at which the veolcity changes with time.

Adrenalin (*see* epinephrine)

adrenotropic having the effect of stimulating normal proliferation of the functioning tissue of the adrenal gland.

adsorption adherence of a substance to the surface of another substance. For example, in the removal of a pigment from a solution by the addition of charcoal and filtration, the pigment is adsorbed or bound to the surfaces of the charcoal particles.

albuminuria the presence of albumin and other proteins in the urine.

alchemy the art practiced by an early group of experimenters with matter whose principal objective was to convert base metals into gold.

alcohol a derivative of a hydrocarbon in which one or more hydrogen atoms are replaced by an OH group.

aldehyde a derivative of a hydrocarbon in which a hydrogen atom is replaced by a

group; the first oxidation product of a primary alcohol.

alkalosis a condition of the body in which there is an increase in the alkalinity of the tissues. This may arise from an excessive administration of alkalies into the body, from a loss of the acid of the gastric juice by prolonged vomiting, or from respiratory overventilation.

allotropism the property peculiar to certain elements of existing in more than one form. For example, carbon exists in the form of diamond, graphite, and charcoal and therefore has the property of allotropism.

amine an organic derivative of ammonia. In an amine one or more hydrogen atoms of the ammonia are replaced by a hydrocarbon radical or radicals.

anabolism the building up, or synthetic chemical processes, taking place in the body after food substances have been absorbed from the alimentary tract; those chemical changes in the body by which food substances are stored or are converted into tissues.

analysis in chemistry a procedure in which a substance is broken down into its constituent parts and then identified.

anhydrous without water.

antidiuretic opposing the flow of urine.

antidote a substance that will prevent the action or combat the effects of a poison.

atheroma a lesion in which there are focal deposits of lipid material in the intima of the arteries.

atherosclerosis one form of arteriosclerosis in which there are degenerative changes leading to deposits of lipid material in the intima of the arteries. These changes bring about a reduction in the size of the lumen of the vessels.

atom the smallest fundamental unit of an element still maintaining the distinguishing features of that element that is able to take part in a chemical change.

atomic mass number the average atomic mass of an element.

atomic mass a value applied to an element that represents the mass of an atom of this element when compared with the mass of an atom of carbon, which has been arbitrarily set as 12.

atomic number a number that shows the position of an element in the periodic table. It is equal to the number of protons in the nucleus of the atom.

Avogadro's law in equal volumes of gases at the same temperature and pressure, there are equal numbers of molecules.

base a substance that produces hydroxyl ions in aqueous solution; a proton acceptor and an electron pair donor.

beriberi a disease caused by a lack of thiamine in the diet. It is characterized by loss of appetite, polyneuritis, lesions in the central nervous system, muscular atrophy, and finally paralysis.

calorie when written with the small letter c, it is the amount of heat required to raise the temperature of 1 g of water from 15° to 16° C. When written with the capital letter C, it is 1000 times as much heat as that specified for the small calorie (kcal).

carbohydrate an aldehyde or ketone derivative of a polyhydroxy alcohol.

catabolism the chemical changes in the body by which stored food materials or tissues are broken down into simpler substances that ultimately become waste products.

catalyst a substance that enters into a chemical reaction altering the rate of that reaction but is itself reformed when the reaction is ended.

centimeter one-hundredth of a meter.

chemistry a science that deals with the nature of matter and the changes that matter undergoes.

chlorophyll the green substance in the leaves of plants.

colloid the term colloid applies to a state of matter in which the particles range in size from 1 to 100 millimicrons. Colloids will not pass readily through membranes made of such materials as parchment, collodion, or cellophane.

combining weight the number of grams of an element that will combine with 8 g of oxygen or with an amount of another element that is chemically equivalent to 8 g of oxygen.

combustion a chemical change accompanied by the evolution of heat and, usually, light. If the change occurs in a confined space where the heat accumulates until a conflagration occurs, the phenomenon is called spontaneous combustion.

compound a pure substance composed of two or more elements firmly united in definite proportions.

crystalloid the term applied to a state of matter in which the particles are less than 1 millimicron in size. Crystalloids pass readily through membranes made of such materials as parchment, collodion, or cellophane.

dehydration the removal of water from a substance.

deliquescence the taking up of moisture from the air at ordinary conditions of temperature and pressure.

density the concentration of matter or the amount of matter per unit volume.

diabetes insipidus a disease caused by injury to the posterior pituitary gland. It is characterized by severe polyuria and marked thirst.

diabetes mellitus a disease caused by insufficient secretion of insulin by the pancreas. It is characterized by an increased blood sugar, sugar in in the urine, and—in the more severe stages—by ketone bodies in the blood and urine and by acidosis.

diabetes, renal a condition in which the blood sugar is normal but glucose is excreted in the urine in abnormal amounts.

diabetogenic having the quality of producing diabetes mellitus.

dialysis the passage of a crystalloid through a membrane that will not permit the passage of a colloid.

digestion the chemical processes in the alimentary tract by which molecules of proteins, carbohydrates, and fats are broken down into molecules that are more soluble and are small enough to pass readily through the intestinal walls into the blood or lymph streams.

disinfectant a substance that has the power to kill small organisms.

distillation the process of purifying a substance by

heating it until it becomes a gas and passing the gas through a cooling chamber in which it becomes a liquid or solid again.

diuretic stimulating the flow of urine.

efflorescence the giving up of water by a substance when exposed to the air at ordinary conditions of temperature, pressure, and moisture.

electrolysis decomposition of a substance by the action of an electric current.

electrolyte a substance that conducts electric current when in solution. Acids, bases, and salts are electrolytes.

electron a structural unit of the atom that is the basic unit of negative electricity. It has a mass of 0.00055 on a scale on which the weight of carbon is 12. It exists in the outer portions of the atom.

element a basic form of matter incapable of being decomposed by chemical means into more simple substances, having distinct chemical and physical characteristics and occupying a specific place in the periodic table of elements.

energy the ability to do work or the capacity to put matter in motion.

enzyme an organic catalyst produced by the living cells of plants or animals but independent of the cell in its action. Enzymes promote chemical reactions of substances associated with living matter.

epinephrine a secretion of the medulla of the adrenal glands—of which the more important physiological effects are acceleration of the heartbeat, raising of the blood pressure, relaxation of the bronchial muscles, and stimulation of the conversion of glycogen to glucose. Epinephrine is the official U.S.P. designation, although it is also called Suprarenin and Adrenalin.

equation in chemistry an abbreviated expression of a chemical change.

erythrocytes the red blood cells. Normal human blood contains 4 to 5 million erythrocytes per cubic millimeter.

ester an organic compound produced by the reaction of an alcohol with an acid.

estradiol one of the female sex hormones. It has the same physiological effects as estrone but is about six times as potent.

estriol one of the female sex hormones, having the same physiological effects as estrone.

estrone one of the female sex hormones. It stimulates the cyclic regeneration of the endometrium of the uterus, produces estrus in lower animals, and has a feminizing effect in the development of sex characteristics.

estrus a condition in the female of lower animals characterized by a sexual urge for mating.

ether an organic oxide. An ether consists of two hydrocarbon radicals joined by an atom of oxygen. Ethyl ether ($C_2H_5-O-C_2H_5$) is the ether used for general anesthesia.

fat an ester of glycerol and fatty acids.

fermentation in general, a chemical reaction liberating gas. More specifically the action of enzymes such as in yeast or in bacteria on sugars, with the formation of ethyl alcohol and carbon dioxide.

food a substance that, when ingested, forms body tissues, yields energy by its oxidation in the body, and aids in the promotion of normal metabolism.

force the cause of motion of matter.

formula an abbreviated expression for a compound.

glomerulonephritis a kidney disease in which there are inflammatory and degenerative changes primarily in the glomeruli.

gonadotropic having the quality of stimulating growth of the gonads or organs of reproduction.

gram a metric unit of weight. The weight of 1 ml of water at a temperature of 4° C.

halogen a member of a family of elements which consists of fluorine, chlorine, bromine, and iodine.

humidity water vapor in the air. Relative humidity means the amount of water vapor in the air as compared with the maximum amount of moisture the air will contain at a given temperature.

hydrate a compound containing one or more molecules of water in loose combination.

hydrocarbon an organic compound containing hydrogen and carbon.

hydrolysis the splitting of a compound through the action of water, with the introduction of the constituents of water into the products formed.

hyperacidity a condition of the stomach in which there occurs a secretion of hydrochloric acid in the gastric juice in excess of the normal amount.

hyperglycemia an increase in the blood sugar above the normal circulating level—approximately above 130 mg of glucose per 100 ml of blood.

hypoacidity a condition of the stomach in which there is less than the normal secretion of hydrochloric acid.

hypoglycemia a blood sugar concentration that is below the normal level, that is, below 65 mg of glucose per 100 ml of blood.

hypothesis an idea that is tentatively proposed as the most reasonable explanation of observed phenomena.

inertia the ability that matter possesses to remain at rest, or in uniform motion, until it is acted upon by some outside force.

indicanuria the presence in the urine of indican, a product resulting from the action of bacteria on the amino acid tryptophan in the alimentary tract.

insulin a protein secreted by the β-islet cells of the islands of Langerhans of the pancreas whose function is to stimulate the metabolism of glucose in the body.

ion an atom or group of atoms that has gained or lost electrons and has formed a negatively or positively charged particle.

ionization the separation of ionic compounds in solution into freely moving positive and negative ions.

isotopes forms of an element that have the same chemical properties but different atomic mass. Isotopes have the same number and arrangement of electrons in their shells but differing numbers of neutrons in their nuclei.

ketone an organic compound containing the elements carbon, hydrogen, and oxygen, the characteristic part of the molecule being a CO group; the first oxidation product of a secondary alcohol.

kilogram the metric unit of weight. A kilogram is the weight of the international prototype for mass, a metal block composed of platinum and iridium, which is kept in the International Bureau of Weights and Measures in Paris. It is equal to the weight of 1 liter of pure water at the temperature of its maximum density (4° C). The kilogram corresponds approximately to 2.2 pounds.

law of definite composition the law that states that a compound always contains the same elements united in the same proportions by weight.

law of multiple proportions when two elements form more than one compound with each other, a simple whole-number relationship exists in the proportions of the element whose amounts are varied.

leukocytes the white cells of the blood. Normal blood contains from 5000 to 10,000 leukocytes per cubic millimeter.

liter a metric unit for the measurement of volume. The volume of 1 kg of water at the temperature of its maximum density (4° C) under a pressure of 1 atmosphere. A liter contains 1000 ml or 1000 cc.

luteotropin a gonadotropic hormone secreted by the anterior pituitary gland. It stimulates the corpora lutea to secrete progesterone and functions also to initiate the secretion of milk in mammals.

mass a quantitative measure of the inertia of an object to having its mobility, or immobility, changed.

matter anything that possesses inertia, has mass, and occupies space.

metabolism the chemical changes that take place in the body after food is absorbed from the alimentary tract. Metabolism includes anabolism and catabolism.

metabolism, basal the metabolic activity of the body, 12 to 15 hours after the last meal, while the individual is awake, comfortably warm, and at complete rest.

meter a standard unit for measuring length. The meter is 1,650,763.73 wavelengths of the orange-red line of krypton −86; it is equal to 39.37 inches.

micron a unit of measurement for organisms or particles of matter of microscopical size. One-millionth of a meter.

milliliter the volume of a cube whose edge is 1 cm in length. One milliliter equals 1 cc.

millimicron one-thousandth of a micron.

molecular mass the average of the masses of the molecules of a substance according to the atomic mass scale.

molecule the smallest electrically neutral particle

of any substance that can maintain a stable existence.

neutralization the union of the hydrogen ions of an acid with the hydroxyl ions of a base to form water, a neutral substance.

neutron a particle of matter existing in the nucleus of the atom. The neutron has the same mass as the proton, but it does not bear an electrical charge.

nucleus the central portion of the atom consisting mainly of protons and neutrons and having a positive charge.

orbit a major energy level outside the nucleus of an atom where electrons can assume stable configurations. An orbit may consist of suborbits. Orbits and suborbits are made up of one or more orbitals.

oribital a volume of space around the nucleus of an atom where there is a high probability of finding an electron.

osmosis the passage of solvent through a semipermeable membrane.

oxidation a chemical reaction in which there is a loss of electrons by an element.

oxide a compound in which oxygen is combined with some other element or elements.

oxytocic stimulating contractions of the uterus.

oxytocin a chemical compound isolated from the posterior pituitary gland that stimulates contraction of smooth muscle. It is especially effective on uterine muscle, having an oxytocic action.

pathology a science that deals with the nature of disease in animals and plants.

pellagra a disease caused by a deficiency of niacin in the diet. It is characterized by dermatitis, diarrhea, and dementia.

periodic law the ordinary physical and chemical properties of the elements are periodic functions of their atomic numbers. (See text for elucidation.)

photosynthesis the production of a chemical compound from simpler substances through the action of light.

precocity the quality of developing maturity more rapidly than usual.

progesterone a hormone produced by the corpus luteum of the ovary, which is essential for the implantation of the fertilized ovum in the wall of the uterus.

proteins compounds composed or carbon, hydrogen, oxygen, nitrogen, and sometimes sulfur or iodine. They are produced by and associated with living matter and structurally are groups of amino acids joined together chiefly through their amino and carboxyl groups.

proton a structural unit of the atom that is the basic unit of positive electricity. It has a mass of 1.0073 on a scale on which the weight of carbon is 12. It exists in the nucleus of the atom.

radical a group of atoms held together by covalent bonds and possessing the ability to react chemically as a unit.

reduction the opposite of oxidation. A chemical change in which there is an addition of electrons to an element.

rickets a disease of infancy and childhood, characterized by a failure in the deposition of calcium phosphate in the bones, which results in various bone malformations such as bowlegs, knock-knees, deformed skull, and enlargements of bones at their junctions with cartilage. It may be caused by a dietary deficiency of calcium or phosphorus, an incorrect ratio of calcium to phosphorus in the diet, a lack of vitamin D in the diet, or inadequate exposure of the body to sunlight.

roentgen that amount of radiation that will produce one electrostatic unit of ions per cubic centimeter of volume.

salt the product, other than water, resulting from the reaction of an acid with a base.

science the systematic observation of facts and of experimental results and the interpretation of the relationships of such observations to each other.

scurvy a disease caused by a deficiency of vitamin C in the diet. It is characterized by hemorrhages in the joints, beneath the skin, and in mucous membranes and periosteum; swollen, bleeding gums; decalcification of the teeth and bones; loss of weight; and pain in the joints upon motion.

shell (See orbit.)

soap a salt of a higher fatty acid. The principal soaps used for cleansing are sodium or potassium salts of fatty acids.

solution a homogenous mixture of two or more substances.

symbol in chemistry an abbreviation of the name of an element.

synthesis in chemistry the putting together of elements to form compounds: for example, the formation of water from hydrogen and oxygen; also the combination of small-molecule compounds to form compounds that have larger molecules.

thyrotropic having the effect of stimulating proliferation of the functioning tissue of the thyroid gland.

thyroxin the internal secretion of the thyroid gland whose principal function is to regulate the rate of oxidations in the body. It also stimulates growth of the young and is concerned in the maintenance of a healthy condition of the skin.

valence a number that represents the combining requirement of an element for hydrogen or other elements of the same combining capacity as hydrogen.

vasopressin a compound isolated from the posterior pituitary gland, which has an antidiuretic effect. It stimulates the reabsorption of water by the kidney tubules. On blood pressure it has a pressor effect in man and a depressor effect in chickens.

velocity the distance an object moves in a given period of time.

vitamin an essential constituent of the diet that does not contribute energy to the body but is necessary for proper metabolism.

weight a characteristic possessed by matter when its mass is acted upon by gravitational attraction.

xerophthalmia a diseased condition of the eyes characterized by dryness, ulceration, and the development of layers of horny tissue on the cornea. It is caused by a deficiency of vitamin A in the diet.

Index

A

Abnormal metabolism of carbohydrates, 289-290
Abnormal respiration, acid-base imbalance due to, 366
Absolute alcohol, 169
Absolute temperature scale, 8
Acceleration, 4
Acetaldehyde, 173
Acetanilide, 195-196
Acetic acid, 113, 176, 178
Acetoacetic acid, 292
Acetone, 174, 292
Acetophenetidin, 196
Acetyl coenzyme A, 291, 292
Acetylcholine, 231
Acetylene, 84, 164-165
Acetylsalicylic acid, 194-195
Achromotrichia, 339
Acid carbonate, 120
Acid phosphate, 120
Acid salt, 119, 120
Acid-base balance, 364-367
  abnormalities in
    due to abnormal respiration, 366
    in body tissues, 366
    causes of, 365-366
    due to kidney failure, 366
  laboratory findings in, 367
Acidification of urine, 357
Acidosis, 113
  diabetic, 293
  nephritic, 366
  respiratory, 366
Acids, 110-114; see also specific acid
  amino; see Amino acids
  aromatic, 193-194
  binary, 113
  definition of, 110
  fatty; see Fatty acids
  inorganic, naming of, 113-114
  keto, uses of, 297
  organic; see Organic acids
  properties of, 111-112
  reaction of amines with, 184

Acids—cont'd
  strong, 112-113
  weak, 112-113
  wounds produced by, antidotes for, 121-122
Acrolein test for fats, 225
Acromegaly, 257
Activators, 250
Activity and energy metabolism, 318
Acyclic organic compounds, 159
Addison's disease, 261, 356
Adenine, 235
Adenosine, 236
Adenosine diphosphate in storage and release of energy, 288, 289
Adenosine phosphate; see Adenylic acid
Adenosine triphosphate in metabolism of carbohydrates, 287-289
Adenylic acid, 236-237
Adrenal cortex, secretions of, 261-262
Adrenal glands, hormones of, 261-263
Adrenal medulla, secretions of, 263
Adrenocorticotropic hormone, 259
Adsorption by charcoal, 152-153
Agar, 207
Age and energy metabolism, 318
Air, oxygen preparation from, 80
Alanine, 210
Albuminoids, 220
Albumins, 220
Alchemists, 2
Alchemy, 2
Alcoholic fermentation, 205
Alcohols, 166-170; see also specific alcohol
  absolute, 169
  aromatic, 191-192
  complex, 170
  denatured, 169-170
  grain; see Ethyl alcohol
  oxidation of, 167-168
  poisoning with, 169
  preparation of, 166
  primary, 166
    oxidation of, 167-168

Alcohols—cont'd
  properties of, 166
  reactions of, 167-168
  secondary, 166
    oxidation of, 168
  tertiary, 167
  wood; see Methyl alcohol
Aldehydes, 171-173
  aromatic, 192-193
  preparation of, 171
  properties of, 171
  as reducing agents, 171-172
Alimentary glucosuria, 290
Alimentary tract
  bacterial synthesis of vitamins in, 342
  disturbances of, causing acid-base imbalance, 366
Aliphatic organic compounds, 159
Alkali, 114; see also Bases
Alkali metals, 63
Alkaline earth metals, 63-64
Alkaline powders, excessive intake of, causing acid-base imbalance, 366
Alkanes, 162-164
Alkenes, 164-165
Alkynes, 164-165
Allotropism, 62
Alpha particles, 22
  deflection of, 23
Alpha rays, 22, 138
  penetrating power of, 138
Amidases, 248
Amines, 182-185
  aromatic, 195-197
  chemical behavior of, 184-185
  definition of, 182
  methyl, preparation of, 183-184
  physical properties of, 184
  primary, 183
  secondary, 183
  tertiary, 183
  toxicity of, 185
Amino acid oxidase, 296
Amino acids, 210
  as blood buffers, 346
  deamination of, 296-297
  degradation of, 296
  essential, 210
    requirements for, 305-307
  importance of, 295
  of oxytocin, 260
  polypeptide chains of, 213
  transamination of, 297-298
  of vasopressin, 260
Aminoacetic acid; see Glycine
Aminopeptidase, role of, in digestion, 277
Ammonia, 67-68, 115
  covalent bonds in, 48
  release of, in deamination of amino acids, 296-297
Ammonia gas, 68
Ammonium hydroxide, 68
Amorphous solid matter, 54
Amyl acetate, 180

Amyl butyrate, 180
Amylase
  pancreatic, role of, in digestion, 278
  salivary, 274-275
Anabolism, 282
  of proteins, 295-296
Anaerobes, 82
Anaerobic metabolism of carbohydrates, diagram of, 283
Analysis, 13-14
Androstane, 198
Androsterone, 267-268
Anemia
  from folic acid deficiency, 340
  nutritional, 315
  pernicious, 342
  from pyridoxine deficiency, 337
Aniline, 195
Animal charcoal, 152
Animal starch; see Glycogen
Anions, 107
  and cations in extracellular fluid, balance between, 361
Anode, 21
Anterior pituitary gland
  relation of, to carbohydrate metabolism, 259-260
  secretions of, 257-260
Anthracene, 198
Antidotes for wounds produced by acid or base, 121-122
Antiovulatory drugs, 267
Aphagia, 295
Appetite, regulation of, by hypothalamus, 295
Arachidonic acid, 224
  need for, 308
Aragon, atomic structure of, 40
Arginine, 211
Aromatic acids, 193-194
Aromatic alcohol, 191-192
Aromatic aldehydes, 192-193
Aromatic amines, 195-197
Aromatic esters, 194-195
Aromatic halogens, 190-191
Aromatic hydrocarbons, 186-190
Aromatic organic compounds, 159
Arrangement, octet, 41
Arsenic, 71
Arsenic oxide, 71
Artificial radioactivity, 140-141
Ascorbic acid, 330-332
  deficiency of, 330-331
Aspirin, 194-195
Astatine, 71
"Atomic cocktail," 149
Atomic energy, 145
Atomic fission, 144-146
Atomic fusion, 146-148
  energy from, calculation of, 147-148
  loss of mass from, 147
Atomic mass, 11
Atomic mass number, 25
Atomic number, 23-24
Atomic pile, 146

Atomic structure, 24-35
  according to quantum theory, 28-35
  rules governing, 24-27
Atomic theory, Dalton's, 20
Atoms, 20-24
  Bohr, 27-28
  decomposition of, 26
  definition of, 10-11
  difference of, from ions, 41-42
Atwater-Rosa-Benedict calorimeter, 316
Avidin, 340
Avogadro's law, 60
Avogadro's number of molecules, 61

B

Bacterial synthesis of vitamins in alimentary tract, 342
Balanced diet, 309-314
  distribution of components in, 309
Barium sulfate, 69
Barometer, 6
  simple, 7
Basal metabolism, 318-320
  determination of, 319-320
Bases, 114-117
  concentrated, 115
  definition of, 114
  dilute, 115
  naming of, 115
  properties of, 114
  reactions of organic acids with, 177
  strong, 115
  weak, 115
  wounds produced by, antidotes for, 121-122
Basic salt, 119
Beeswax, 232
Beet sugar; see Sucrose
Bending of cathode ray, 21
Bends, 66-67
Benedict's test
  for aldehydes, 172
  of carbohydrates, 204
Benzaldehyde, 192-193
Benzene, 186-189
  homologues of, 189-190
  preparation of, 187-188
  properties and reactions of, 188-189
  side chains of, 189
  source of, 187
Benzenecarbonal; see Benzaldehyde
Benzenecarboxylic acid; see Benzoic acid
Benzoic acid, 193-194
Benzoic acid liver function test, 193-194
Benzyl alcohol, 192
Beriberi, 332, 333
Beryllium, structure of, 31, 33
Beta rays, 22
  penetrating power of, 138
Beta-oxidation of fatty acids, 290-292
Betatron, 140
Bicarbonates as blood buffers, 346
Bichloride of mercury, 64

Bile, 273, 279-280
  composition of, 279
  functions of, 280
  role of, in digestion, 278
  storage of, 280
  in urine, 360
Bilirubin, 280
Binary acids, 113
Binary compounds, name of, 50-52
  Stock system of, 52
Biology, 2
  uses of radioactive isotopes in, 148-149
Biotin, 339-340
Bismuth, 65-66
Bismuth salts, 65-66
Bismuth subcarbonate, 65
Bismuth subnitrate, 65, 66
Biuret test of proteins, 219
Black phosphorus, 70
Black tongue, 334
Blindness, night, 323
Blood, 344-352
  buffers in, 346-347, 364
  changes in, due to osmosis, 134-135
  chemistry of, application of, to disease, 347-349
  coagulation of, 344-345
  composition of, 345
  functions of, 344
  pH of, and respiration, 350
  transportation of carbon dioxide by, 350-352
  transportation of oxygen by, 349-350
  in urine, 360
Blood cells
  crenation of, 134, 135
  hemolysis of, 134
  red, 344
  white, 344
Blood glucose (blood sugar), 282-284
  need for, 307
  sources of, 307-308
Blood plasma, electrolyte composition of, 360-361
Body, water in
  amount of, 361
  uses of, 97
Bohr atom, 27-28
Boiling point
  of liquid, 57, 58
  of water, 94
Bomb calorimeter, 316
Bombardment reaction in producing artificial radioactivity, 141
Bonding, 47-50
Bonds, 47-50
  covalent, formation of, 48-49
  electrovalent, formation of, 47-48
  hydrogen, in water, 93
  polar, 50
Borax, 71
Boric acid, 71
Boron, 71
  structure of, 31, 33, 34

Bowman's capsule, 356
Boyle-Charles gas law, 60
Boyle's law, 60
Breeding of fissionable material, 146
Bromine, 71
  atomic composition of, 72
  properties of, 72
Brownian movement, 56-57
Buffers, 123-125
  blood, 346-347, 364
  definition of, 123
  uses of, 125
Burning, 81
Butanoic acid; see Butyric acid
Butter, 228
Butyric acid, 178

C

Calciferol, 325-327
Calcium, 63-64
  dietary need for, 309-310
  metabolism of, 299-300
Calcium arsenate, 71
Calcium carbonate, 64, 156
Calcium fluoride, 71
Calcium hydroxide, 64
Calcium ions, functions of, 299-300
Calcium paracaseinate, 275
Calcium phosphate, 64
Calcium salts, 64
Calcium sulfate, 64
Calomel, 64
Caloric value of common foods, 317
Calorie, 8
Calorimeters, 316
Calorimetry, 315-318
Cane sugar; see Sucrose
Capsule
  Bowman's, 356
  glomerular, 356
Carbocyclic compounds, 199
Carbocyclic organic compounds, 159
Carbohydrases, 247-248
Carbohydrates, 201-208
  Benedict's test for, 204
  chemical properties of, 204-205
  classification of, 202-203
  definition of, 201
  digestion of, in intestinal tract, 277-278
  Fehling's test of, 204
  fermentation of, 205
  hydrolysis of, 205
  metabolism of; see Metabolism of carbohydrates
  in nutrition, 307-308
  occurrence of, 201
  origin of, 201
  oxidation of, 205
  physical properties of, 204
  reducing substances of, 204
  stability of, 204
  structure of, 203-204

Carbolic acid, 191, 192
Carbon, 151-153
  compounds of, inorganic, 153-156
  fundamental principles about, 157-159
  and hydrogen in methane, 48
  importance of, 151
  occurrence of, 151
  properties of, 151
  radioactive, uses of, 148-149
  structure of, 33, 34
Carbon 14, uses of, 148-149
Carbon cycle in nature, 201-202
Carbon dioxide, 154-156
  transportation of, by blood, 350-352
Carbon dioxide fire extinguisher, 154-155
Carbon monoxide, 153-154
Carbon tetrachloride, 165
Carbonated water, 155
Carbonic acid, 155
Carbonic anhydrase, 248
Carboxyl, 176
Carboxypeptidase, role of, in digestion, 277
Carnauba wax, 232
Carotene, dietary need for, 325
Catabolism, 282
  of proteins, 296-298
Catalases, 248
Catalyst
  definition of, 15
  enzymes as, 246
  negative, 15
  positive, 15
Cathode, 21
Cathode ray, 21-22
  bending of, 21
Cathode-ray tube, 21-22
Cations, 106-107
  and anions in extracellular fluid, balance between, 361
Caustic, 114; see also Bases
Cells; see Blood cells
Cellulose, 207
Cellulose acetate, 181
Cellulose nitrate, 181
Celsius temperature scale, 8
Centigrade temperature scale, 8
Cerebrosides, 222
Charcoal
  adsorption by, 152-153
  animal, 152
  wood, 152
Charles' law, 60
Cheilosis, 335-336
Chemical bonds; see Bonds
Chemical change, 12-19; see also Chemical reactions
  definition of, 12
Chemical energy, 145
Chemical reactions, 10, 38-43
  forward, 14
  reverse, 14
  reversible, 14-15
  types of, 13-14

Chemical reactivity, 38-43
Chemical symbol, 9
Chemical theory, relation of noble gases to, 39-41
Chemistry
  blood, application of, to diagnosis of disease, 347-349
  divisions of, 2
  history of, 2
  importance of, 2
  inorganic, 2
  organic, 156-161
    definition of, 2
    fundamental principles of, 157-159
  as science, 1-2
Chloride, metabolism of, 301-302
Chloride ions, functions of, 301-302
Chloride shift, 351
Chlorine
  atomic composition of, 72
  properties of, 72
  uses of, 73
Chlorobenzene, 191
Chloroethane; see Ethyl chloride
Chloroform, 165
Cholecystokinin, 269, 280
Cholesterol, 229-230
Choline, 230-231, 337-338
  deficiency of, 338
  need for, 294
Cholinesterase, 231
Chromoproteins, 220
Chyle in metabolism of fats, 290
Chylomicrons in metabolism of fats, 290
Chyme, 276
Chymotrypsin in digestion, 277
Chymotrypsinogen, 277
Citric acid, 178-179
Citric acid cycle in metabolism of carbohydrates, 285-287
  occurrence of, 287
Classification
  of carbohydrates, 202-203
  of elements, periodic, 36-38
  of enzymes, 247-248
  of lipids, 222-223
  of organic compounds, 159-160
  of proteins, 220, 221
Clotting, blood, 344-345
Coagulated proteins, 220
Coagulation
  blood, 344-345
  of proteins, 219
Coal, formation of, 153
Cobalt, dietary need for, 312
Codon, genetic, 243-244
Coefficient, phenol, 192
Coenzyme A, 249-250
Coenzymes, 249-250
Colloid goiter, 254
Colloids, 127-130
  definition of, 127
  hydrophilic, 129
  lyophilic, 129
  lyophobic, 129

Colloids—cont'd
  protective, 129
  solvated, 128, 129
  surface area of, 128-129
Colloidal solution, 127
  phases of, 129
Color tests of proteins, 219, 221
Combination, 13
Combining masses, 17-18
Combustion, 81
  spontaneous, 82
Complex alcohols, 170
Composition, definite, law of, 16-17
Compound lipids, 222
Compounds
  binary, naming of, 50-52
    Stock system of, 52
  definition of, 9-10
  halogen, 165-166
  inorganic carbon, 153-156
  organic; see Organic compounds
Concentrated bases, 115
Concentrated solution, 99
Concentration
  hydrogen ion, 111
  of solutions, 99-103
Condensation point of gas, 57
Conjugated proteins, 220
Conservation of mass and energy, laws of, 15-16
Constant
  equilibrium, 88-90
  ionization, 108
Continuous phase of colloidal solution, 129
Convulsions from pyridoxine deficiency, 336
Coordination number, 55
Copper, dietary need for, 310
Copper arsenite, 71
Copper sulfate, crystallized, 96
Corpuscle, malpighian, 356
Cortex, adrenal, secretions of, 261-262
Corticosterone, 262
Counter, Geiger-Müller, 148
Covalence of carbon, 157
Covalent bonds, formation of, 48-49
Crenation of blood cells, 134, 135
Cretinism, 253-254
Critical temperatures, 59
  calculation involving, 109-110
Crystalline solid matter, 54
Crystallization, water of, 95-96
Crystallography, 54
Crystalloids, 127-130
Crystals, 55
Curie, 149
Cyanocobalamin, 341-342
Cyclohexane, 188, 199
Cyclopentanoperhydrophenanthrene, 198
Cyclopropane, 199
Cyclotron, 140
Cystine, 210
Cytidine, 236
Cytidine phosphate; see Cytidylic acid

Cytidylic acid, 236-237
Cytosine, 234

D

Dakin's solution, 73
Dalton's atomic theory, 20
Dating, radiocarbon, 141-142
Deamination of amino acids, 296-297
Decarboxylase, 248
Decomposition, 13-14
  of atoms, 26
  double, 14
Definite composition, law of, 16-17
Deflection of alpha particles, 23
Degradation of amino acids, 296
7-Dehydrocholesterol, 327
Dehydrogenases, 248
Democritus' theory of atom, 20
Denaturation of proteins, 214
Denatured alcohol, 169-170
Deoxyribonucleic acid
  molecule of, 237-239, 240-241
    replication of, 241
  sections of chain of, 240
Deoxyribose, 235, 236
Derived lipids, 222
Derived proteins, 220
Dermatitis
  from biotin deficiency, 339
  from niacin deficiency, 334
  from pantothenic acid deficiency, 338-339
  from pyridoxine deficiency, 336
  from riboflavin deficiency, 335
Detergents, 227
Dextrans, 207
Dextrins, 206
Dextrose; see Glucose
Diabetes, renal, 289, 359
Diabetes insipidus, 261, 356
Diabetes mellitus, 256, 289
Diabetic acidosis, 293
Diabetogenic factor, 259-260
Diagnex test of gastric function, 276
Diagnosis of disease
  application of blood chemistry to, 347-349
  use of radioactive isotopes in, 149
Dialysis, 130-131
Dialyzing apparatus, 131
Diamond, 151-152
Diarrhea causing acid-base imbalance, 366
Diastole, 167
Diet, balanced, 309-314
  distribution of components in, 309
Diffusion, 130
  of gases, 59
  of liquids, 57
  of solids, 56
Digestion, 272-280
  carbohydrate, in intestinal tract, 277-278
  definition of, 272
  emulsification of fats during, 278
  fat, in intestinal tract, 278

Digestion—cont'd
  gastric, 275-276
  in intestinal tract, 277-278
  protein, in intestinal tract, 277
  salivary, 274-275
  summary of, 279
Digestive system, diagram of, 274
Digestive tract, supplementary action of enzymes in, 278
Dihydrogen carbonate, 120
Dihydrogen salt, 119
2,3-Dihydroxybutanedioic acid; see Tartaric acid
Dilute bases, 115
Dilute solution, 99
Dipeptidase in digestion, 277
Diphenylamine, 195
Direct calorimeter, 316
Disaccharides, 204, 206
Disease
  Addison's, 261, 356
  diagnosis of, application of blood chemistry to, 347-349
  diagnosis and treatment of, uses of radioactive isotopes in, 149
Disodium hydrogen phosphate, 119
Disodium phosphate, 70, 119
Dispersed phase of colloidal solution, 129
Displacement, double, 14
Dissociation, 106
Distillation of water, 94
Disulfide linkage of polypeptide chains, 214
Divinyl ether, 182
Double decomposition, 14
Double displacement, 14
Double helix of deoxyribonucleic acid, 237-239
Drugs, antiovulatory, 267
Duodenum, diagram of, 256
Dwarfism, 257

E

Earth metals, alkaline, 63-64
Einstein equation, 147
Electricity, study of, 21
Electrodes, 91
Electrolysis
  diagram showing, 107
  of water, 91
    apparatus used for, 92
Electrolyte balance, 360-361
Electrolyte composition of blood plasma, 360, 361
Electrolytes, 106
  ions of, 106
Electrophoresis, 130
Electrostatic field, radioactive particles in, 139
Electrovalent bonds, formation of, 47-48
Electromotive series, 65, 66
Electron acceptors, 43
Electron donors, 43
Electron structure of noble gases, 39
Electronegativity, 66
Electrons, 21-22
  energy of, quantization of, 27
  energy levels of, 30

386 *Index*

Electrons—cont'd
  excited states of, 27
  ground state of, 27
  positive, 141
  valence, 35
Elements
  abundance of, in earth's crust, 79
  definition of, 9
  discovery of, based on periodic table, 38
  early concepts of, 8-9
  metallic, 43
  periodic classification of, 36-38
  serial number of, 38
  simplest structure of, 31
  valence of, 44-45, 51
Embden-Meyerhof pathway
  diagram of, 283
  in metabolism of carbohydrates, 284-285
Emulsification of fats, 226
  during digestion, 278
Emulsifying agent, 129
Emulsion, 129
Enamel, mottled, 312
End point of titration, 116
Endocrine organs, 251; *see also* Hormones
Endothermic reaction, 39
Energy
  atomic, 145
  from atomic fusion, calculation of, 147-148
  chemical, 145
  conservation of, laws of, 15-16
  definition of, 5
  of electron, quantization of, 27
  kinetic, 5
  matter and, measurement of, 5-8
  potential, 5
  storage and release of, 287-289
Energy change, 39
Energy levels, 26-27
  of electron, 30
Energy metabolism, 315-320
  factors affecting rate of, 318
English system of measurement, 5-6
Enterocrinin, 269
Enterogastrone, 269
Enterokinase, 277
Enzymes, 246-250
  as catalysts, 246
  classification of, 247-248
  definition of, 246
  hydrolytic, 247
  importance of, 249
  mechanism of action of, 248-249
  nomenclature for, 247
  optimal temperature for, 246-247
  oxidation-reduction, 248
  properties of, 246-247
  substrate of, 246
  supplementary action of, in digestive tract, 278
Epinephrine
  effects of, 263
  in metabolism of carbohydrates, 284

Epsom salts, 69
Equation, 13; *see also* Formula
  definition of, 13
Einstein, 147
Equilibrium, 14-15
Equilibrium constant, 88-90
Equivalence point of titration, 116
Erythrocytes, 344
  crenated, 134, 135
Essential amino acids, 210
  requirements of
    by man, 307
    in rats, 305-306
Essential fatty acids, 224
  in nutrition, 308
Esterases, 247
  simple, 247
Esters, 179-181
  aromatic, 194-195
  definition of, 179
  important for their odors, 180-181
Estradiol, 264
Estrane, 198
Estriol, 264
Estrone, 264
Estrus, 264
Ethane, 163
Ethanedioic acid; *see* Oxalic acid
Ethanoic acid; *see* Formic acid
Ethanol; *see* Ethyl alcohol
Ethereal salts; *see* Esters
Ethers, 181-182
Ethoxyethane; *see* Ethyl ether
Ethyl acetate, 180
Ethyl alcohol, 167, 169
Ethyl butanoate; *see* Ethyl butyrate
Ethyl butyrate, 180
Ethyl chloride, 165-166
Ethyl ethanoate, 180
Ethyl ether, 181, 182
Ethyl formate, 180
Ethyl methanoate; *see* Ethyl formate
Ethylene, 164
Ethylene glycol, 170
17α-Ethyl-19-nortestosterone, 267
Evaporation, 57
Exchange, ion, 97
Excited states of electron, 27
Excretion, water, 363
Exothermic reaction, 39
Extracellular fluid, balance between anions and cations in, 361
Extracellular fluid compartment, fluid volume of, measurement of, 364
Extracellular water, 362
Eye, disturbances of, from riboflavin deficiency, 335-336

F

Fahrenheit temperature scale, 8
Families, natural, on periodic table of elements, 38
Fats, 181, 223-228
  acrolein test for, 225

Fats—cont'd
  chemical behavior of, 225-226
  definition of, 223
  digestion of, in intestinal tract, 278
  emulsification of, 226
    during digestion, 278
  energy value of, 228
  functions of, 228
  as heat insulation, 228
  hydrolysis of, 225-226
  iodine number of, 227-228
  metabolism of, 290-295
  neutral, 222
  in nutrition, 308
  occurrence of, 223-224
  oxidation of, 225
  physical properties of, 225
  as protection against mechanical injury, 228
  as reserve food supply, 228
  saponification of, 225-226
  saponification number of, 227
  as vitamin carriers, 228
Fat-soluble vitamins, 323-329
Fatty acids
  beta-oxidation of, 290-292
  essential, 224
    in nutrition, 308
  further oxidation of, 292
  unsaturated, 224
Fatty liver, 293-294, 338
Fehling's test
  for aldehydes, 172
  of carbohydrates, 204
Female sex hormones 264-267
Fermentation
  alcoholic, 205
  of carbohydrates, 205
Ferric iron, 300
Ferritin, functions of, 301
Ferrous iron, functions of, 300-301
Fibrous protein molecule, 214, 216
Field, electrostatic, radioactive particles in, 139
Film, x-ray, 144
Fire extinguisher, carbon dioxide, 154-155
Fission, atomic, 144-146
Flask, volumetric, 100
Fluid compartments, 361-362
  extracellular, fluid volume of, measurement of, 364
Fluid volume of extracellular compartment, measurement of, 364
Fluoride for prevention of caries, 312
Fluorine, 71
  atomic composition of, 72
  dietary need for, 311-312
  properties of, 72
  structure of, 34, 35
  uses of, 73
  and xenon, reaction of, 39-40
Fluorocarbons, 166
Fluoroscopy, 144
Fluorosis, 312
Fluorspar, 71

Folic acid, 340-341
  deficiency of, 340
Folinic acid, 341
Follicle-stimulating hormone, 258
Follicular hyperkeratosis, 324
Follicular phase of menstrual cycle, 266
Food and energy metabolism, 318
Force, 4
Formaldehyde, 172-173
Formalin, 173
Formic acid, 176, 177-178
Formula
  definition of, 10
  valence in writing, 46
Formula mass, 12
Forward chemical reactions, 14
  speed of, 89
Fraction, mole, 102-103
Freezing point
  of solid, 56
  of water, 94
Fructose, 206
  structure of, 203
Furan, 198
Fusion
  atomic, 146-148
    energy from, calculation of, 147-148
    loss of mass from, 147
  heat of, 56

G

Galactose, 206
  structure of, 203
Gallbladder, storage of bile in, 280
Gamma rays, 138
Gas laws, 60
Gases, 59-62
  condensation point of, 57
  diffusion of, 59
  inert; see Noble gases
  noble
    electron structure of, 39
    relation to chemical theory, 39-41
  pressure of, 59
Gastric analysis, 276
Gastric digestion, 275-276
Gastric juice, 272, 275
  summary of role of, in digestion, 279
Gastrin, 268-269, 273
Gastrointestinal hormones, 268-269
Geiger-Müller counter, 148
Geiger-Müller tube, 148
Gel, 129
Gelatin, phases of, 129
Genes, 240
Genetic code, 239, 242-244
Genetic codons, 243-244
Genetic information
  storage of, 240-241
  transmission of, 241-242
Genetic mutation, 244
Genetics, 240-244

388 *Index*

Giantism, 257
Glacial acetic acid, 178
Glands
  adrenal, hormones of, 261-263
  parathyroid, hormones of, 255
  pituitary; *see* Pituitary gland
  salivary, diagram of, 273
  thyroid, hormones of, 251-254
Glandular secretions and energy metabolism, 318
Glauber's salt, 69
Globular protein molecule, 214, 215
Globulins, 220
Glomerulus, 355
  capsule of, 356
Glucocorticoids, 262
Gluconeogenesis, 262, 308
Glucose, 205-206
  as blood sugar, 282-284
  structure of, 203
  in urine, tests for, 204
α-Glucose, 203, 204
β-Glucose, 203, 204
Glucosuria, 359
  alimentary, 290
Glutamic acid, 211
Glutelins, 220
Glycerin, 170
Glycerol, 170
  metabolism of, 290
Glyceryl trinitrate, 170, 180
Glycine, 210
Glycogen, 207
Glycol, 170
Glycolipids, 222, 231
Glycolysis
  diagram of, 283
  in metabolism of carbohydrates, 284-285
Glycolytic sequence in metabolism of carbohydrates, occurrence of, 287
Glycoproteins, 220
Glycosuria, causes of, 289-290
Goiter, 254
Gonadotropic hormones, 258-259
Grain alcohol; *see* Ethyl alcohol
Gram-molecule, 61
Graphite, 152
Grass tetany, 310
Graying of hair from pantothenic acid deficiency, 338, 339
Ground state of electron, 27
Growth hormone, 257
Guanine, 235
Gypsum, 64, 96

H

Hair, graying of, from pantothenic acid deficiency, 338, 339
Half-cell reaction, 86
Half-life of radioactive materials, 140
Half-reaction, 86
Halides, 165-166

Halogens, 71-73, 165-166
  aromatic, 190-191
  atomic composition of, 72-73
  properties and periodic relationships of, 71-72
Hard water, 96-97
Heat
  of fusion, 56
  measurement of, 8
  of vaporization, 58
Helium, 31, 32
Hemolysis of blood cells, 134
Hemosiderin, 301
Hemosiderosis, 301
Heparin, 345-346
Heterocyclic compounds, 198-199
Heterocyclic organic compounds, 159
Hexoses, 202
High-energy phosphate bond, 288
High-energy phosphate compounds, 288
Histidine, 211
Histones, 220
Homogenization, 129
Homologous series of hydrocarbons, 162
Homologues, 162
Hopkins-Cole test for proteins, 221
Hormonal control of water movement, 363
Hormones, 251-269
  of adrenal glands, 261-263
  adrenocorticotropic, 259
  follicle-stimulating, 258
  gastrointestinal, 268-269
  gonadotropic, 258-259
  growth, 257
  luteinizing, 258-259, 266
  of pancreas, 255-257
  of parathyroid glands, 255
  of pituitary gland, 257-261
  sex
    female, 264-267
    male, 267-268
  somatotropic, 257
  of thymus, 263
  of thyroid gland, 251-254
  thyrotropic, 259
Hydrates, 95-96
Hydrocarbons, 162-165
  aromatic, 186-190
  International Union of Pure and Applied Chemistry nomenclature for, 186-187
  homologous series of, 162
  polynuclear, 197-198
  saturated, 162-164
  unsaturated, 164-165
Hydrochloric acid, 111, 113
  in digestion, 275
  and zinc, hydrogen production from, 75, 76
Hydrofluoric acid, 73
Hydrogen, 75-78
  and carbon in methane, 48
  chemical changes involving, 41
  chemical properties of, 77-78

Hydrogen—cont'd
  and nitrogen in ammonia, 48
  occurrence of, 75
  and oxygen in water, 48-49
  physical properties of, 77
  preparation of, 75-76, 77
  structure of, 31, 32
  uses of, 78
Hydrogen bond
  in polypeptide chains, 214
  in water, 93
Hydrogen carbonate, 120
Hydrogen ion concentration, 111
Hydrogen peroxide
  enzymes in decomposition of, 249
  preparation of, 97
  properties of, 97-98
  uses of, 98
Hydrogen salt, 119
Hydrogen sulfide, 69-70
Hydrolases, 247
Hydrolysis
  of carbohydrates, 205
  of fats, 225-226
  of nucleic acids, 237
  of proteins, 219
  of water, 95
Hydrolytic enzymes, 247
Hydronium ion, 110, 111
Hydrophilic colloid, 129
Hydrostatic pressure and water movement, 362
Hydroxides, 114
1-Hydroxybutanedioic acid; see Malic acid
β-Hydroxybutyric acid, 292
Hydroxyl ion, 114
2-Hydroxy-1,2,3-propanetricarboxylic acid; see Citric acid
2-Hydroxypropanoic acid; see Lactic acid
Hyperacidity, 113
Hyperglycemia, 255
Hyperkeratosis, follicular, 324
Hyperthyroidism, 254
Hypertonic solution, 134
Hypoacidity, 113
Hypoglycemia, 257
Hypothalamus, appetite regulation by, 295
Hypotonic solutions, 134

I

Imbibition, 128
Imidazole, 198
Indican in urine, 359-360
Indicators, 111-112
Indirect calorimeters, 316
Indole, 198
Inert gases; see Nobel gases
Inertia, 4
Inhibitors, 250
Inorganic acids, naming of, 113-114
Inorganic carbon compounds, 153-156
Inorganic chemistry, 2

Inorganic salts, metabolism of, 298-302
Insoluble salts, 120-121
Insulin, 214, 255-257
  in metabolism of carbohydrates, 284
Intermediary metabolism of carbohydrates 284-287
International Union of Pure and Applied Chemistry system of nomenclature
  for hydrocarbons, 186-187
  for organic compounds, 160
Intestinal juice, 273
  summary of role of, in digestion, 279
Intestinal tract, digestion in
  of carbohydrates, 277-278
  of fats, 278
  of proteins, 277
Intracellular water, 362
Inulin, 206-207
Inulin clearance test, 207
Invert sugar, 206
Iodine, 71
  atomic composition of, 72
  dietary need for, 311
  properties of, 72
  protein-bound, 252
  uses of, 73
Iodine 131, uses of, 149
Iodine number of fats, 227-228
Iodoform, 165
Iodothyronines, 251
Ion exchange, 97
Ion-exchange resin, 97
Ionic crystals, 55
Ionization, 106-108
  definition of, 106
  diagram showing, 107
  nonaqueous, 108
Ionization constant, 108
Ions, 41-43
  definition of, 41
  difference of, from atoms, 41-42
  of electrolytes, 106
  protein, 216-218
Iron, 64
  dietary need for, 310
  ferric, 300
  ferrous, 300-301
  metabolism of, 300-301
Iron carbonate, 64
Iron oxide, 64
Isoamyl acetate, 180
Isoamyl butyrate, 180
Isoamyl nitrite, 179-180
Isoelectric point, 129-130
  of proteins, 214, 216-218
Isomorphs, 54
Isosmotic solutions, 134
Isotonic solutions, 134
Isotopes, 25-26
  importance of, 26
  of oxygen, 80

Isotopes—cont'd
  radioactive
    labeling with, 148
    uses of, in biology and medicine, 148-149

J

Juice
  gastric, 272, 275
    summary of role of, in digestion, 279
  intestinal, 273
    summary of role of, in digestion, 279
  pancreatic, 272-273
    summary of role of, in digestion, 279

K

Kelvin temperature scale, 8
Keratinization caused by lack of vitamin A, 324
Keto acids, 297
α-Ketoglutaric acid in transamination, 297
Ketone bodies
  metabolism of, 292-293
  in urine, 359
Ketones, 173-174
Ketosis causing acid-base imbalance, 366
Kidney, failure of, acid-base imbalance due to, 366
Kinetic energy, 5
Krebs cycle in metabolism of carbohydrates, 285-287
  occurrence of, 287

L

Labeling with radioactive isotopes, 148
Lactase in digestion, 278
Lactic acid, 178
Lactobacillus casei factor; see Folic acid
Lactose, 204, 206
Lanolin, 231-232
Lard, 228
Law
  Avogadro's, 60
  Boyle's, 60
  Charles', 60
  of conservation of mass and energy, 15-16
  of definite composition, 16-17
  gas, 60
  of multiple proportions, 18-19
  of partial pressures, 62
  periodic, 38
Lead arsenate, 71
Lecithin, 230
  need for, 294
Lecithoproteins, 220
"Leg weakness," 311
Length, measurement of, 5
Lepidolite, 63
Leucine, 210
Leukocytes, 344
Levulose; see Fructose
Life, relation of oxygen to, 82-83
Linkage, peptide, 211-214
Linoleic acid, 224
  need for, 308

Linolenic acid, 224
  need for, 308
Lipases, 247
  in digestion, 275
Lipids, 222-232
  classification of, 222-223
  compound, 222
  derived, 222
  properties of, 222
  simple, 222
Lipotropic agents, need for, 294
Liquefaction, 56
Liquid oxygen, 84
Liquids, 56-59
  boiling point of, 57, 58
  diffusion of, 57
  surface tension of, 57
  viscosity of, 57
Lithium, 63
  chemical changes involving, 41
  structure of, 31, 32
Lithium aluminum hydride, 63
Lithium boron hydride, 63
Lithium salts, 63
Litmus paper, 112
Liver
  fatty, 293-294, 338
  in metabolism of carbohydrates, 284
Liver function test, benzoic acid, 193-194
Logarithms, 369-371
  four-place, 372
Luteal phase of menstrual cycle, 266
Luteinizing hormone, 258-259, 266
Luteotropin, 259, 266
Lyases, 248
Lyophilic colloid, 129
Lyophobic colloid, 129
Lysine, 211

M

Magnesium, 64
  dietary need for, 310
  metabolism of, 300
Magnesium ions, functions of, 300
Magnesium salts, 64
Magnesium sulfate, 69
Male sex hormones, 267-268
Malic acid, 179
Malpighian corpuscle, 356
Maltase in digestion, 278
Maltose, 204, 206
Manganese, dietary need for, 311
Mass
  atomic, 11
  combining, 17-18
  conservation of, laws of, 15-16
  definition of, 4
  formula for, 12
  loss of, from atomic fusion, 147
  molecular, 61
    definition of, 11-12
    determination of, 61

Mass number, atomic, 25
Matter, 4-19
　chemical concepts regarding, 8-12
　definition of, 4
　and energy, measurement of, 5-8
　gaseous state of; see Gases
　liquid; see Liquids
　physical states of, 54-62
　solid; see Solids
Measurement, units of, 6
　of radioactivity, 149-150
Medicine, uses of radioactive isotopes in, 148-149
Medulla, adrenal, secretions of, 263
Melting point of solid, 56
Membranes, permeability of, 132-133
Menadione, 328-329
　deficiency of, 329
Menstrual cycle, 265-267
Mercuric chloride, 64
Mercuric oxide, oxygen preparation from, 78-79
Mercurous chloride, 64
Mercury, 64
　bichloride of, 64
　poisoning from, proteins as antidote for, 219
Messenger ribonucleic acid, 241, 242
Metabolism
　basal, 318-320
　　determination of, 319-320
　of calcium, 299-300
　of carbohydrates, 282-290
　　abnormal, 289-290
　　absorption and distribution in, 282
　　adenosine triphosphate in, 287-289
　　anaerobic, diagram of, 283
　　blood sugar and, 282-284
　　citric acid cycle in, 285-287
　　effects of adrenal cortex secretions on, 262
　　glycolysis of, 284-285
　　glycolytic sequence in, occurrence of, 287
　　intermediary, 284-287
　　liver in, 282-290
　　relation of anterior pituitary gland to, 259-260
　of chloride, 301-302
　definition of, 282
　energy, 315-320
　　factors affecting rate of, 318
　of fats, 290-295
　　absorption and distribution in, 290
　　glycerol, 290
　of inorganic salts, 298-302
　of iron, 300-301
　of ketone bodies, 292-293
　of magnesium, 300
　of phosphorus, 302
　of potassium, 299
　of proteins, 295-298
　　effects of adrenal cortex secretions on, 262
　of sodium, 298
Metallic elements, 43
Metallic solids, 55-56
Metalloids, 43, 71

Metals, 43, 55-56, 63-66
　alkali, 63
　earth, alkaline, 63-64
　heavy, precipitation of proteins by, 218-219
　reaction of organic acids with, 177
Metaproteins, 220
Meta-xylene, 190
Meter, standardization of, 5
Methane, 163
　covalent bonds in, 48
　formation of, 162
　molecule of, 157
　preparation of, 163
Methanoic acid; see Formic acid
Methanol; see Methyl alcohol
Methionine, need for, 294
Methoxyflurane, 182
Methoxymethane; see Methyl ether
Methyl alcohol, 167, 168-169
Methyl amine, preparation of, 183-184
Methyl ether, 181
Methyl propyl ether, 182
Methyl salicylate, 181, 194
γ-Methylbutyl butanoate; see Isoamyl butyrate
γ-Methylbutyl ethanoate; see Isoamyl acetate
γ-Methylbutyl nitrite; see Isoamyl nitrite
Metropryl; see Methyl propyl ether
Metric system, 5-6
Microcurie, 149-150
Milk, 314-315
Millicurie, 149-150
Milliequivalent, 103-104
Milligrams percent, 104
Milliosmole, 133
Millon's test for proteins, 219
Mixture, 9-10
Molal solution, 102
Molar solutions, 100-101
Mole, 61
Mole fraction, 102-103
Molecular crystals, 55
Molecular mass, 61
　definition of, 11-12
　determination of, 61
Molecules
　Avogadro's number of, 61
　definition of, 11
　deoxyribonucleic acid, 237-239, 240-241
　　replication of, 241
　methane, 157
　nucleic acid, 237-239
　polar, of water, 93
　protein
　　large, 218
　　shape of, 214
　ribonucleic acid, 237
Molybdenum, dietary need for, 312
Monohydrogen phosphate, 120
Monosaccharides, 202-203, 205-206
Monosodium phosphate, 119
Mottled enamel, 312
Movement, Brownian, 56-57

Mucin in digestion, 276
Multiple proportions, law of, 18-19
Mutation, genetic, 244
Myxedema, 253-254

N

Naming
  of bases, 115
  of binary compounds, 50-52
    Stock system of, 52
  of inorganic acids, 113-114
  of organic compounds, 160-161
  of salts, 118-120
Naphthalene, 198
Natural families on periodic table of elements, 38
Nature
  carbon cycle in, 201-202
  oxygen cycle in, 83
Negative catalyst, 15
Neon
  atomic structure of, 40
  structure of, 34, 35
Neothyl; see Methyl propyl ether
Nephritic acidosis, 366
Nephron, structure of, 355-356
Neutral fats, 222
Neutral salt, 119
Neutralization, 115-116
  preparation of salts by, 117, 118
  in treatment of wounds produced by acids or bases, 121-122
Neutron, 24
Niacin, 334-335
Niacinamide, 334, 335
Night blindness, 323
Nitrates, 67
Nitric acid, 67
Nitrogen, 66-67
  and hydrogen in ammonia, 48
  structure of, 34
Nitrogen dioxide, 67
Nitrogen equilibrium, proteins and, 304-305
Nitroglycerin, 170
Noble gases
  electron structure of, 39
  relation of, to chemical theory, 39-41
Nonaqueous ionizations, 108
Nonelectrolytes, 106
Nonmetals, 43, 66-73
Norepinephrine, effects of, 263
Norite eluate factor; see Folic acid
Normal solutions, 101-102
19-Norprogesterone, 267
Nuclear reactor, 146
Nucleases, 248
Nucleic acid, 233-239
  components of, 234-236
  hydrolysis of, 237
  molecule of, 237-239
  structure of, 237-239
  sugars in, 235-236

Nucleoproteins, 220, 233-234
Nucleosides, 236
Nucleotides, 236-237
Number
  atomic, 23-24
  Avogadro's, 61
  coordination, 55
  iodine, of fats, 227-228
  mass, atomic, 25
  oxidation, assignment of, 86
  saponification, of fats, 227
  serial, of elements, 38
Nutrition, 304-320
  essential fatty acids in, 308
  fats in, 308
  proteins in, 304-307
Nutritional anemia, 315
Nyctalopia, 323

O

Obesity, 294-295
Obligatory water loss, 363
Octet arrangement, 41
1-Octanol ethanoate; see Octyl acetate
Octyl acetate, 180
Octyl butyrate, 180
Odor of urine, 358
Oils, 181
Oleic acid, 224
Oleomargarine, 228
Olive oil, 228
Orbital, 28, 29
Orbits, 26-27
Organic acids, 176-179
  definition of, 176
  examples of, 176
  preparation of, 176-177
  properties and reactions of, 177
Organic chemistry, 156-161
  definition of, 2
  fundamental principles of, 157-159
Organic compounds, 159
  classification of, 159-160
  International Union of Pure and Applied chemistry system of nomenclature for, 160
  naming of, 160-161
Organs, endocrine, 251; see also Hormones
Ortho-hydroxybenzoic acid; see Salicylic acid
Ortho-sulfobenzoic acid imide; see Saccharin
Orthy-xylene, 190
Osmolarity, 133
Osmosis, 131-135
  blood changes due to, 134-135
  importance of, 133
Osmotic pressure, 131-132
  and water movement, 362-363
Oxalic acid, 178
Oxidases, 248
Oxidation, 81, 85-86
  of alcohols, 167-168
  of carbohydrates, 205
  of fats, 225

Oxidation—cont'd
  of fatty acids
    beta, 290-292
    further, 292
  of proteins, 219
  and reduction, 87
Oxidation numbers, assignment of, 86
Oxidation-reduction enzyme, 248
Oxidation-reduction reaction, 87
Oxidative phosphorylation, 288
Oxides, 81-82
Oxonium ion, 168
Oxyacids, 113-114
Oxygen, 78-85
  chemical changes involving, 41
  chemical properties of, 80-81
  in decay, 84
  as disinfectant, 84
  and hydrogen in water, 48-49
  isotopes of, 80
  liquid, 84
  occurrence of, 78
  physical properties of, 80
  preparation of, 78-80
  relation of, to life, 82-83
  structure of, 34
  transportation of, by blood, 349-350
  uses of, 83-84
Oxygen cycle in nature, 83
Oxygen therapy, 83
Oxyhemoglobin, 209, 300
  formation of, 350
Oxytocin, 260-261
Ozone, 84-85

P

Pancreas
  diagram of, 256
  hormones, of 255-257
Pancreatic juice, 272-273
  summary of role of, in digestion, 279
Pancreozymin, 269, 273
Pantothenic acid, 338-339
  deficiency of, 339
Para-aminobenzenesulfonamide; see Sulfanilamide
Paracasein, 275
Paraffins, 163
Paraformaldehyde, 172
Paraldehyde, 173
Parathyroid extract, 255
Parathyroid glands, hormones from, 255
Para-xylene, 190
Paris green, 71
Partial pressure, law of, 62
Particles
  alpha, 22
    deflection of, 23
  radioactive, in electrostatic field, 139
  subatomic, 20-24
Pasteurization of milk, 315

Pectic acid, 207
Pectin, 207-208
Pellagra, 334-335
Penetrating power of alpha, beta, and gamma rays, 138
1-Pentanol butanoate, 180
1-Pentanol ethanoate; see Amyl acetate
Penthrane; see Methoxyflurane
Pepsin in digestion, 275
Pepsinogen in digestion, 275
Peptidases, 248
Peptide linkage, 211-214
Peptides, 220
Paptones, 220
Percent solutions, 100
Periodic classification of elements, 36-38
Periodic law, 38
Peristalsis, 276
Permeability of membranes, 132-133
Pernicious anemia, 342
Perosis, 311
Peroxidases, 248
Petalite, 63
pH, 122-123
  blood, and respiration, 350
  definition of, 122
  of urine, 358
pH system, diagrammatic representation of, 124
Phase, 59
Phase diagram for water, 58, 59
Phenacetin, 196
Phenanthrene, 198
Phenol coefficient, 192
Phenols, 191-192
Phenomenon, Tyndall, 128
Phenyl radical, 187
Phenyl salicylate, 181, 195
Phenylacetic acid, 193
Phenylpropionic acid, 193
Phosphatases, 247
Phosphate bond, high-energy, 288
Phosphate compounds, high-energy, 288
Phosphates, 70
  as blood buffers, 347
Phosphatides; see Phospholipids
Phospholipids, 222, 230-231
Phosphoproteins, 220
Phosphoric acid, 70
Phosphorus, 70
  dietary need for, 310
  metabolism of, 302
Phosphorus 32, uses of, 149
Phosphorus pentoxide, 70
Phosphorylation, oxidative, 288
Photosynthesis, 83, 201
Physical change, 12-19
  definition of, 12
Physical states of matter, 54-62
Physics, 1
Physiological salt solution, 133-134
Pile, atomic, 146

Pituitary gland
  anterior
    in carbohydrate metabolism, 259-260
    secretions of, 257-260
  hormones of, 257-261
  posterior, secretions of, 260-261
Plasma, 344
  blood, electrolyte composition of, 360, 361
Plasma bicarbonate/carbonic acid system, 364-365
Plasma volume of water, measurement of, 363-364
Plaster of Paris, 96
Platelets, 344
Poisoning
  alcohol, 169
  carbon monoxide, 154
  mercury, proteins as antidote for, 219
  "ptomaine," 185
Polar bonds, 50
Polar water molecule, 93
Poliovirus molecules, 215
Polymerization of formaldehyde, 172
Polynuclear hydrocarbons, 197-198
Polypeptide chains
  of amino acid, 213
  disulfide linkage of, 214
  hydrogen bond in, 214
Polypeptides, 214
Polyphagia, 295
Polysaccharidases, 248
Polysaccharides, 204, 206-208
Polyuridylic acid, 243
Positive catalyst, 15
Positive electrons, 141
Positrons, 141
Posterior pituitary gland, secretions of, 260-261
Potassium, 63
  dietary need for, 309
  metabolism of, 299
Potassium bisulfate, 119
Potassium chlorate, oxygen preparation from, 79-80
Potassium chloride, 118
Potassium hydrogen phosphate, 119
Potassium iodide, uses of, 73
Potassium ions, functions of, 299
Potassium salts, 63
Potential energy, 5
Precipitation of proteins by heavy metals, 218-219
Pregnancy tests, 267
Pressure
  effect of
    on solubility, 99
    on solution, 108
  gas, 59
  hydrostatic, and water movement, 362
  measurement of, 6, 8
  osmotic, 131-132
    and water movement, 362-363
  partial, law of, 62
  and respiration, 350
  and temperature, standard conditions of, 60
  vapor, 57
Primary alcohols, 166
  oxidation of, 167-168

Primary amine, 183
Procarboxypeptidase, 277
Product, solubility, 120, 121
Progesterone, 264, 265
Progestins, synthetic, 267
Prolamines, 220
Prontosil, 197
Propane, 163
Propanol; see Propyl alcohol
Proportions, multiple, law of, 18-19
Propyl alcohol, 167
Protamines, 220
Proteans, 220
Proteases, 248
Protective colloid, 129
Protein, 209-221
  anabolism of, 295-296
  as antidote for mercury poisoning, 219
  behavior of, in solution, 214, 216-218
  biological value of, 305
  biuret test of, 219
  as blood buffer, 346
  and carbon dioxide transport, 351
  catabolism of, 296-298
  chemical nature of, 209-211
  classification of, 220, 221
  coagulated, 220
  coagulation of, 219
  color tests of, 219, 221
  conjugated, 220
  definition of, 211
  denaturation of, 214
  derived, 220
  digestion of, in intestinal tract, 277
  as energy producers, 221
  functions of, 221
  Hopkins-Cole test for, 221
  human requirement for, 307
  hydrolysis of, 219
  ions of, 216-218
  isoelectric points of, 214, 216-218
  metabolism of, 295-298
    effects of adrenal cortex secretions on, 262
  Millon's test for, 219
  molecules of
    large, 218
    shape of, 214
  and nitrogen equilibrium, 304-305
  in nutrition, 304-307
  occurrence of, 209
  origin of, 209
  oxidation of, 219
  precipitation of, by heavy metals, 218-219
  properties of, 218-221
  reaction of, 218
  simple, 220
  solubility of, 218
  structure of, 211-214
  as tissue builders, 221
  in urine, 358-359
  xanthoproteic test for, 219, 221
Proteinases, 248
Protein-bound iodine, 252

Proteinuria, 359
Proteoses, 220
Proton, 22-23
Pteroylglutamic acid, 341
Ptomaine poisoning, 185
Ptyalin, 274-275
Pure water, 94
Purines, 235
Pyran, 198
Pyridine, 198
Pyridoxamine, 337
Pyridoxine, 336-337
Pyridoxol, 337
Pyrimidines, 234
Pyrrole, 198

Q

Quantitative analysis of urine, 357-360
Quantization of energy of electron, 27
Quantum theory, atomic structure according to, 28-35
Quaternary ammonium salt, 183

R

Radical, 43, 162
  of aromatic hydrocarbons, 187
  phenyl, 187
  valence of, 45-46, 61
Radioactive carbon, uses of, 148-149
Radioactive isotopes
  labeling with, 148
  uses of, in biology and medicine, 148-149
Radioactive materials, half-life of, 140
Radioactive particles in electrostatic field, 139
Radioactivity, 137, 138
  artificial, 140-141
  units of measurement of, 149-150
Radiocarbon, 141-142
Radiochemistry, 137-150
Radiography, 143-144
Radioiodine, uses of, 149
Radiophosphorus, uses of, 149
Radium, 137-142
  uses of, 137-138
Rays
  alpha, 22
    penetrating power of, 138
  beta, 22
    penetrating power of, 138
  cathode, 21-22
    bending of, 21
  gamma, 138
  roentgen; see X-rays
Reactions
  of alcohols, 167-168
  bombardment, in producing artificial radioactivity, 141
  chemical; see Chemical reactions
  endothermic, 39
  exothermic, 39
  forward, speed of, 89
  fusion, 146-148
  half-cell, 86
  oxidation-reduction, 87

Reactions—cont'd
  redox, 87
    balancing, 87-88
  reverse, speed of, 89
Reactivity, chemical, 38-43
Reactor, nuclear, 146
Red cells, 344
  crenated, 134, 135
Red phosphorus, 70
Redox reaction, 87
  balancing, 87-88
Reducing agents, 78
  aldehydes as, 171-172
Reduction, 86
  and oxidation, 87
Renal diabetes, 289, 359
Rennin in digestion, 275-276
Replacement, single, 14
Replication of deoxyribonucleic acid molecule, 241
Research, use of radioactive isotopes in, 148-149
Resin, ion-exchange, 97
Respiration, 349-352
  abnormal, acid-base imbalance due to, 366
  blood pH and, 350
  pressure and, 350
  temperature and, 350
Respiratory acidosis, 366
Retinol, 323-325
  deficiency of, 323-324
Reverse chemical reactions, 14
  speed of, 89
Reversible chemical reactions, 14-15
Riboflavin, 35-36
Ribonucleic acid
  messenger, 241, 242
  molecule of, 237
  soluble, 241, 242
  transfer, 241-242
Ribose, 235
Rickets, 325, 327
  symptoms of, 326
Ringer's solution, 133-134
Roentgen, 150
Roentgen rays; see X-rays

S

Saccharidases, 247-248
Saccharin, 196
Saccharose; see Sucrose
Salicylic acid, 194
Saliva, 272
  in digestion, 274-275
  summary of role of, 279
Salivary amylase, 274-275
Salivary digestion, 274-275
Salivary glands, diagram of, 273
Salivation, stimulation of, 273
Salt solution, physiological, 133-134
Salts, 117-121
  acid, 119, 120
  balance of, effects of adrenal cortex secretions on, 261-262
  basic, 119

Salts—cont'd
  dihydrogen, 119
  ethereal; see Esters
  hydrogen, 119
  inorganic, metabolism of, 298-302
  insoluble, 120-121
  naming of, 118-120
  neutral, 119
  preparation of, 117-118
  quaternary ammonium, 183
Saponification of fats, 225-226
Saponification number of fats, 227
Saturated carbon compound, 158
Saturated hydrocarbons, 162-164
Saturated solutions, 99
Scales, temperature, 7, 8
Schrödinger's theory of atomic structure, 28
Science
  chemistry as, 1-2
  definition of, 1
Scurvy, 330-331
Secondary alcohols, 166
  oxidation of, 168
Secondary amine, 183
Secretin, 269, 273
Secretions
  of adrenal cortex, 261-262
  of adrenal medulla, 263
  of anterior pituitary gland, 257-260
  glandular, and energy metabolism, 318
  of posterior pituitary gland, 260-261
  of urine, 355-357
    mechanism of, 356-357
Sediment, urine, 358
Selenium, dietary need for, 312
Serial number of elements, 38
Serum glutamic oxalacetic transaminase, 297, 298
Serum glutamic pyruvic transaminase, 297, 298
Sex hormones
  female, 264-267
  male, 267-268
Side chains of benzene, 189
Simple esterases, 247
Simple goiter, 254
Simple lipids, 222
Simple proteins, 220
Single replacement, 14
Single substitution, 14
Size and energy metabolism, 318
Soaps, 225, 226
Sodium, 63
  chemical changes involving, 41
  dietary need for, 309
  metabolism, 298
Sodium acetate, 118
Sodium bicarbonate, 119, 156
Sodium bromide, uses of, 73
Sodium carbonate, 155-156
Sodium chloride, 71
  uses of, 73
Sodium chlorite, 118
Sodium cyclamate, 199

Sodium dihydrogen phosphate, 119
Sodium hydrogen carbonate, 119
Sodium iodate, 71
Sodium iodide, 71
Sodium ions, functions of, 298
Sodium nitrate, 66
Sodium perchlorate, 118
Sodium salts, 63
Sodium sulfate, 69, 118
Sodium sulfite, 118
Soft water, 96-97
Sol, 129
Solids
  amorphous, 54
  crystalline, 54
  diffusion of, 56
  freezing point of, 56
  melting point of, 56
  metallic, 55-56
Solubility, 98-99
Solubility product, 120, 121
Soluble ribonucleic acid, 241-242
Solute, 98
  nature of, 99
Solutions, 98-103
  behavior of proteins in, 214, 216-218
  colloidal, 127
    phases of, 129
  concentrated, 99
  concentration of, 99-103
  definition of, 98
  dilute, 99
  effect of, on solution, 108-110
  effect of solute on, 108-110
  hypertonic, 134
  hypotonic, 134
  isosmotic, 134
  isotonic, 134
  molal, 102
  molar, 100-101
  normal, 101-102
  percent, 100
  phase diagram of, 109
  pressure and, 108
  Ringer's, 133-134
  salt, physiological, 133-134
  saturated, 99
  supersaturated, 99
  temperature and, 108-109
Solvated colloidal particle, 128, 129
Solvent, 98
  nature of, 99
  phase diagram for, 109
Solvent power of water, 93-94
Somatotropic hormone, 257
Sorbitol, 170
Sörensen scale of pH, 122, 123
Souring of milk, 315
Sperm oil, 232
Spermaceti, 232
Spontaneous combustion, 82
Standard conditions of temperature and pressure, 60

Starch, 206
  animal; see Glycogen
Steapsin in digestion, 278
Sterility, prevention of, by vitamin E, 327-328
Steroids, 230
Sterols, 229-230
Stock system of naming of binary compounds, 52
Stomach
  digestion in, 275-276
  function of, clinical testing of, 276
Storage
  of genetic information, 240-241
  and release of energy, 287-289
Strong acids, 112-113
Strong bases, 115
Structure
  atomic; see Atomic structure
  electron, of noble gases, 39
Subatomic particles, 20-24
Sublimation, 56
Substitution, single, 14
Substrate, enzyme, 246
Succus entericus, 273
Sucrase in digestion, 278
Sucrose, 204, 206
Sugar
  beet; see Sucrose
  blood, 282-284
  cane; see Sucrose
  invert, 206
  in nucleic acid, 235-236
  in urine, 359
Sugar threshold, 289
Sulfadiazine, 197
Sulfamerazine, 197
Sulfamethazine, 197
Sulfas, 196-197
Sulfolipids, 222
Sulfonamides, 196-197
Sulfur, 68-70
Sulfur dioxide, 68-69
Sulfur trioxide, 69
Sulfuric acid, 69
Supersaturated solutions, 99
Surface tension of liquids, 57
Symbol, chemical, 9
Synthesis, 13
  bacterial, of vitamins in alimentary tract, 342
  of water, 92
Synthetic progestins, 267

T

Table of elements, periodic, 36-37
Tallow, 228
Tartaric acid, 179
Temperatures
  critical, 59
    calculations involving, 109-110
  effect of
    on solubility, 99
    on solution, 108-109
  measurement of, 8

Temperatures—cont'd
  and pressure, standard conditions of, 60
  and respiration, 350
Temperature scales, 7, 8
Tension, surface, of liquids, 57
Tertiary alcohol, 166-167
Tertiary amine, 183
Testosterone, 267-268
Tests
  acrolein, for fats, 225
  Benedict's
    for aldehydes, 172
    of carbohydrates, 204
  benzoic acid liver function, 193-194
  biuret, of proteins, 219
  color, of proteins, 219, 221
  Fehling's
    for aldehydes, 172
    of carbohydrates, 204
  for glucose in urine, 204
  Hopkins-Cole, for proteins, 221
  inulin clearance, 207
  Millon's, for proteins, 219
  pregnancy, 267
  Tollens', for aldehydes, 171
  xanthoproteic, for proteins, 219, 221
Tetany, grass, 310
Tetrachloromethane; see Carbon tetrachloride
Theory
  atomic, Dalton's, 20
  chemical, relation of noble gases to, 39-41
  quantum, atomic structure according to, 28-35
Therapy, oxygen, 83
Thiamine, 332-333
  deficiency of, 332
Thiazole, 198
Thiophene, 198
Threonine, 210
Thromboplastin, 345
Thymine, 234
Thymus, hormones of, 263
Thyroglobulin, 252-253
Thyroid gland
  hormones of, 251-254
  testing of, with radioiodine, 149
Thyrotropic hormone, 259
Thyroxin, 251-252
Tissues, body, acid-base imbalance in, 366
Titration, 116-117
Tobacco mosaic virus, 216
α-Tocopherol, 327-328
Tollens' test for aldehydes, 171
Toluene, 190
Tongue, black, 334
Torr, 8
Total body water, measurement of, 363
Transaminases, 297-298
Transamination of amino acids, 297-298
Transfer ribonucleic acid, 241-242
Transferases, 248
Transferrin, 301
Transmission of genetic information, 241-242

Transportation
  of carbon dioxide by blood, 350-352
  of oxygen by blood, 349-350
Treatment of disease, uses of radioactive isotopes in, 149
Tributyrin, 223
Trichloromethane; see Chloroform
Triiodomethane; see Iodoform
Triphenylamine, 195
Tryphilite, 63
Trypsin in digestion, 277
Trypsinogen, 277
Tryptophan, 210
Tube
  cathode-ray, 21-22
  Geiger-Müller, 148
  x-ray, 142, 143
Tubules of nephron, 355, 356
Tyndall phenomenon, 128
Tyrosine, 210

U

Units of measurement, 6
  of radioactivity, 149-150
Unsaturated carbon compounds, 158
Unsaturated fatty acids, 224
Unsaturated hydrocarbons, 164-165
Uracil, 234
Uranium, structure of, 34-35
Uranium 235, fission of, 144-146
Uranium 238, fission of, 144-146
Uranium disintegration series, 139
Urine, 354-360
  acidification of, 357
  amount of, 357-358
  bile in, 360
  blood in, 360
  color of, 358
  composition of, 354
  glucose in, tests for, 204
  indican in, 359-360
  ketone bodies in, 359
  odor of, 358
  pH of, 358
  proteins in, 358-359
  quantitative analysis of, 357-360
  secretion of, 355-357
    mechanism of, 356-357
  sediment of, 358
  specific gravity of, 358
  sugar in, 359
  volume excreted, 357-358

V

Valence, 43-47
  elemental, 44-45
  of elements and radicals, 51
  in formula writing, 46
  mechanisms underlying, 47
  of radicals, 45-46
Valence electrons, 35
Vanillin, 193
Vapor pressure, 57

Vaporization, heat of, 58
Vasopressin, 260
Velocity, 4
Vinegar, 178
Vinethane; see Divinyl ether
Viruses, 233
  tobacco mosaic, 216
Viscosity of liquid, 57
Vitamins, 322-342
  A, 323-325
    deficiency of, 323-324
  B_1, 332-333
  B_2, 335-336
  B_6; see Pyridoxine
  B_{12}, 341-342
    dietary need for, 312
  B_c; see Folic acid
  B complex, 333-342
  bacterial synthesis of, in alimentary tract, 342
  C, 330-332
    deficiency of, 330-331
  D, 325-327
  D_2, 327
  D_3, 327
  dietary need for, 312, 314
  E, 327-328
  fat-soluble, 323-329
  H; see Biotin
  K, 328-329
  K_1, 329
  K_2, 329
  K_3, 329
  M; see Folic acid
  water-soluble, 330-333
Volume
  measurement of, 6
  of urine excreted, 357-358
Volumetric flask, 100
Vomiting causing acid-base imbalance, 366

W

Water, 91-97
  absorption of, 362
  activity of, 95
  analysis of, 91-92
  in body, amount of, 361
  boiling point of, 94
  carbonated, 155
  chemical behavior of, 95
  color of, 92-93
  composition of, 91-92
  covalent bonds in, 48-49
  of crystallization, 95-96
  density changes in, 94-95
  distillation of, 94
  distribution of, 362-363
  electrolysis of, 91
    apparatus for, 92
  excretion of, 363
  extracellular, 362
  freezing point of, 94
  hard, 96-97

Water—cont'd
    hormonal control of movement of, 363
    hydrogen bonds in, 93
    hydrolysis of, 95
    intake of, 362
    intracellular, 362
    occurrence of, 91
    odor of, 92-93
    phase diagram for, 58, 59
    physical properties of, 92-95
    plasma volume of, measurement of, 363-364
    polar molecules of, 93
    pressure effects on movement of, 362-363
    pure, 94
    reaction of amine with, 184
    soft, 96-97
    solvent power of, 93-94
    stability of, 95
    synthesis of, 92
    taste of, 92-93
    total body, measurement of, 363
    uses of, in human body, 97
    volume change in, 94-95
Water balance, 361-364
    importance of, 363
Water-soluble vitamins, 330-333
Waxes, 222, 231-232
Weak acids, 112-113
Weak bases, 115
Weakness, leg, 311

Weight
    definition of, 4-5
    measurement of, 6
White cells, 344
White phosphorus, 70
Wood alcohol; *see* Methyl alcohol
Wood charcoal, 152
Wounds produced by acids or bases, antidotes for, 121-122

X

Xanthoproteic test for proteins, 219, 221
Xenon and fluorine, reaction of, 39-40
Xerophthalmia, 323-324
X-ray film, 144
X-ray tube, 142, 143
X-rays, 24, 142-144
    discovery of, 142
    generation of, 142-143
    nature of, and application, 142-144
    in radiography, 143-144
Xylenes, 190

Z

Zinc
    dietary need for, 311
    and hydrochloric acid, hydrogen production from, 75, 76
Zwitterions, 216

ATOMIC WEIGHTS OF THE ELEMENTS*

Element	Symbol	Atomic number	Atomic weight	Element	Symbol	Atomic number	Atomic weight
Actinium	Ac	89	(227)†	Mercury	Hg	80	200.59
Aluminum	Al	13	26.9815	Molybdenum	Mo	42	95.94
Americium	Am	95	(243)	Neodymium	Nd	60	144.24
Antimony	Sb	51	121.75	Neon	Ne	10	20.179
Argon	Ar	18	39.948	Neptunium	Np	93	237.0482
Arsenic	As	33	74.9216	Nickel	Ni	28	58.71
Astatine	At	85	(210)	Niobium	Nb	41	92.9064
Barium	Ba	56	137.34	Nitrogen	N	7	14.0067
Berkelium	Bk	97	(249)	Nobelium	No	102	(254)
Beryllium	Be	4	9.0122	Osmium	Os	76	190.2
Bismuth	Bi	83	208.981	Oxygen	O	8	15.9994
Boron	B	5	10.811	Palladium	Pd	46	106.4
Bromine	Br	35	79.904	Phosphorus	P	15	30.9738
Cadmium	Cd	48	112.40	Platinum	Pt	78	195.09
Calcium	Ca	20	40.08	Plutonium	Pu	94	(242)
Californium	Cf	98	(251)	Polonium	Po	84	(210)
Carbon	C	6	12.01115	Potassium	K	19	39.102
Cerium	Ce	58	140.12	Praseodymium	Pr	59	140.9077
Cesium	Cs	55	132.9055	Promethium	Pm	61	(147)
Chlorine	Cl	17	35.453	Protactinium	Pa	91	231.0359
Chromium	Cr	24	51.996	Radium	Ra	88	226.0254
Cobalt	Co	27	58.9332	Radon	Rn	86	(222)
Copper	Cu	29	63.546	Rhenium	Re	75	186.2
Curium	Cm	96	(247)	Rhodium	Rh	45	102.906
Dysprosium	Dy	66	162.50	Rubidium	Rb	37	85.4678
Einsteinium	Es	99	(254)	Ruthenium	Ru	44	101.07